Orthopaedic Emergencies

This material is not intended to be, and should not be considered, a substitute for medical or other professional advice. Treatment for the conditions described in this material is highly dependent on the individual circumstances. And, while this material is designed to offer accurate information with respect to the subject matter covered and to be current as of the time it was written, research and knowledge about medical and health issues is constantly evolving, and dose schedules for medications are being revised continually, with new side effects recognized and accounted for regularly. Readers must therefore always check the product information and clinical procedures with the most up-to-date published product information and data sheets provided by the manufacturers and the most recent codes of conduct and safety regulation. The publisher and the authors make no representations or warranties to readers, express or implied, as to the accuracy or completeness of this material. Without limiting the foregoing, the publisher and the authors make no representations or warranties as to the accuracy or efficacy of the drug dosages mentioned in the material. The authors and the publisher do not accept, and expressly disclaim, any responsibility for any liability, loss, or risk that may be claimed or incurred as a consequence of the use and/or application of any of the contents of this material.

Orthopaedic Emergencies

Casey Jo Humbyrd, MD

Resident
Department of Orthopaedic Surgery
Johns Hopkins Hospital
Baltimore, MD

Benjamin Petre, MD

Administrative Chief Resident
Department of Orthopaedic Surgery
Johns Hopkins Hospital
Baltimore, MD

Arjun S. Chanmugam, MD, MBA

Associate Professor
Department of Emergency Medicine
Johns Hopkins University School of Medicine
Baltimore, MD

Dawn M. LaPorte, MD

Associate Professor
Department of Orthopaedic Surgery
Johns Hopkins University School of Medicine
Baltimore, MD

OXFORD
UNIVERSITY PRESS

OXFORD
UNIVERSITY PRESS

Oxford University Press, Inc., publishes works that further
Oxford University's objective of excellence
in research, scholarship, and education.

Oxford New York
Auckland Cape Town Dar es Salaam Hong Kong Karachi
Kuala Lumpur Madrid Melbourne Mexico City Nairobi
New Delhi Shanghai Taipei Toronto

With offices in
Argentina Austria Brazil Chile Czech Republic France Greece
Guatemala Hungary Italy Japan Poland Portugal Singapore
South Korea Switzerland Thailand Turkey Ukraine Vietnam

Copyright © 2012 by Oxford University Press, Inc.

Published by Oxford University Press, Inc.
198 Madison Avenue, New York, New York 10016
www.oup.com

Oxford is a registered trademark of Oxford University Press.

Oxford is a registered trademark of Oxford University Press
All rights reserved. No part of this publication may be reproduced,
stored in a retrieval system, or transmitted, in any form or by any means,
electronic, mechanical, photocopying, recording, or otherwise,
without the prior permission of Oxford University Press.

Library of Congress Cataloging-in-Publication Data
Orthopaedic emergencies / Casey Jo Humbyrd … [et al.].
p. ; cm.
Includes index.
ISBN 978–0–19–973574–7 (pbk. : alk. paper)
I. Humbyrd, Casey Jo.
[DNLM: 1. Musculoskeletal System–injuries–Handbooks. 2. Orthopaedic Procedures methods–Handbooks. 3. Emergency Treatment–methods–Handbooks. WE 39]
617.47044—dc23
2011039726

9 8 7 6 5 4 3 2 1
Printed in the United States of America
on acid-free paper

Contents

Series Preface

Emergency physicians care for patients with any condition that may be encountered in an emergency department. This requires that they know about a vast number of emergencies, some common and many rare. Physicians who have trained in any of the subspecialties—cardiology, neurology, OBGYN and many others—have narrowed their fields of study, allowing their patients to benefit accordingly. The Oxford University Press *Emergencies* series has combined the very best of these two knowledge bases, and the result is the unique product you are now holding. Each handbook is authored by an emergency physician and a sub-specialist, allowing the reader instant access to years of expertise in a rapid access patient-centered format. Together with evidence-based recommendations, you will have access to their tricks of the trade, and the combined expertise and approaches of a sub-specialist and an emergency physician.

Patients in the emergency department often have quite different needs and require different testing from those with a similar emergency who are inpatients. These stem from different priorities; in the emergency department the focus is on quickly diagnosing an undifferentiated condition. An emergency occurring to an inpatient may also need to be newly diagnosed, but usually the information available is more complete, and the emphasis can be on a more focused and in-depth evaluation. The authors of each *Handbook* have produced a guide for you wherever the patient is encountered, whether in an outpatient clinic, urgent care, emergency department or on the wards.

A special thanks should be extended to Andrea Seils, Senior Editor for Medicine at Oxford University Press for her vision in bringing this series to press. Andrea is aware of how new electronic media have impacted the learning process for physician-assistants, medical students, residents and fellows, and at the same time she is a firm believer in the value of the printed word. This series contains the proof that such a combination is still possible in the rapidly changing world of information technology.

Over the last twenty years, the Oxford Handbooks have become an indispensible tool for those in all stages of training throughout the world. This new series will, I am sure, quickly grow to become the standard reference for those who need to help their patients when faced with an emergency.

Jeremy Brown, MD
Series Editor
Associate Professor of Emergency Medicine
The George Washington University Medical Center

Preface

This book is intended to be a rapid reference guide to the approach and initial management of the more common orthopaedic injuries. It aims to provide basic information about the initial management of musculoskeletal (MSK) injuries in general, including reduction, splinting, and casting techniques for specific fractures and soft tissue injuries. It is meant to be a "how-to" guide to the most basic of orthopaedic procedures and management. This book shares knowledge and "tricks of the trade" typically acquired through the apprenticeship model whereby new practitioners learn procedures and techniques from more experienced individuals. It will guide practitioners in how to avoid the most common pitfalls in casting and splinting and ideally limit the morbidity of inexpertly applied splints and casts which can include skin breakdown, impaired fracture healing, permanent loss of joint motion and compartment syndrome. The intended audience is emergency physicians, orthopaedic residents, family practice physicians, and other primary care providers as well as medical students and midlevel providers. Emergency medical services (EMS) personnel and other frontline heath care providers who need a quick access to key information will find this book to be a handy reference.

This book is not intended to be a reference for definitive treatment of injuries, although a brief discussion of definitive treatment is provided where appropriate. Many of the injuries listed here will need further nonoperative or operative management for long-term healing. Much of orthopaedic evaluation and treatment can be provided on an outpatient basis, and our hope is to convey how to best manage these outpatients versus those that require admission. There are certain procedures and injuries are that are beyond the scope of this reference but are described so as to improve time to recognition and involvement of orthopaedic surgeons. In addition, there are true, but rare, orthopaedic emergencies that require immediate orthopaedic specialist intervention as a life- or limb-saving procedure and these are briefly described.

In summary, the goal of this text is to help the reader properly identify orthopaedic injuries, provide guidance for initial management, distinguish those patients who can be appropriately treated as outpatients, and improve recognition and stabilization of those patients who require urgent and emergent orthopaedic consultation. It will aid in communication between practitioners providing the initial management and the final treating physician.

This book would not have been possible without the help of many people. Dr. Humbyrd thanks her father, Dr. Dan Humbyrd, for introducing and guiding her in the field of orthopaedics and her husband, Kent Grasso, for his continued love and support. Dr. Petre would like to thank his loving and caring family, Kristen, Grace and Hannah for their continued support. Dr. Chanmugam would like to extend thanks to his family, Karen, Sydney, William, Nathan, Sarah and Tamara for their wonderful support and terrific inspiration as well his both his parents, Malathi and Jayarajan who instilled the love of learning in their children. Dr. Laporte would like to thank her wonderful family, Paul, Sydney, Zachary, and Cooper, for their patience and for always making her smile, and her parents, Jere and Michael, for their unwavering faith and encouragement. As a group, we would like to thank our editor, Andrea Seils, for her advice and guidance. Many thanks to Derek Papp, MD for contributing the chapter on pelvic injuries and to Melinda J. Ortmann, PharmD, BCPS for her review and feedback on the analgesia chapter. Lastly, we thank the members of West Bay Orthopaedic Associates, Warwick, RI for their review of the manuscript and guidance. Their insight has been invaluable in making sure the manuscript is widely applicable to audiences in all practice settings.

Chapter 1

Introduction

Purpose of the Book

Improve Recognition and Management of the Most Common Fractures and Dislocations

This book is intended to be a rapid reference guide to the approach and initial management of the more common orthopaedic injuries. It aims to provide basic information about the initial management of musculoskeletal (MSK) injuries in general, including reduction, splinting, and casting techniques for specific fractures and soft tissue injuries. It is meant to be a "how-to" guide to the most basic of orthopaedic procedures and management. The intended audience is emergency physicians, orthopaedic residents, family practice physicians, and other primary care providers as well as medical students and midlevel providers. Emergency medical services (EMS) personnel and other frontline heath care providers who need a quick access to key information will find this book to be a handy reference.

This book is not intended to be a reference for definitive treatment of injuries, although a brief discussion of definitive treatment is provided where appropriate. Many of the injuries listed here will need further nonoperative or operative management for long-term healing. Much of orthopaedic evaluation and treatment can be provided on an outpatient basis, and our hope is to convey how to best manage these outpatients versus those that require admission. There are certain procedures and injuries are that are beyond the scope of this reference but are described so as to improve time to recognition and time to involvement of orthopaedic surgeons. In addition, there are true, but rare, orthopaedic emergencies that require immediate orthopaedic specialist intervention as a life- or limb-saving procedure and these are briefly described.

In summary, the goal of this text is to help the reader properly identify orthopaedic injuries, to provide guidance for initial management, to identify those patients who can be appropriately treated as outpatients, and to improve recognition and stabilization of those patients who require urgent and emergent orthopaedic consultation.

Guide to Effective Immobilization

Nonsurgical Management of Injuries

The musculoskeletal system is one of the few systems in the body where major injuries to the system can be overcome, regenerated, and healed solely by placing the two broken or injured parts in proximity and stabilizing them in place. An important concept to note when caring for injuries is that bones can repair and remodel, and tendons and ligaments can heal, but cartilage damage is often irreversible. Healing and recovery takes time (in children, as little as a few weeks), while at the other end of the spectrum, adults with comorbidities can take many months to heal. The key to healing is bringing the injured ends together and holding them there long enough to allow for healing to occur. The main point is to provide appropriate stabilization because motion may prevent healing in many circumstances.

In order to limit motion of a broken bone, three things must happen. The bone itself must be immobilized, the joint above the bone must be immobilized, and the joint below the bone must be immobilized. To illustrate this, picture a tibia broken directly in the center of the tibial shaft. If the knee is not immobilized, extension of the knee requires the quadriceps to act through the patellar tendon, pulling up on the tibia, but only the proximal fragment of the fracture. The distal fragment will be left behind. The same is true when the hamstrings are activated to flex the knee. Activation of the knee through the quadriceps or the hamstrings creates motion at the fracture site, which often is painful, but more importantly, the motion decreases the chance of effective healing.

In contrast to immobilization of bones, immobilization of a joint to allow for healing requires that the bone above the joint not move in relation to the one below. However, the joints not involved in the extremity usually can all move without jeopardizing healing. To illustrate this, picture a soft tissue injury to the knee. As long as the femur and tibia stay immobilized, the patient can move his or her hip, ankle, foot, and toes. This is why a knee immobilizer is an effective treatment for soft tissue injuries about the knee.

Importance of Immobilization Prior to Definitive Surgical Treatment

There are circumstances in which surgical intervention will be required to treat an injury, as there is no effective nonoperative treatment that can be tried first. In these circumstances the question is often asked: "If surgery is needed, why put the patient through the process of reduction and splinting in the meantime?" This question seems to have logical merits on the surface. There are a few extremely important reasons to insist on reduction and splinting, or at the very least, just simple splinting, of almost all injuries.

The first reason and arguably most important, is for soft tissue rest. A fracture or injury to the MSK system causes swelling and inflammation at the time of injury. Continued abnormal motion of a joint or motion at a fracture leads to ongoing swelling and soft tissue damage, which is a major contributing factor to failure of surgery and compartment syndromes; both are disastrous complications. Inadequate reductions or splinting in improper positions can also lead to contractures, articular cartilage death, soft tissue compromise, nerve compression, and other irreparable complications while the patient awaits surgery. Finally, there may be extenuating circumstances that prevent definitive treatment (certain cardiovascular events or associated injuries that prevent surgery, or the patient may have difficulty in timely follow-up for definitive care.) Therefore, the initial treatment needs to be as good as possible. The key point is to provide the necessary stabilization in all situations, for several reasons, not the least of which is that the initial stabilization may end up being the only treatment.

Circumstances Requiring Orthopaedic Consultation

Operative Interventions

This book is meant for frontline providers seeing the patients in emergency and or primary care settings. The operative treatment of traumatic injuries encompasses volumes of textbooks and can be found elsewhere.

Life or Limb-Threatening Injuries

True orthopaedic emergencies are rare. There are only a handful of orthopaedic injuries that have an immediate risk of loss of life or loss of limb. These need orthopaedic involvement as soon as possible. Recognition of these injuries is critical, but treatment will not be discussed in this text. These injuries include:

- Open-book pelvis injury with hemodynamic instability
- Spinal injury with any neurologic change including cauda equina syndrome
- Compartment syndrome of any compartment
- Septic arthritis, especially if the patient is acutely septic and unstable from the septic arthritis
- Injuries resulting in neurologic manifestations or vascular compromise.

Open Fractures

The acute treatment of open fractures has been evolving in the past decade and will likely continue to evolve long after this text. The current accepted practice is as follows:

- Open fractures with an opening < 1 cm and no exposed bone:
    - Appropriate antibiotics and consideration of tetanus booster immunization with timely irrigation and debridement either in the emergency department (ED) or operating room (OR) depending upon age of the patient, and orthopaedist's evaluation of the wound character. Appropriate treatment of the fracture.

- Open fractures with wounds 1–10 cm:
  - Appropriate antibiotics and consideration of tetanus booster immunization
  - Thorough irrigation and debridement in the ED with temporary stabilization of the fracture. Formal irrigation and debridement in the operating room with definitive or temporary fixation of the fracture in 6–12 hours.
- Open fractures >10 cm, grossly contaminated, vascular compromise, or large soft tissue defect:
  - Immediate tetanus and appropriate antibiotics. Cursory irrigation and debridement with temporary stabilization and formal treatment in the OR within usually 1–2 hours from injury.

Soft Tissue Injuries

Pyogenic flexor tenosynovitis of the hand may be caused by a number of pathophysiological processes, but for the purposes of this text, it should be considered an infectious process requiring immediate antibiotics and orthopaedic consultation for urgent surgical debridement. It is best characterized by the original description provided by Kanavel:

1. Finger held in slight flexion
2. Fusiform swelling
3. Tenderness along the flexor tendon sheath
4. Pain with passive extension of the digit

Infections

Septic arthritis, which is an infection of a joint, requires antibiotics and orthopaedic consultation and prompt intervention to prevent permanent damage to the joint and to prevent progression to systemic infection.

Importance of Imaging

Radiographs

Of the utmost importance to the diagnosis and treatment of musculoskeletal injuries is imaging. This is the stethoscope of MSK injuries; it is the electrocardiogram (EKG) equivalent for bone and soft tissue. Imaging is becoming more complex every year as there are more tools and techniques available all the time. Often, however, it is the classic radiograph that gives the most cost-effective and rapid information. Throughout training and practice, there will always be resistance to reordering radiographs, getting additional studies, including a joint above and a joint below, or obtaining tests with "less information" when a more advanced imaging modality is already available. We cannot emphasize enough the need for the proper imaging and the importance of radiographs.

Using Imaging to Plan Your Management

Radiographs

Plain radiography is often the diagnostic modality of choice. It is an invaluable tool for the orthopaedist and primary practitioner. Plain radiographs are inexpensive to obtain, have a very low morbidity for the patient, are fast, are widely available, and often provide all necessary information without a more involved test. Almost every injury listed in this text will have associated radiographs that are necessary for diagnosing the injury or ruling out other injuries with similar presentations. One can never settle for inadequate or incomplete imaging series as every radiograph in a series is designed to tell the practitioner very specific information. Both for the care of the patient and for medical-legal reasons, inadequate or incomplete series should be repeated without hesitation.

Computed Tomography. CT scanning is now widely available in many emergency centers. It is often a necessary test in orthopaedics but rarely a first-line imaging study. Computed tomography when compared with plain radiography is excellent for showing bony detail but is often less desirable for showing the "whole picture." There are some injuries that require CT scanning for complete diagnosis and treatment and these will be identified.

As a general rule, fractures involving articular surfaces should have CT scans after the reduction is complete and the injury is stabilized. CT scans are helpful for both assessment of reduction and preoperative planning. Due to the complexity of the bones involved, spinal pathology and pelvic pathology generally require CT scans.

Magnetic Resonance Imaging. MRI certainly has its role in orthopaedic diagnostic testing. It is an excellent modality to look for occult bony injuries and soft tissue injuries. However, many of these injuries can also be diagnosed by physical exam, and advanced imaging modalities can be delayed or performed on an outpatient basis.

There are many other useful diagnostic imaging tools available, including bone scans, tagged white cell scans, myelograms, contrast-enhanced arthrograms, and so forth. These will be reviewed where appropriate. They are rarely first-line diagnostic tools.

Postreduction Radiographs—the Need for Post-Splint/Cast Radiographs

Just as diagnostic imaging is crucial to the diagnosis and treatment of orthopaedic injuries, postprocedure radiographs are as important. An incomplete or poor reduction, an errant fold in a splint or cast, or an inadequate procedure can cause significant harm to patients. These often cannot be seen with the unaided human eye. Even if the reduction "felt good" and the limb looks well aligned, traumatic injuries are often unstable, and one needs to evaluate and document that the procedure was successful. Postreduction films are necessary to confirm that any splint or cast that was applied will not cause harm and is holding appropriate reduction. Without the use of postprocedure radiographs, the practitioner is sending patients out based on the guess that their intervention has helped.

Follow-up Imaging

For injures when an occult fracture is suspected but cannot be demonstrated on radiograph, it is recommended that the affected limb be splinted. The injury should be reimaged in 10–14 days when an occult fracture is better identified on a radiograph. In other words, a negative radiograph is a true negative only when it is taken 10–14 days after the injury.

Describing Fractures

Recommended Radiographic View for Specific Injuries

Just as describing an EKG to colleagues has its own vernacular, so does describing radiographs of musculoskeletal injury. For example, saying "normal sinus rhythm with ST elevations in the anterior leads" provides much more information to the person you are communicating to than does "the squiggly lines all look the same except sometimes the medium-sized bumps after the really big ones are higher." The following scheme will allow one to communicate efficiently with orthopaedic colleagues. Although a full orthopaedic glossary is presented in Chapter 10 of this book, here are some basic words used to describe fractures:

- Open versus closed. Any full-thickness skin defect in the zone of injury is considered open. Please note that a superficial skin injury near the musculoskeletal injury does not constitute an open fracture (e.g., road rash). An open fracture requires a full-thickness skin defect.
  - "I have a patient with a closed..."
- Identify the bone of interest and the location of the fracture in the bone:
  - "...distal third tibia fracture..."
- Describe the type of fracture using appropriate words. (Please note a full orthopaedic glossary is presented in Chapter 10 of this book.)
  - Simple: One fracture line
  - Comminuted: Multiple fracture lines
  - Butterfly: A triangle or wedge that has broken off at the site of the fracture
  - Segmental: Fractures above and below a segment of a bone such that it is free floating

- Displacement: The amount of translation one segment has compared with the other. Fractures can be nondisplaced (often difficult to see on radiograph), minimally displaced, or displaced. Displaced fractures can be described by the percent of the width of the bone in question.
- Angulation: The degree and direction of the distal segment compared with the proximal segment
- Apex: The point of a fracture
- Direction of the fracture line: Transverse, oblique, spiral, buckle, and so forth
- Intra-articular: Extending into or involving a joint
- Extra-articular: Outside of the joint
  - "...that is comminuted, extra-articular, angulated 30 degrees with the apex anterior and displaced 100%."
- Subluxation: Partial dislocation; loss of normal joint congruity
- Dislocation: Complete incongruity of a joint
- Certain fractures have epynomic classification schemes that are widely used and understood. Because of this, they can provide a large amount of information with a small amount of description. Where appropriate, these will be described in this book.

Pediatric Fractures

Classification of Pediatric Physeal Fractures (Salter-Harris)

Because children have growth plates and fractures can occur in or through them, a special descriptive classification system has been developed to describe physeal fractures, these can be seen graphically in Figure 1.1:

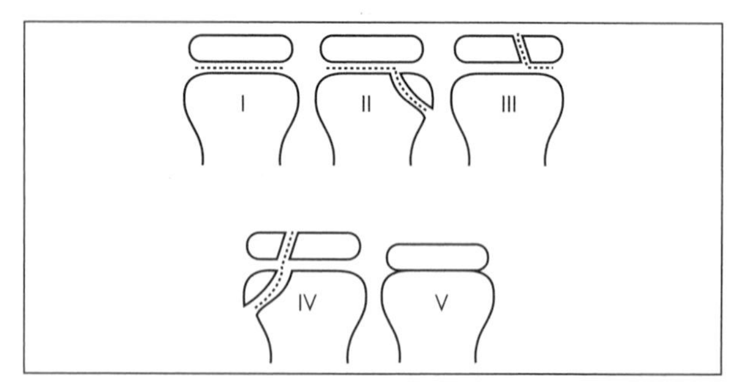

Figure 1.1 Diagram of Salter-Harris classification.

- Type I: Physeal separation
- Type II: Traverses the physis and exits through the metaphysis
- Type III: Traverses the physis and exits through the epiphysis
- Type IV: Passes through the epiphysis, physis, and metaphysis
- Type V: Crush injury to the physis

SALTR: A simple mnemonic to help remember the Salter-Harris fracture classification.

- I-S = Slipped (or straight across). Fracture of the cartilage of the physis (growth plate)
- II-A = Above. The fracture lies above the physis.
- III-L = Lower. The fracture is below the physis in the epiphysis.
- IV-T = Through. The fracture is through the metaphysis, physis, and epiphysis.
- V-R = Rammed (crushed). The physis has been crushed.

Chapter 2

Basic Techniques

Basic Techniques in Splinting and Casting

When to Splint versus When to Cast

This is a common question that does not have a perfect answer. Compared with splints, casts are more stable and more rugged and can be designed to be weight bearing. The disadvantage of a cast is that its circumferential nature does not allow for expansion secondary to swelling. An improperly placed cast can cause a compartment syndrome and in the worst case scenario can lead to limb loss.

In general, splinting is preferable to casting, especially if there is concern for future swelling—which is the case in most circumstances. In young children (under 4 years of age), it is important to provide as much structure as necessary, which generally means casting. Although these young children need more substantial immobilization, if concerns exist about swelling, it may be more appropriate to cast and then bivalve the cast with a cast saw and subsequently wrap the cast with an elastic wrap.

Here are some guidelines regarding splints and casts:

- If the patient will need an operation—splint
- If the patient is an adult and the injury is acute (less than a week old or still very swollen)—splint.
- If the patient is a child and the immobilization will be the definitive treatment—cast and bivalve in the acute setting, cast alone in the chronic setting.

General Principles of Immobilization—How to Avoid the Major Pitfalls

There are a few major pitfalls and complications associated with casting and splinting. We will review them individually and how to avoid them.

Compartment Syndrome: Often caused by the swelling from the injury, this can be accelerated by a bad cast or splint.

- Never use a material that won't expand in circumferential wraps (Kling or Kerlix should always be avoided).
- Never cast an acute adult fracture; use a splint that will allow for expansion
- Elevate once complete.
- Never wrap any layer, especially ace wraps, tightly.
- The more layers circumferentially placed with any material around a limb, the tighter and less expansile the entire construct becomes. Minimize unnecessary layers!

Pressure Sores or Cast Ulcers: These occur when the cast is improperly padded and pressure is placed chronically in one place. Pressure sores/cast ulcers result in skin breakdown if left unchecked. This can happen anywhere, but they usually occur in a place where there is motion and a lack of padding, typically at the very top or very bottom of the cast, and bony prominences (heel, patella). Ways to avoid cast problems are listed below.

- Pad bony prominences well.
- Add extra padding at the very top and very bottom of casts.
- Never allow the rigid casting/splinting material to go beyond the padding
- Avoid moving a joint when the padding/rigid material is being applied. Movement may result in folds. Folds cause pressure and pressure causes unnecessary skin breakdown.
- Place the proper molds to minimize the cast/splint motion and slippage

"A Bad Cast": Often people with less experience find themselves in a position where the postcast/splint radiographs look less than ideal due to positioning, folds, reduction, and so on. (This can happen to anyone!) Bad casts should always be redone. Reductions of articular surfaces or physes should be limited to three attempts to minimize shear on these sensitive structures. To avoid the "bad cast":

- Be prepared! Have more than enough materials at the bedside every time. That way when you accidentally drop a crucial material in the middle of the procedure, you do not have to start over.
- Get help. Many of these procedures are more than a one-person job. Getting a trustworthy extra pair of hands or two is crucial to doing it right the first time
- Exert the necessary force (don't wimp out.) Even with the best analgesia, these procedures can cause some discomfort and pain. This discomfort is usually quick and temporary. Doing it right the first time despite patient discomfort will prevent the patient from having to go through it all again. Explaining the uncomfortable steps beforehand so they know what is coming will often make the patient more tolerant and helpful during the painful parts. Once the bones are in the best alignment, the patient will feel better. Attention to analgesia prior to starting the procedure can minimize or eliminate discomfort.
- Get the proper positioning before you start. Avoid moving during casting or splinting as it causes the dreaded folds. See above.

Three-Point Mold (see Figure 2.1)

The mold of a cast or a splint determines the position of the fracture. A proper mold requires three points of applied force (the mold), and the force should be applied to counteract the forces displacing the fracture. The concept of a three-point mold is analogous to a seesaw, in which the fulcrum of the seesaw is balanced on either side by forces in the opposite direction.

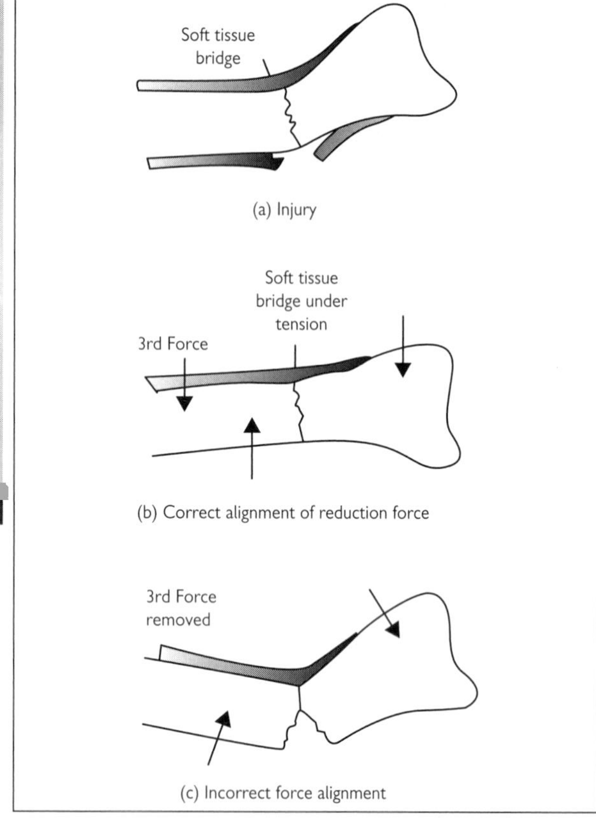

Figure 2.1 Diagram of a three-point mold with a fracture.

Techniques for Patient Padding: Sandwich Technique or Overwrap Technique

There are two general ways to place padding between the patient and the rigid material. The "sandwich" is a one-step method that essentially pads the rigid material first and then is placed on the patient. The second technique is the "overwrap" method, which is a two-step process. The first step pads the patient, then the rigid material is added. Casts always use overwrap; splints can use either.

The Sandwich Technique (see Figure 2.2)

1. The rigid splinting material is measured out, and the proper number of layers are created and laid out on a table.

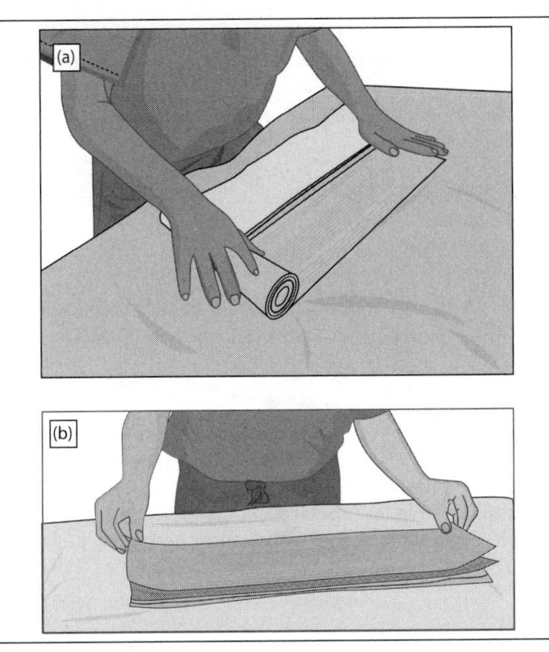

Figure 2.2 Sandwich technique. a. The sandwich technique requires three layers to be rolled out: Sof-Roll, followed by plaster, followed by Sof-Roll. b. The sandwich is assembled by stacking the three layers of the sandwich.

2. The padding material is then laid out on top of the splinting material and layered to give the appropriate padding. The padding should be slightly larger in all dimensions (wider and longer) than the rigid material to make sure all aspects of the rigid material are covered.
3. A final single layer of padding is measured off to cover the outside of the plaster to prevent the elastic wrap from sticking to the plaster.
4. The plaster or fiberglass is then activated with water and is placed between the padding that will be touching the patient and the single covering layer. Then, the entire construct is placed on the patient.
5. The construct is then held in place typically with an elastic wrap, and the proper mold is introduced until the rigid material has hardened.

The Overwrap Technique (see Figure 2.3)

1. The rigid splinting material is premeasured or casting material is at hand.
2. Position the extremity into position of immobilization NOW and do not move again.
3. Wrap the padding layers circumferentially around the extremity, making sure to go above and below where you plan to end the rigid material.
4. Use a 50–50 wrap where each layer overlaps the last by 50% of its width.
5. Take care to provide adequate padding to bony prominences.
6. Add the rigid material in either a splint or casting fashion and secure with elastic wraps.
7. Apply proper mold.

Padding the Splint—Importance and Techniques

Proper padding is a large part of the art of immobilization. Too little padding, especially at bony prominences, causes serious skin breakdown and can lead to infections and other serious complications. However, the converse of too much padding will prevent the rigid layers from holding the extremity immobilized and can lead to nonunion.

- Splints that are only to temporarily hold immobilization while awaiting definitive procedures should always get extra padding as holding the perfect reduction is less important than preventing skin breakdown, which will delay surgery.

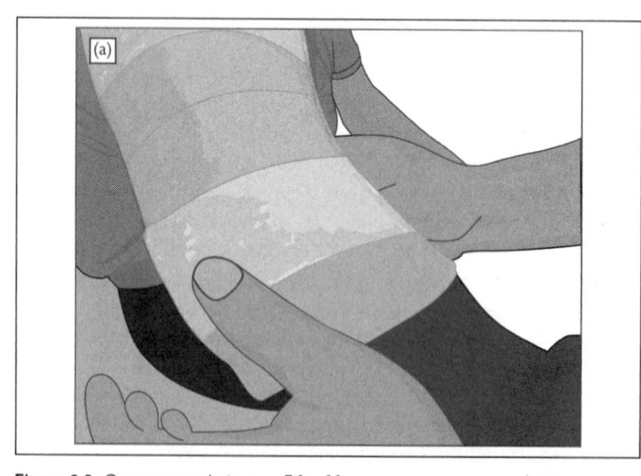

Figure 2.3 Overwrap technique. a. Fifty–fifty overwrap.

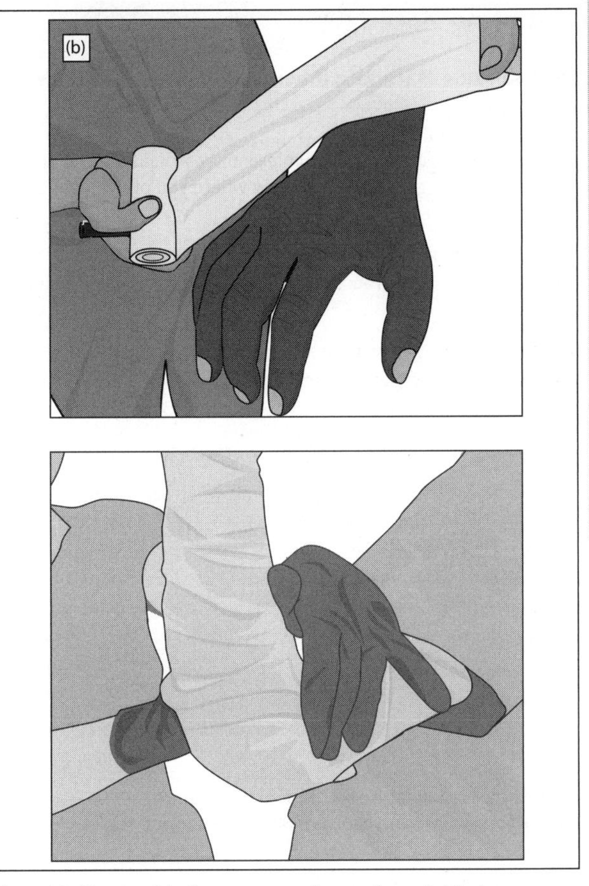

Figure 2.3 (Continued) b. Correct way to roll out an Ace. c. Padding bony prominence; applying extra padding to the elbow.

- Splints and casts that are holding reductions and immobilization should have just slightly more that the minimum padding at the areas that are crucial to maintain the reduction while other areas are padded more to prevent pressure sores.
- Bony prominences such as the olecranon or heel are often the problem locations and should get addressed with specific padding directed at preventing pressure sores.

Splinting Materials

Immobilizing Materials: Fiberglass, Plaster, Elastic Wrap

Fiberglass

- Extremely rigid. When laminated together can create a lightweight strong construct.
- Less pliable than plaster
- Mild-moderately exothermic when curing
- Can be water resistant
- Comes in rolls and premade "sandwiches" with padding built in

Plaster

- Extremely rigid when enough layers laminated together
- Heavier than fiberglass for similar rigidity
- Extremely pliable when wet, can be made into almost any shape and allow for odd bony prominences
- Not water resistant
- Moderately to significantly exothermic when setting and has been known to cause burns if used improperly. Using cool water is the best way to prevent this.
- Comes in rolls or sheets

Elastic Wraps

- Poor rigidity
- Extremely lightweight
- No curing necessary, therefore no heat
- Useful for creating gentle pressure, holding other materials in place, or to create a gentle decrease in the range of motion at a joint

Padding Materials: Sof-Roll, Kerlix, Gauze, Bulky Jones Cotton, Army Basic Dressing Pads

Sof-Roll

- Very soft, thin padding material
- Tears easily
- Can be stacked or rolled to create multiple layers increasing the padding
- Comes in various sizes
- Some stretch to allow for swelling

Kerlix

- Thinner than Sof-Roll and less padding per layer
- Extremely strong in axial stretch
- Does not allow for stretch from swelling

- Comes in various sizes
- Not recommended for most padding applications, preferred as a positioning tool

Gauze

- Fairly soft thin padding material
- Limited in sizes, usually in sheets only
- Little to no stretch
- Strong, resists tearing

Bulky Jones Cotton

- Extremely soft, thick padding
- Thickness allows for swelling/prevents pressure sores.
- Thickness decreases ability to hold reduction.
- Tears easily

Army Basic Dressing (ABDs—Sometimes Called Abdominal Pads)

- Thick, well-padded gauze
- Limited sizes
- Strong, resists tearing
- Excellent for padding bony prominences

Patient Management

How to Care for Your Cast or Splint

- The vast majority of casts and splints are NOT waterproof and should be kept dry at all times.
- Because they are not waterproof, cleaning them is difficult, and therefore the cast or splint should be kept clean and protected at all times.
- Garbage bags, newspaper bags, Saran Wrap, and commercially available "cast bags" can all be used to keep a cast or splint dry when bathing.
- The majority of casts and splints are non-weight-bearing and will deteriorate or break quickly if subjected to weight-bearing stress.
- Casts or splints should never be rewrapped by untrained practitioners.
- New pain or worsening pain in casts or splints is always concerning and should be evaluated by a qualified practitioner.

Standard Discharge Instructions

With any splint or cast, the patients and their families need to be counseled on key points. Discharge instructions are absolutely vital to ensure the integrity of the splint or cast:

1. Keep clean and dry
2. Call or return to provider for pressure sores, any new numbness, tingling, or weakness. Also return to provider if there is an increase in pain or if pain is not controlled by medication.
3. Elevate
4. If the cast or splint becomes loose, NEVER rewrap at home, return to be evaluated by a provider. There are case reports of parents rewrapping bandages or splints causing ischemia and amputations of their child's limbs.

After Splinting/Casting Emergencies (i.e., Compartment Syndrome)

Compartment syndrome (see Chapter 7, "Orthopaedic Emergencies and Urgencies") can be caused by an improperly placed cast or splint or excess swelling in a cast or splint. This is a limb-threatening surgical emergency. All patients and family members should get educated about the signs and symptoms of compartment syndrome before being discharged from any medical center with a new cast or splint. They need to understand that they must return to a medical facility at once if these signs or symptoms occur. If a patient or family cannot comprehend this or does not have some ability to return, the patient should be kept for 24–48 hours for observation.

Cast Removal Technique, Including Bivalving of Casts and Wedging

Splints can be removed by using a sharp pair of trauma shears or cast shears (scissors with a protected blunt end on one side). The layers of a splint should be sequentially cut off starting with the outer layer and progressing inward toward the patient's skin. It is easiest to work between the slabs of rigid material.

Casts are more difficult to remove as they are circumferential, and therefore there is no space between slabs in which to insert scissors. Therefore, a cast saw is usually required to cut the plaster or fiberglass.

Technique (see Figure 2.4)

1. Talk to the patient and family about the cast saw: how it works, how it is designed to not cut skin, that it is extremely loud, and how it sometimes gets hot when it has been on for a short period of time.
2. Check that the cast saw blade is sharp and free of defects.

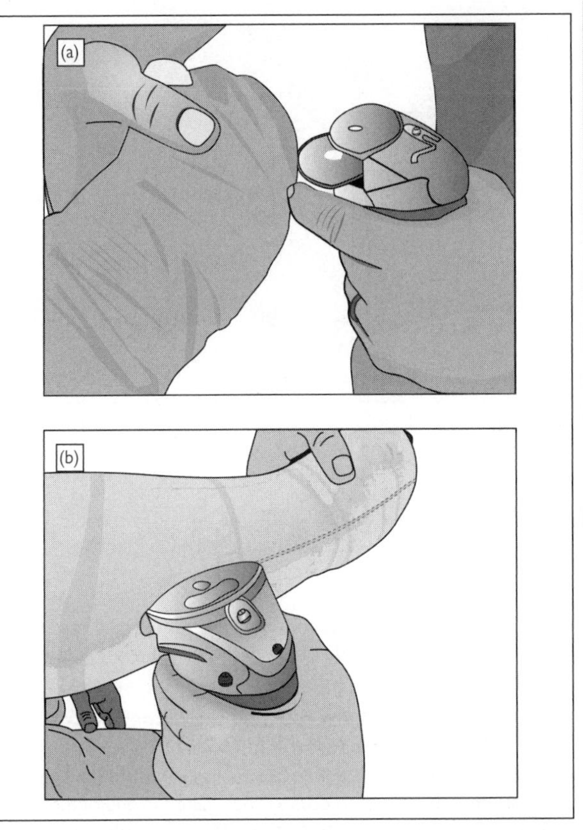

Figure 2.4 Cast removal. a. Positioning the hand to use the cast saw. b. Cutting the cast in a straight line.

3. Check that the cast saw blade is securely attached to the saw as the nut that secures the blade often loosens.
4. NEVER use a saw that is in any disrepair.
5. Plan the cuts before beginning. You should never work in a concavity such as in the antecubital fossa when the arm is bent. Working in a concavity greatly increases your chance of injuring a patient.
6. The cast saw is designed to move directly in and out perpendicular to the cast.

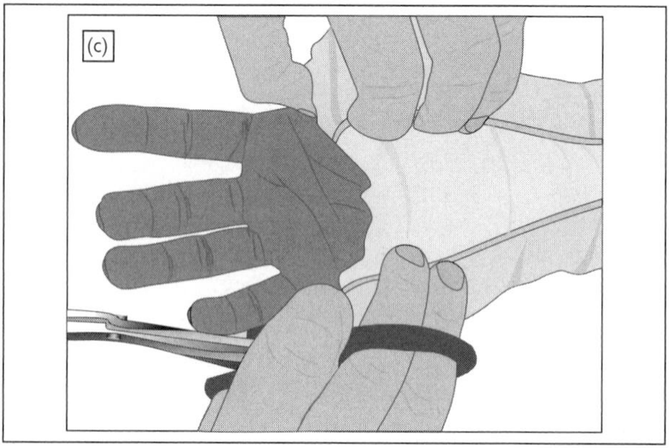

Figure 2.4 (Continued) c. Splitting open a bivalved cast.

7. Important: The saw should NEVER be moved longitudinally up or down the cast while still inside the cast. Remove the blade completely, move down slightly, and then repeat cut.
8. The saw blade heats up. Check the blade frequently to ensure it is not too hot, as the blade can burn the skin. If it is hot, stop and allow blade to cool.

Patient Positioning: How to Hang the Upper or Lower Extremity Using Kerlix (see Figure 2.5)

- Double a length of Kerlix and pass the loop through the second webspace from palmar to dorsal. Then, take the end of the loop at end of Kerlix dorsally and pull over the index and middle fingers (toes) and pull taut.
- Secure free ends to a stable IV pole or weighted IV pole.
- We discourage the use of ceiling-mounted IV poles as they are not strong enough to hold an extremity, and cannot move with the patient if the stretcher needs to be moved.

Splinting and Casting the Upper Extremity

Coaptation Splint

Gear List

- Six inch stockinette (3× length of Sof-Roll)
- Two to three 4 inch plaster rolls

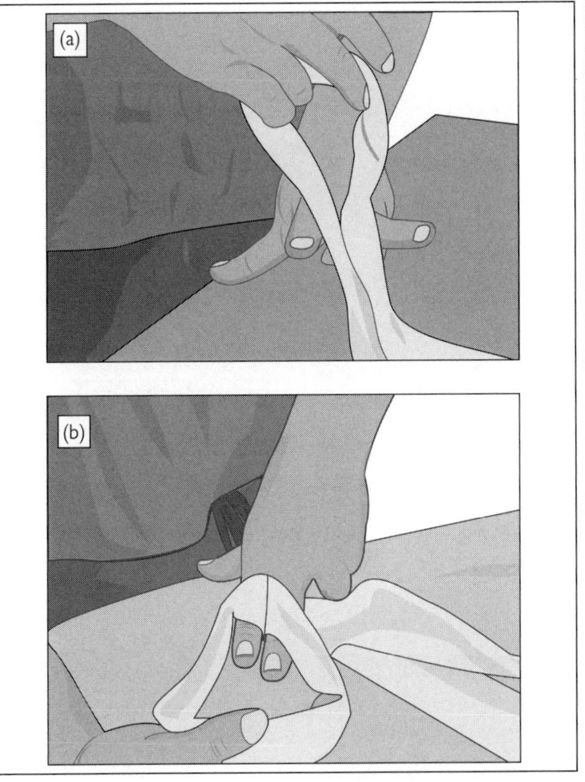

Figure 2.5 Patient positioning—how to hang an arm using Kerlix. a. Putting Kerlix between the fingers to be suspended. b. Pulling Kerlix loop over the fingers to be suspended.

- Three 6 inch Sof-Rolls
- Two 6 inch ace wraps

Patient Positioning

- Position the patient sitting up on a stretcher or a chair and have an assistant support the wrist to maintain the arm in 45 degrees abduction at the shoulder, elbow bent to 90 degrees. When applying the splint, the assistant can remove the upper hand and assist with supporting the splint in the axilla. The lower hand is maintained on the wrist. If an assistant is not available, the patient can support their hand on a Mayo stand or on their

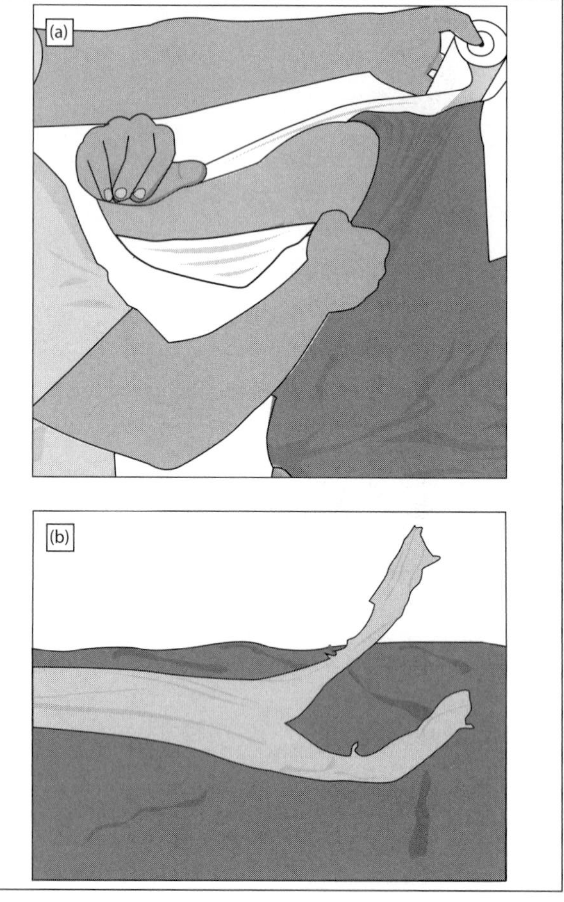

Figure 2.6 Coaptation splint. a. Measuring the splint's length. b. Cutting the stockinette.

abdomen. The goal is to have the elbow at 90 degrees and arm in some abduction. The splint needs to extend into the axilla (with appropriate padding).

Preparing the Splint

- Measure length of plaster from AC joint around lateral arm and then around elbow and up the medial arm to the axilla.
- Roll out 10 layers of plaster to this length

Figure 2.6 (Continued) c. Tying the second axilla knot.

- Measure Sof-Roll to extend 1 inch beyond plaster at both ends (six layers minimum on skin side and two on outer side of plaster).
- Measure stockinette to be 3× length of Sof-Roll and split proximal and distal 1/3 with scissors to make two "tails" on both sides (should be enough stockinette to cover entire Sof-Roll).

Technique (Sandwich Technique of Splinting) (see Figure 2.6)

1. Dunk plaster in cool water.
2. Squeeze out excess water.
3. Stretch and smooth plaster for laminating.
4. Plaster onto premeasured padding.
5. Apply top layer Sof-Roll.
6. Insert sandwich into stockinette.
7. Place splint on the medial aspect of arm as high into axilla as possible and, if available, have assistant hold splint in axilla.
8. Place stockinette over the shoulder and tie into place using the tails of the stockinette (tight enough to hold plaster into axilla).

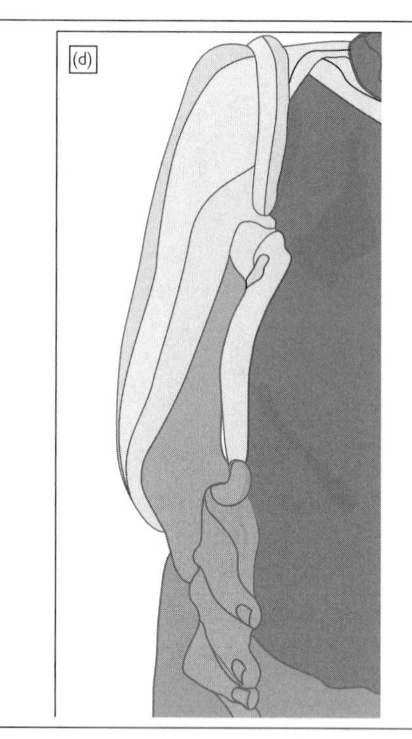

Figure 2.6 (Continued) d. Splint applied and tied, before molding and application of Ace wrap.

9. Wrap around elbow and up lateral arm so end lies on top of existing knot in stockinette and then tie second knot in axilla stockinette over the shoulder stockinette.
10. Wrap ends of the shoulder stockinette around to the contralateral axilla and secure.
11. Apply Ace wraps along length of arm.
12. Mold splint to fracture.
13. Apply sling and swathe/shoulder immobilizer.
14. Remove knot at contralateral axilla and check to ensure the knot in the ipsilateral axilla has not gotten too tight with the addition of the many layers.

Pearls and Pitfalls

Pearls

- Patients should be wearing only gown (no undergarments).
- Always measure off the good arm.

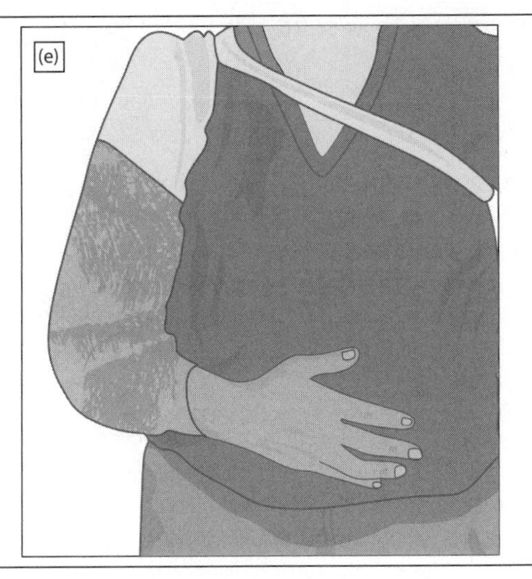

Figure 2.6 (Continued) e. Completed splint.

- Mold based on fracture
- Place ABD pad or extra padding in axilla.
- Because fractures tend towards a varus (distal segment has inward angulation) position, the mold should place the elbow in abducted position compared to the proximal humerus.
- Weight of splint is important in achieving fracture reduction and therefore the elbow in the sling should never be higher than the contralateral elbow (do not make sling too tight).
- Do not need anatomic reduction—reduction will continue to improve with proper splinting

Pitfalls

- Axilla portion of the splint is always lower than you think it is, so the splint must extend all the way into the axilla; otherwise it will slide down and inevitably lever against fracture site.
- Superior portion of splint must extend to the lateral border of the neck, at least to the AC joint, to immobilize the shoulder.
- The axilla knot should be tied no more lateral than the AC joint.

Sugar Tong Forearm Splint

Gear List

- Four 3 inch or 4 inch plaster rolls (plaster should be approximately the width of a closed fist).

- Three Sof-Rolls (go up in size [width]) of Sof-Roll compared to plaster) → two for padding, one for wrap.
- Kerlix or set of finger traps
- Three 4 inch Ace wraps
- IV pole
- Weight (for distal radius reduction) for traction
- One 1L IV bag (for distal radius only)

Patient Positioning (see Figure 2.7)

- Hang hand from IV pole either with Kerlix or finger traps (patient can be seated or supine).
- The elbow should be at 90 degrees.
- Specific fractures require traction, and this should be applied prior to splint application, and traction radiographs obtained before splinting (more detailed instructions on reduction maneuvers unique to particular fractures are covered in Chapter 3).

Preparing the Splint

- Measure the width of the plaster to approximate the width from the first webspace (between the thumb and second finger) to the ulnar side of the palm.
- Measure the length of the plaster from the distal palmar crease around the elbow to the MCP (metacarpophalangeal) joints at the dorsum of the hand.
- Roll out 10–12 layers of plaster.
- Measure Sof-Roll to be 1 inch longer than the plaster at both ends:
  - Sof-Roll should have six layers for the skin side.
  - Sof-Roll should have two layers for the outer surface.

Technique

1. Reduction if indicated (see Chapter 3, "Upper Extremity Injuries," under specific fractures)
2. Dunk plaster in cool water.
3. Squeeze out excess water.
4. Stretch and smooth plaster for laminating.
5. Plaster onto premeasured padding.
6. Apply top layer Sof-Roll.
7. Lay splint from distal palmar crease along volar forearm and around elbow and over dorsal forearm and hand to the MCP joints.
8. Wrap Sof-Roll circumferentially around splint at wrist to hold in place (optional).
9. Apply Ace wraps to secure plaster.
10. Mold plaster according to the injury. Traditionally, a 3-point mold is utilized. The mold for dorsally displaced distal radius fractures is demonstrated in diagram 2.7e and 2.7g.
11. Apply sling.

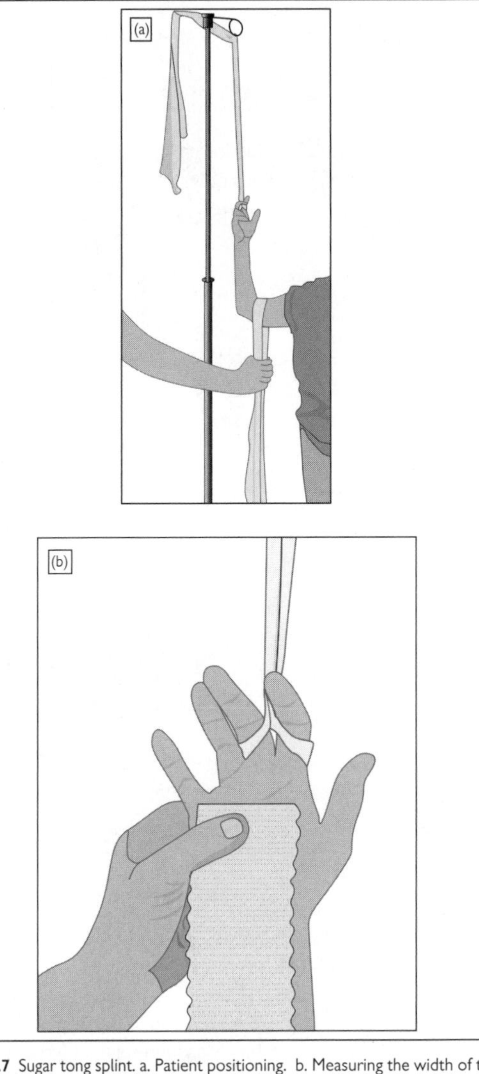

Figure 2.7 Sugar tong splint. a. Patient positioning. b. Measuring the width of the plaster.

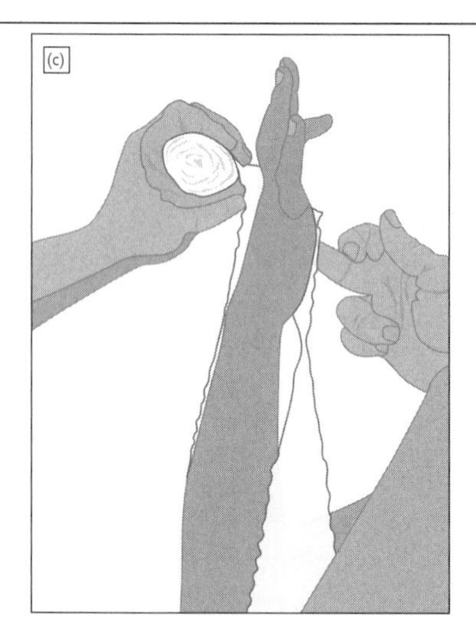

Figure 2.7 (Continued) c. Measuring the length of the plaster. d. Initial splint fixation with Sof-Roll.

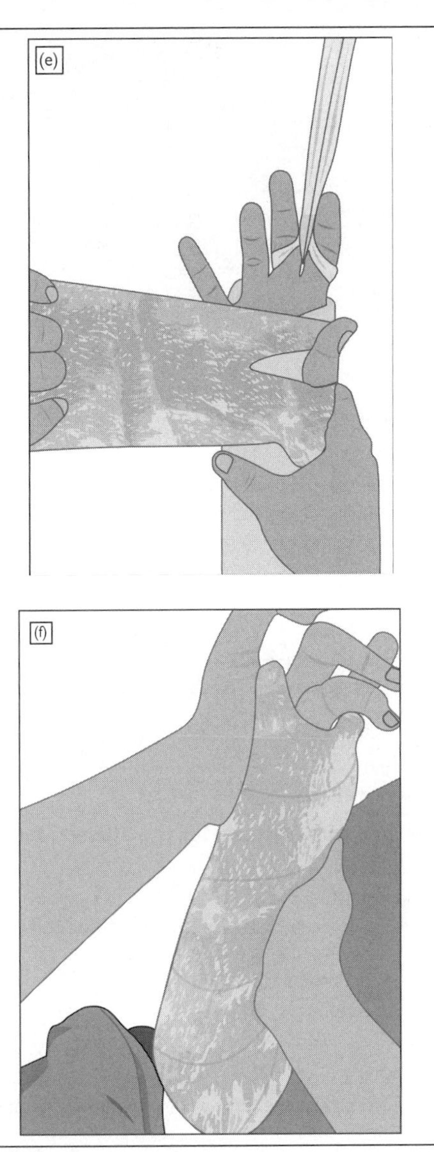

Figure 2.7 (Continued) e. Applying the Ace wrap. f. Traditional three-point mold for sugar tong.

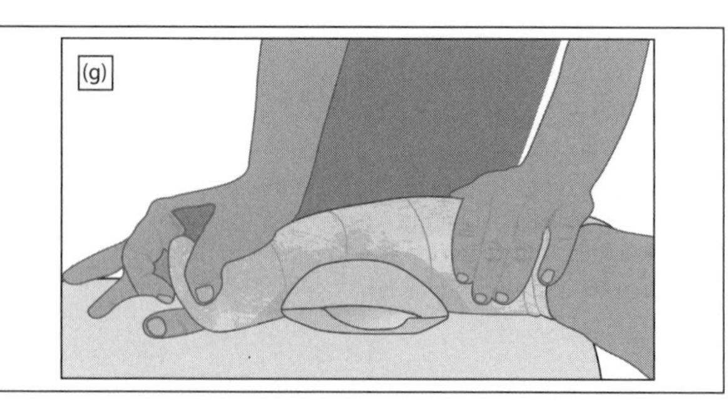

Figure 2.7 (Continued) g. Modified mold for sugar tong using 1 L IV bag for a fulcrum.

Pearls and Pitfalls

Pearls

- Hang the arm with the IV pole at mid- or low height so that there is room to increase height of IV pole.
- Can use 1 liter IV bag as fulcrum for three-point molding
- Sufficient pressure must be applied to achieve adequate molding.

Pitfalls

- The plaster often shortens by 1–2 cm. It is better to measure slightly long and remove excess than too short and have to redo the entire splint.
- Patients must be able to make a fist when complete; if the splint is too high in the palm, it will block finger range of motion (ROM) and cause stiffness.

Posterior Slab with Buttress for the Elbow

Gear List

- Three 4 inch plaster rolls
- Three Sof-Rolls (go up in size [width] of Sof-Roll compared to plaster) → two for padding, one for wrap
- Kerlix or set of finger traps
- Three 4 inch Ace wraps
- IV pole

Patient Positioning

- Hang hand from IV pole either with Kerlix or finger traps (patient can be seated or supine).
- The elbow should be at 90 degrees.
- Specific fractures require traction, and this should be applied prior to splint application, and traction radiographs obtained

before splinting (More detailed instructions on the need for traction and traction radiographs unique to particular fractures are covered in Chapter 3.).

Preparing the Splint

- Measure the plaster from the distal palmar crease at the ulnar aspect of the hand along the ulnar side of the forearm across the elbow and up the posterior arm to just short of the axilla.
- Roll out 10–12 layers of plaster.
- Measure Sof-Roll to be 1 inch longer than the plaster at both ends:
  - Sof-Roll should have six layers for the skin side.
  - Sof-Roll should have two layers for the outer surface.
- Roll out 8–10 layers of plaster measuring ½ length of posterior splint plaster.

Technique (see Figure 2.8)

1. Reduction if indicated (see Chapter 3, "Upper Extremity Injuries," under specific fracture)
2. Dunk plaster in cool water.
3. Squeeze out excess water.
4. Stretch and smooth plaster for laminating.
5. Plaster onto premeasured padding.
6. Apply top layer Sof-Roll.
7. Apply the plaster from the distal palmar crease at the ulnar aspect of the hand along the ulnar side of the forearm across the elbow and up the posterior arm to just short of the axilla.
8. Lay the shorter segment of plaster across the medial elbow (from the proximal forearm to the midlateral humerus).
9. Overwrap with Sof-Roll to secure in place.
10. Apply Ace wraps to secure plaster.
11. Apply sling.

Pearls and Pitfalls

Pearls

- For all elbow splints: Elbow at 90 degrees and forearm in neutral anatomic position (with the thumb pointing up in the air).

Pitfalls

- Take care to pad well with attention to elbow/bony prominences as this is often a location prone to skin breakdown.

Volar Resting Splint

Gear List

- Two 3 or 4 inch plaster rolls
- Two Sof-Rolls
- One or two 4 inch Ace wraps.

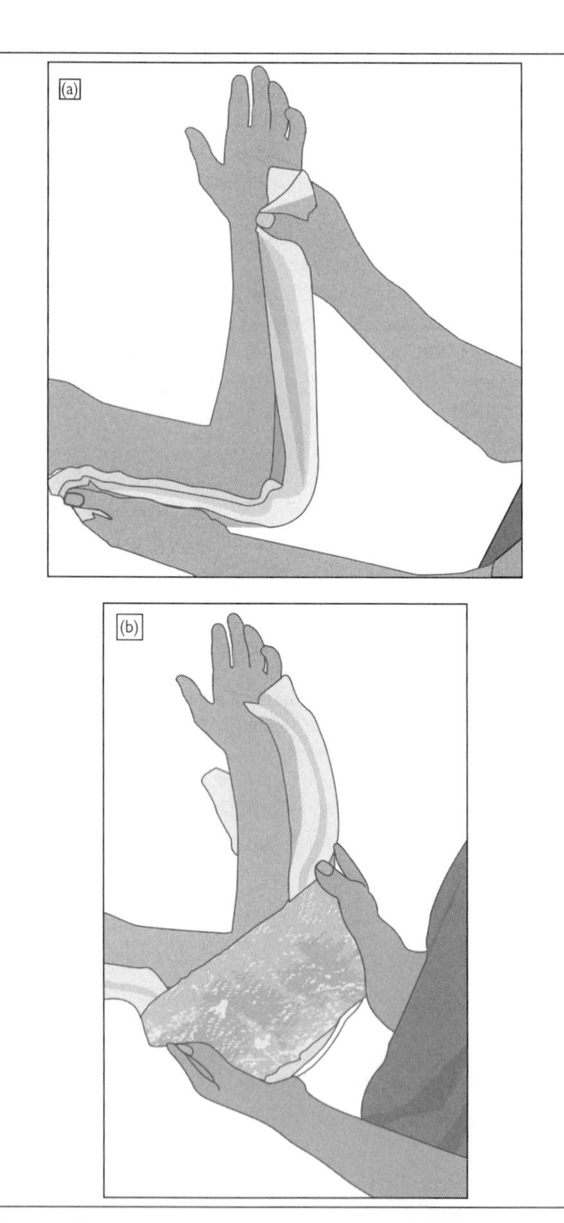

Figure 2.8 Posterior slab with buttress splint. a. Applying posterior slab. b. Applying buttress component.

Patient Positioning

- Patient seated with elbow resting on a Mayo stand or table
- The patient's hand will be positioned in a normal resting posture. Resting posture is the position that is comfortable for the patient and may be different depending upon the patient and his or her ailment.

Preparing the Splint (see Figure 2.9)

- Measure plaster from just distal to palmar tips of fingers along palm to proximal forearm. Stop the plaster distal to the mobile wad.
- Roll out 10–12 layers of plaster.
- Measure Sof-Roll to be 1 inch longer than the plaster at both ends.
- Soft roll should have six layers for the skin side
- Soft roll should have two layers for the outer surface.

Analgesia (see Chapter 8)

- Digital or metacarpal block and/or systemic

Technique

1. Dunk plaster in cool water.
2. Squeeze out excess water.

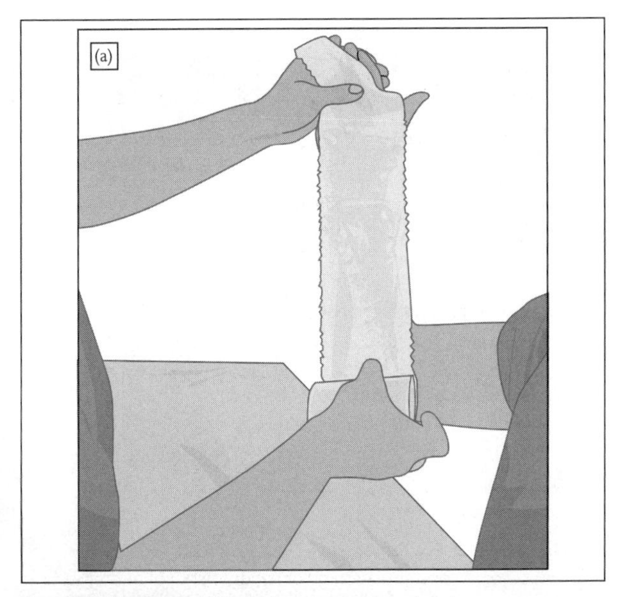

Figure 2.9 Volar Resting Splint. a. Measuring the plaster length.

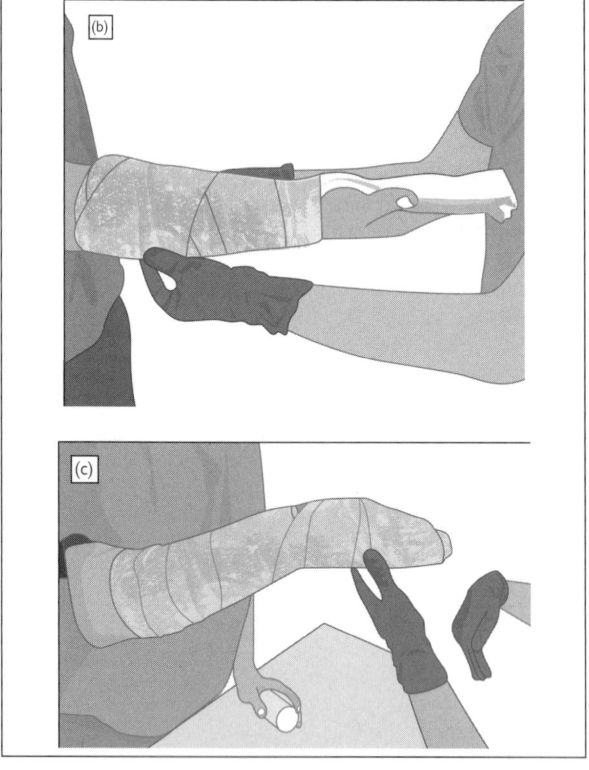

Figure 2.9 (Continued) b. Applying the splint and holding it in place using Ace. c. Resting position for a volar resting splint.

3. Stretch and smooth plaster for laminating.
4. Plaster onto premeasured padding.
5. Apply top layer Sof-Roll.
6. Apply the plaster from just distal to tips of fingers palmarly along palm to proximal forearm.
7. Apply Ace wrap(s) to secure plaster.
8. Mold splint to position wrist in a comfortable resting position.
9. Apply sling.

Pearls and Pitfalls

Pearls

- Do not worry about positioning of hand/wrist until plaster is secured in place.

- Construct in a way that it can be easily removed and reapplied.

Pitfalls

- This splint is primarily for soft tissue rest and should not be utilized if a reduction needs to be maintained.

Intrinsic Plus Splint

Gear List

- Two 3 or 4 inch plaster rolls
- Two Sof-Roll
- One or two 4 inch Ace wraps.

Patient Positioning (see Figure 2.10)

- Patient seated with elbow resting on a Mayo stand or table
- The patient's hand will be positioned with MCP joints flexed to 90 degrees and PIP/DIP (proximal interphalangeal/distal interphalangeal) joints fully extended. The wrist should be held in approximately 30 degrees of dorsiflexion.

Preparing the Splint

- Measure plaster from just distal to palmar tips of fingers along palm to proximal forearm.
- Roll out 10–12 layers of plaster.
- Measure Sof-Roll to be 1 inch longer than the plaster at both ends.
  - Soft roll should have six layers for the skin side.
  - Soft roll should have two layers for the outer surface.

Analgesia (see Chapter 8, Analgesia—Refer to Appropriate Page)

- Digital or metacarpal block and/or systemic

Technique

1. Dunk plaster in cool water.
2. Squeeze out excess water.
3. Stretch and smooth plaster for laminating.
4. Plaster onto premeasured padding.
5. Apply top layer Sof-Roll.
6. Apply the plaster from just distal to tips of the splinted fingers palmarly along palm to proximal forearm.
7. Apply Ace wrap(s) to secure plaster.
8. Mold splint to position wrist in 30 degrees dorsiflexion and flex MCPs to 90 degrees (IPs [interphalangeals] in extension).
9. Apply sling.

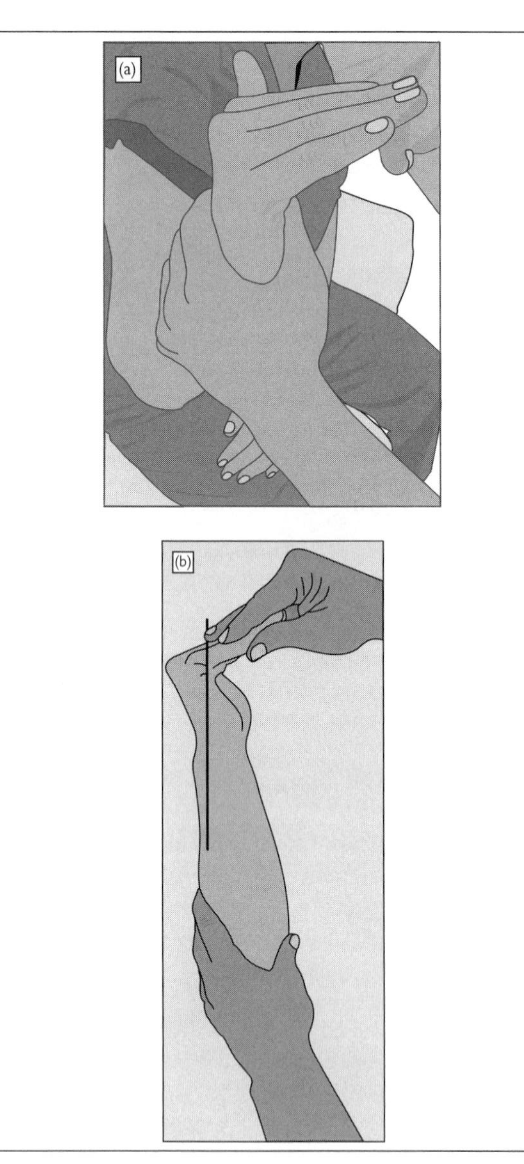

Figure 2.10 Intrinsic plus splint. a. Demonstrating MCP flexion and visible tip of the thumb. b. Demonstrating dorsiflexion of the wrist past neutral where neutral is demonstrated by a straight line.

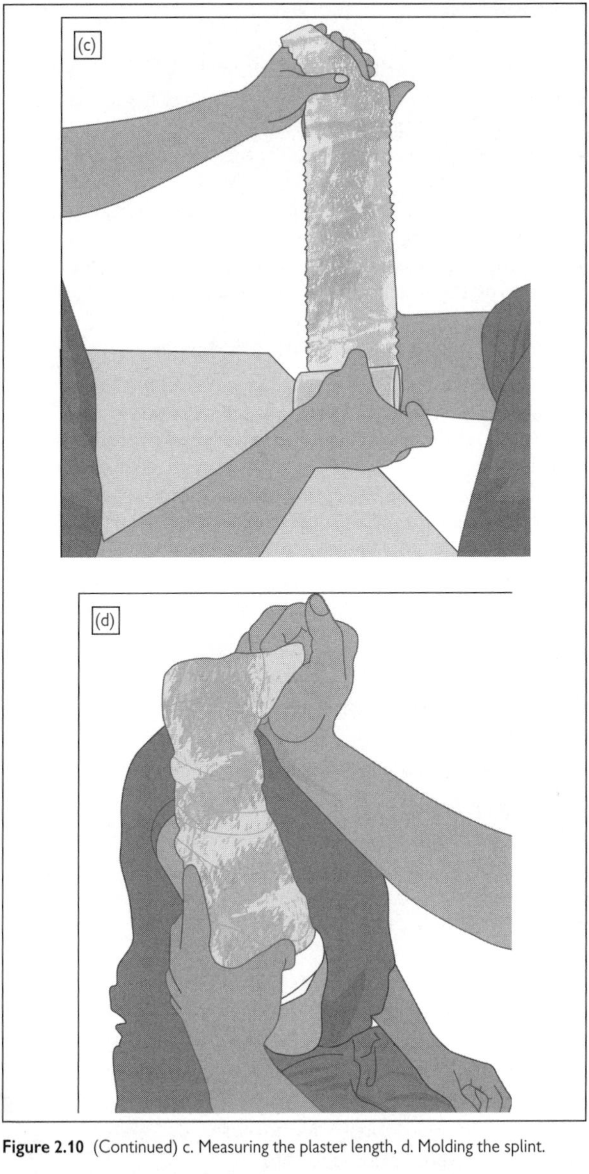

Figure 2.10 (Continued) c. Measuring the plaster length, d. Molding the splint.

Pearls and Pitfalls

Pearls

- Having a knowledgeable assistant is helpful.
- If the tip of thumb is not visible over dorsum of fingers, need more MCP (metacarpophalangeal) flexion
- Do not worry about positioning of hand/wrist until plaster is secured in place.
- Wrist should be in 30 degrees dorsiflexion and MCP joints 90 degrees of flexion.
- For associated proximal/middle phalanx fractures, buddy-tape the affected digit to an adjacent digit before splinting to help maintain reduction.
- Can use thumb to contour/mold plaster into the palm.

Pitfalls

- Obtaining 90 degree MCP flexion and IP extension is very difficult. Using cold water can give extra time to obtain idealized positioning.
- Pushing down on the PIP, up and back on the fingertips with one hand while holding the wrist dorsiflexion and hand position with the other is the best method.

Ulnar Gutter Splint

Gear List

- Two 3 or 4 inch plaster rolls
- Two Sof-Roll
- One or two 4 inch Ace wraps

Patient Positioning

- Patient seated with elbow resting on a Mayo stand or table

Preparing the Splint

- Measure plaster from just distal to palmar tip of fourth finger along the palm to proximal forearm.
- Roll out 10–12 layers of plaster.
- Measure Sof-Roll to be 1 inch longer than the plaster at both ends:
    - Sof-Roll should have six layers for the skin side.
    - Sof-Roll should have two layers for the outer surface.

Technique (see Figure 2.11)

1. Dunk plaster in cool water.
2. Squeeze out excess water.
3. Stretch and smooth plaster for laminating.
4. Plaster onto premeasured padding.
5. Apply top layer Sof-Roll.

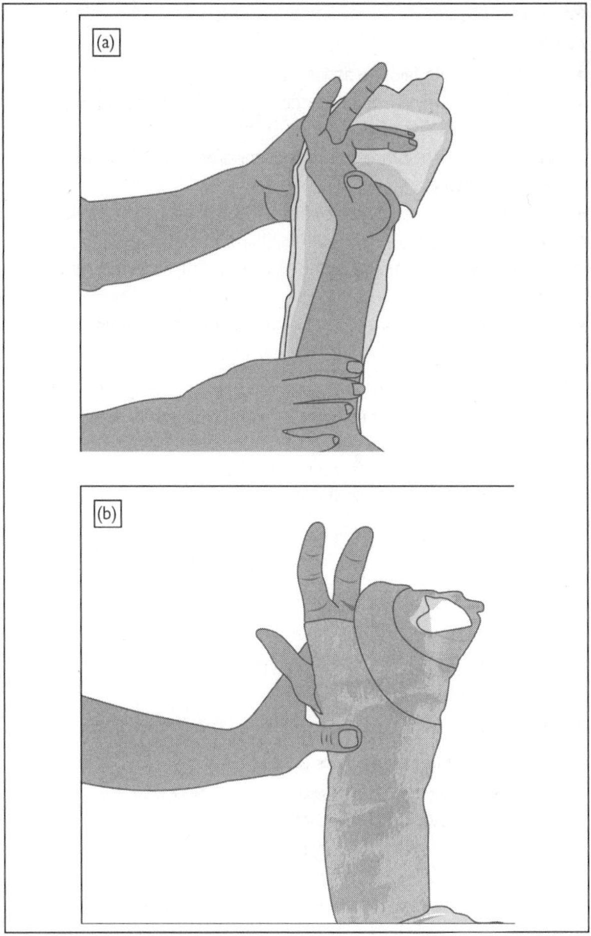

(a)

(b)

Figure 2.11 Ulnar gutter splint. a. Applying the measured ulnar gutter splint. b. Finished ulnar gutter splint.

6. Apply the plaster from just distal to tips of small and ring fingers palmarly along palm to proximal forearm.
7. Apply Sof-Roll and Ace wrap(s) to secure plaster.
8. Mold splint to position wrist in 30 degrees dorsiflexion and flex MCPs to 90 degrees (IPs in extension).
9. Apply sling.

Pearls and Pitfalls

Pearls

- Having a knowledgeable assistant is helpful.
- If tip of thumb not visible over dorsum of the splinted fingers, need more MCP flexion
- Do not worry about positioning of hand/wrist until plaster secured in place.
- Wrist in 30 degrees dorsiflexion and MCP joints 90 degrees
- For associated proximal/middle phalanx fractures, buddy-tape affected digit to adjacent digit before splinting to help maintain reduction.
- Can use thumb to contour/mold plaster into the palm.

Pitfalls

- The hand position is the same as the intrinsic plus splint and like an intrinsic plus splint, obtaining the correct position is difficult. Using cold water can give extra time to obtain idealized positioning. Check the X-ray after casting to make sure there is adequate padding and acceptable positioning.

Radial Gutter Splint

Gear List

- Two 3 or 4 inch plaster rolls
- Two Sof-Roll
- One or two 4 inch Ace wraps

Patient Positioning

- Patient seated with elbow resting on a Mayo stand or table

Preparing the Splint (see Figure 2.12)

- Measure plaster from just distal to palmar tip of the long finger along palm to proximal forearm.
- Roll out 10–12 layers of plaster.
- Measure Sof-Roll to be 1 inch longer than the plaster at both ends (should have six layers for the skin side and two layers for the outer surface).
- Cut triangle out of the plaster and sof-roll to go around the thumb.

Technique (see Figure 2.12)

1. Dunk plaster in cool water.
2. Squeeze out excess water.
3. Stretch and smooth plaster for laminating.
4. Plaster onto premeasured padding.
5. Apply top layer of Sof-Roll.
6. Apply the plaster from just distal to tips of index and middle fingers palmarly or dorsally to the proximal forearm, immobilizing the radial hand and digits.

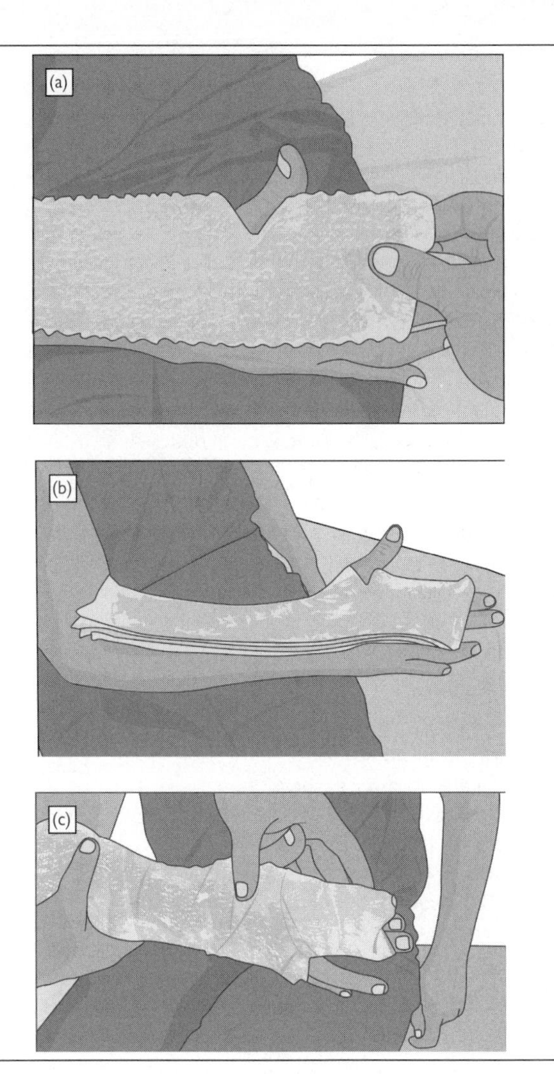

Figure 2.12 Radial gutter splint without thumb spica. a. Cutout for thumb. b. Applying the splint. c. Finished splint with Ace wrap.

7. Apply Ace wrap(s) to secure plaster.
8. Mold splint to position wrist in 30 degrees dorsiflexion and flex MCPs to 90 degrees (interphalangeal joints in extension).

9. Thumb should be freely mobile.
10. Apply sling.

Pearls and Pitfalls

Pearls

- Having a knowledgeable assistant is helpful.
- Do not worry about positioning of hand/wrist until plaster secured in place.
- Wrist in 30 degrees dorsiflexion and MCP joints 90 degrees
- For associated proximal/middle phalanx fractures, buddy-tape affected digit to adjacent digit before splinting to help maintain reduction.
- Can use thumb to contour/mold plaster at distal metacarpals

Pitfalls

- Maintain wrist in neutral to 30 degrees of dorsiflexion.
- The hand position is the same as the intrinsic plus splint and like an intrinsic plus splint, obtaining the correct position is difficult. Using cold water can give extra time to obtain idealized positioning.

Thumb Spica Splint with or without Radial Gutter

Gear List

- Four 3 or 4 inch plaster rolls
- Two Sof-Rolls
- One or two 4 inch Ace wraps

Patient Positioning

- Patient seated with elbow resting on a Mayo stand or table

Preparing the Splint

- Radial gutter: Measure as above.
- Thumb spica: Measure from tip of thumb to midforearm.
- Roll out 10–12 layers of plaster for each.
- Measure Sof-Roll to be 1 inch longer than the plaster at both ends:
  - Sof-Roll should have six layers for the skin side.
  - Sof-Roll should have two layers for the outer surface.

Technique (see Figure 2.13)

1. Dunk plaster in cool water.
2. Squeeze out excess water.
3. Stretch and smooth plaster for laminating.
4. Plaster onto premeasured padding.
5. Apply top layer Sof-Roll.
6. Apply first set of plaster from just distal to tips of index and middle fingers palmarly along palm to proximal forearm.
7. Apply second set of plaster to thumb, extending from dorsum of distal thumb to mid forearm.

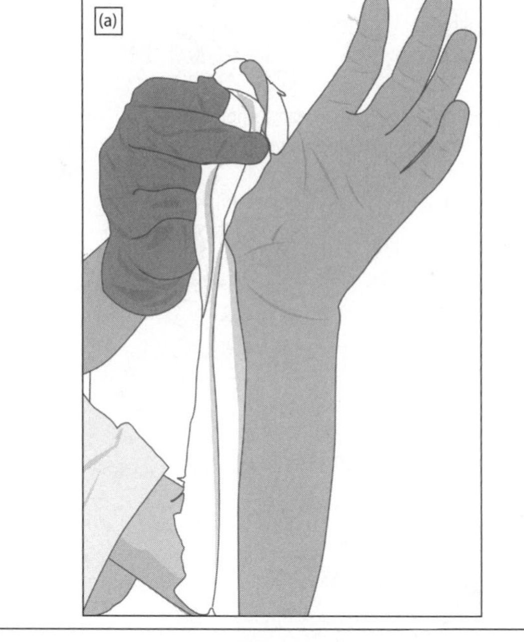

Figure 2.13 Radial gutter splint with thumb spica. a. Applying thumb spica to the hand.

8. Apply ACE wrap(s) to secure plaster.
9. If the radial gutter is utilized, the hand should be positioned with the wrist in 30 degrees of dorsiflexion and the finger MCPs should be flexed to 90 degrees with the IPs in extension. The thumb should have its IP in extension and MCP (metacarpophalangeal) in 0–30 degrees of flexion.
10. Apply sling.

Pearls and Pitfalls

Pearls

- Having knowledgeable assistant is helpful.
- Do not worry about positioning of hand/wrist until plaster secured in place.
- If radial gutter component is utilized, make sure the wrist is in 30 degrees of dorsiflexion and the MCP joints are in 90 degrees of flexion.
- For proximal/middle phalanx fractures, can buddy-tape to adjacent digit before splinting to help maintain reduction

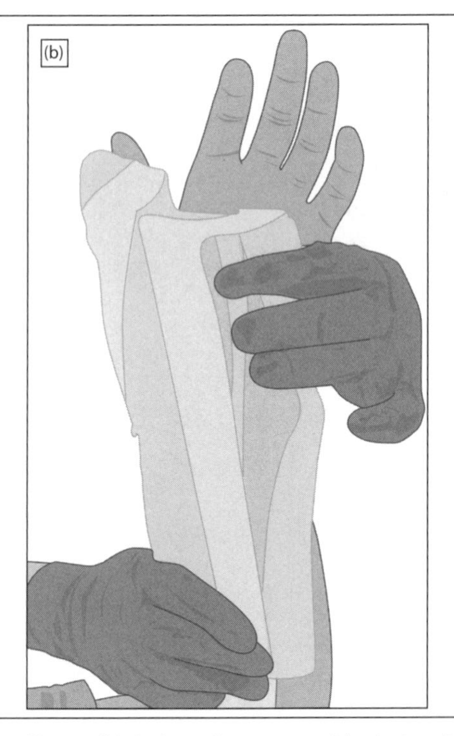

Figure 2.13 (Continued) b. Applying volar component of thumb spica splint.

- Thumb spicas can be applied individually with or without the volar component for isolated thumb injuries.

Pitfalls

- Do not let the plaster around the thumb touch circumferentially.
- As the splint sets, there is a tendency for the thumb spica component to slide proximal relative to the remainder of the splint. If the spica component slides past the interphalangeal joint of the thumb, it is no longer providing the necessary immobilization.
- Maintain wrist in neutral to 30 degrees of dorsiflexion.

Short Arm Cast (SAC)

Gear List

- Three inch stockinette (length of hand and forearm)
- Two rolls 2–3 inch fiberglass
- Two Sof-Roll

Patient Positioning

- Patient seated with elbow resting on a Mayo stand or table

Preparing the Cast

- Measure and cut section of stockinette to extend over hand and forearm.
- Cut hole for thumb in stockinette.
- Apply Sof-Roll over stockinette with each wrap covering 50% of the previous wrap; wrap from distal palmar crease to proximal forearm, leaving 1 cm stockinette uncovered both proximally and distally.

Technique (see Figure 2.14)

1. Place stockinette over hand and forearm (should extend distally over base of fingers).
2. Cut hole for thumb in stockinette.
3. Apply Sof-Roll over stockinette with each wrap covering 50% of the previous wrap; wrap from distal palmar crease to proximal forearm, leaving 1 cm stockinette uncovered both proximally and distally. The Sof-Roll will extend approximately 1 cm past the edge of the fiberglass to protect the skin from the fiberglass.
4. Dip fiberglass in water and squeeze out extra water and apply fiberglass over Sof-Roll, leaving a 1 cm margin of Sof-Roll uncovered both proximally and distally; each fiberglass wrap should cover 50% of previous wrap.
5. Fold excess stockinette over edge of fiberglass to create a cuff (both proximally and distally) and secure with additional one wrap of fiberglass, taking care to ensure that fiberglass does not touch patient's skin.
6. Laminate the edges of fiberglass by rubbing Surgilube or soap across seams until smooth.
7. Bivalve the cast after it dries for most acute fractures, usually on palmar and dorsal aspects of cast.

Pearls and Pitfalls

Pearls

- Bivalve cast to allow for swelling after acute fractures; avoid concavities
- Repeat neurovascular exam after casting.
- Use smaller size Sof-Roll and fiberglass around the hand, such as 2 inch rolls. It is easier to handle and less likely to be touching the skin when completed.

Pitfalls

- Do not allow the fiberglass to touch patient's skin. The first webspace (between the thumb and index finger) is a notoriously challenging region to cast, and fiberglass will often be touching

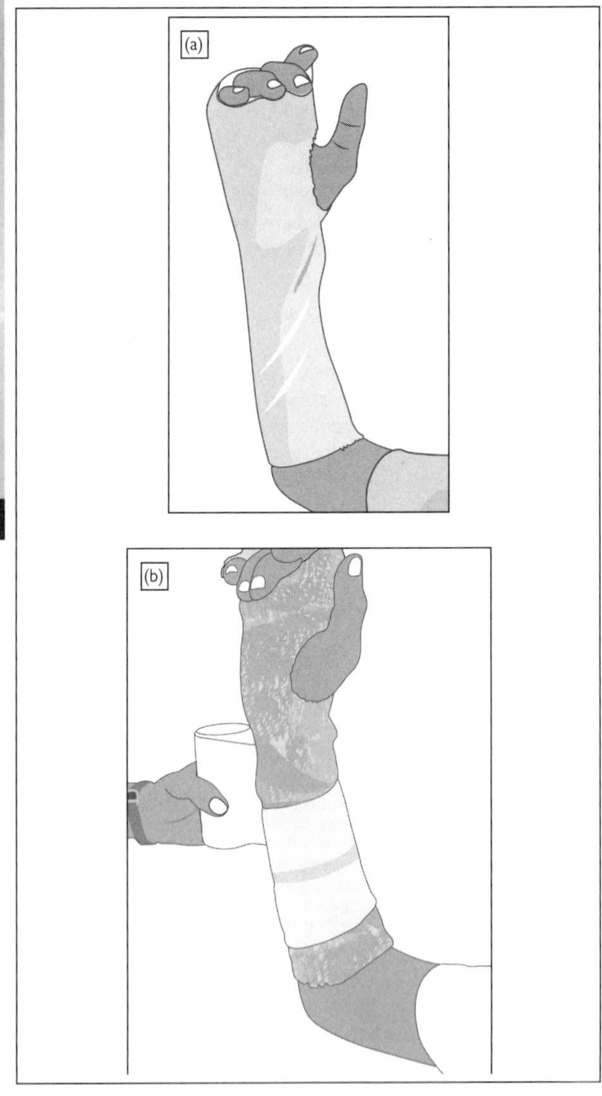

Figure 2.14 Short Arm Cast. a. Cutting hole for thumb. b. Wrapping Sof-Roll 50–50.

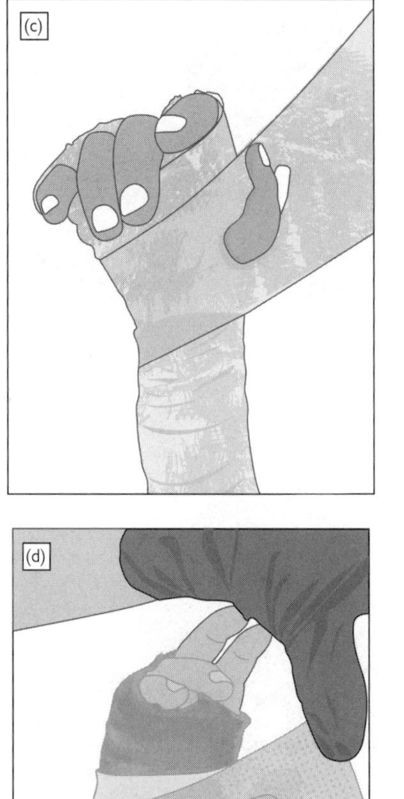

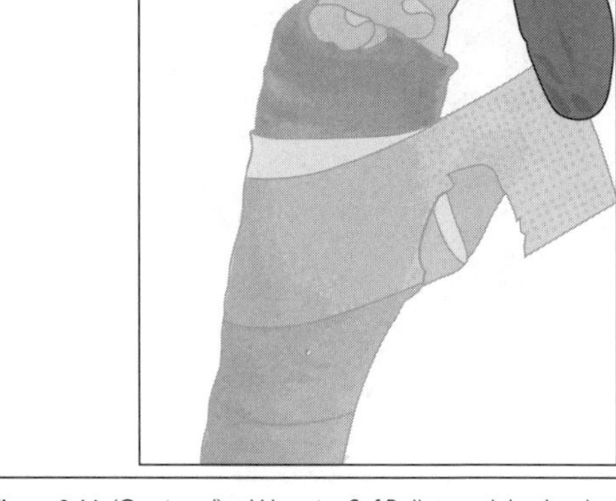

Figure 2.14 (Continued) c. Wrapping Sof-Roll around the thumb. d. Fiberglass around the thumb.

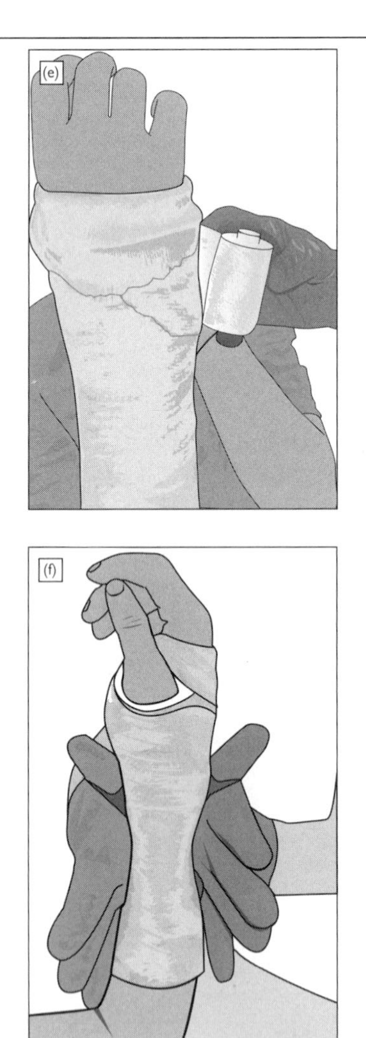

Figure 2.14 (Continued) e. Folding down the stockinette around the thumb and hand. f. Finished SAC and molding the cast.

the skin. Always check every cast to make sure the skin is protected.

- Do not apply fiberglass distal to the metacarpophalangeal joints. The patient should be able to grip when cast is completed. Check the X-ray after casting to make sure there is adequate padding and acceptable positioning.

Long Arm Cast

Gear List

- Stockinette appropriate for size
- Three rolls fiberglass
- Three Sof-Rolls

Patient Positioning

- Patient supine with shoulder just off the edge of bed and index and middle fingers in finger traps suspended from IV pole

Preparing the Cast

- Measure and cut section of stockinette to extend over hand and forearm above elbow to level of axilla.
- Place stockinette on hand up to axilla.
- Cut hole for thumb in stockinette.
- Apply Sof-Roll over stockinette with each wrap covering 50% of the previous wrap; wrap from distal palmar crease to upper arm, leaving 1 cm stockinette uncovered both proximally and distally.

Analgesia (see Chapter 8, Analgesia—Refer to Appropriate Page)

- Hematoma block and/or systemic

Technique (see Figure 2.15)

1. Measure and cut section of stockinette to extend over hand and forearm above elbow to level of axilla.
2. Cut hole for thumb in stockinette.
3. Position elbow in appropriate flexion and forearm rotation (based on fracture type); standard long arm splint is elbow 90 degrees and forearm in neutral rotation (thumb in same plane as arm).
4. Apply Sof-Roll over stockinette with each wrap covering 50% of the previous wrap; wrap from distal palmar crease to upper arm, leaving 1 cm stockinette uncovered both proximally and distally.
5. Add extra padding at posterior elbow but avoid bunching at elbow.
6. Dip fiberglass in water and squeeze out extra water and apply fiberglass over Sof-Roll, leaving a 1 cm margin of Sof-Roll uncovered both proximally and distally; each fiberglass wrap should cover 50% of previous wrap.
7. Fold excess stockinette over edge of fiberglass to create a cuff (both proximally and distally) and secure with additional wrap of fiberglass, taking care to ensure that fiberglass does not touch patient's skin.

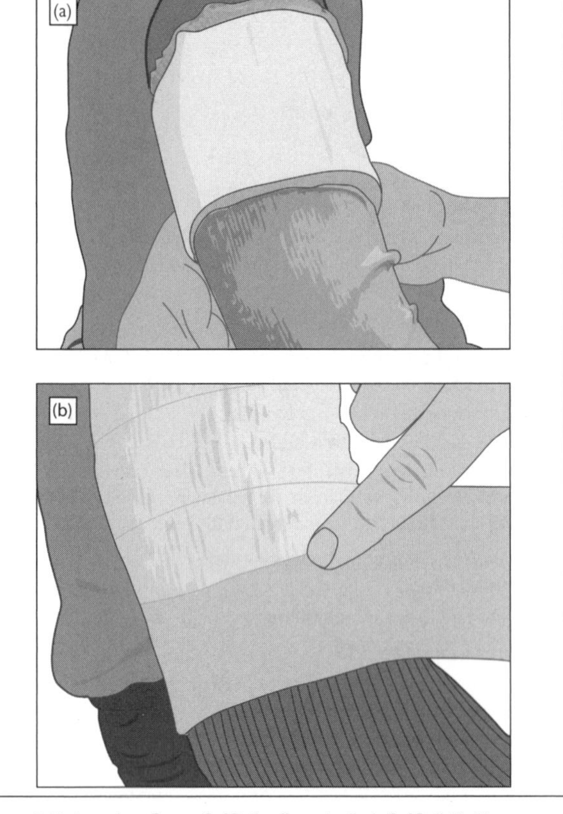

Figure 2.15 Long Arm Cast. a. Sof-Roll cuff proximally. b. Sof-Roll 50–50 wrap.

8. Laminate the edges of fiberglass by rubbing Surgilube or soap across seams until smooth.
9. Apply interosseous mold distally and mold proximally, medially, and laterally above condyles. Having a mold above the condyles will prevent slipping of cast over the condyles.
10. Recommend bivalving cast after dry for most acute fractures.

Pearls and Pitfalls

Pearls

- Maintain elbow at appropriate flexion and rotation (determined by fracture type) throughout casting process

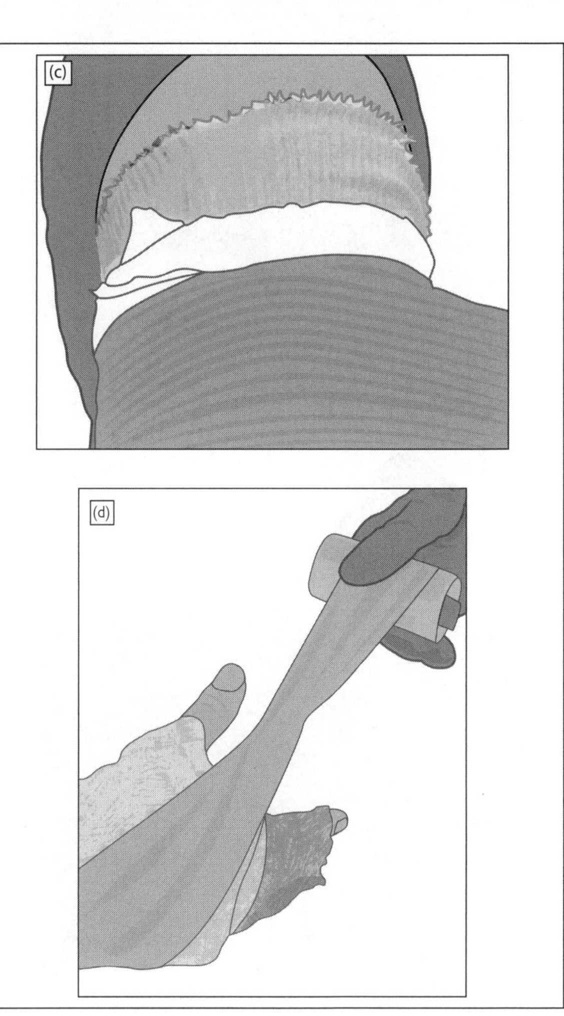

Figure 2.15 (Continued) c. Rolling fiberglass. d. Rolling fiberglass between the thumb and index finger.

(e.g., flexed for extension type supracondylar fracture, extended for flexion type).

- Always use 50–50 overwrap technique.
- Ensure appropriate mold both proximally and distally, especially in children, to minimize mobility of cast; distal mold should be an interosseous mold.

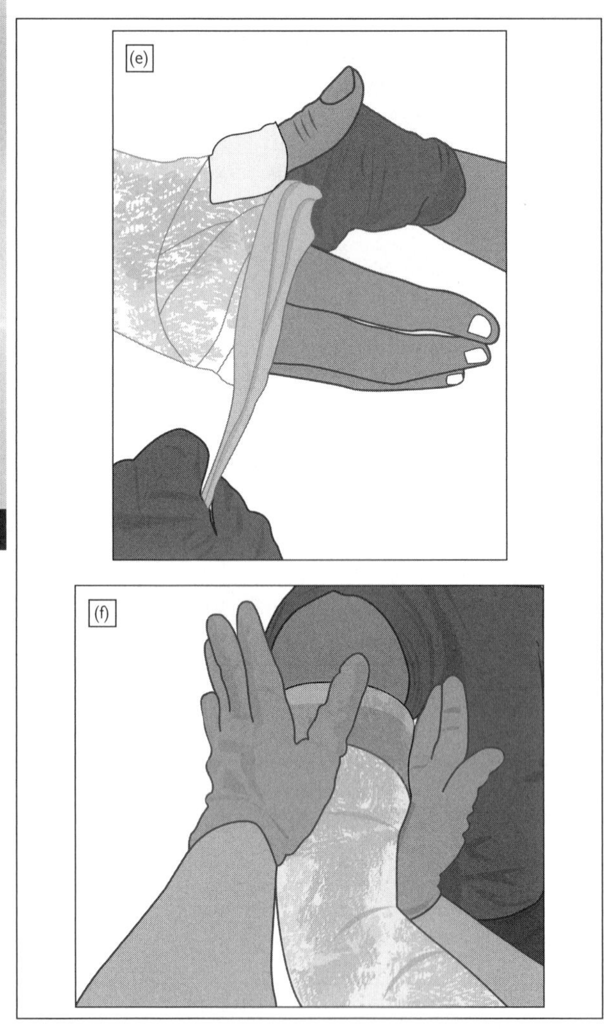

Figure 2.15 (Continued) e. Rolling down the stockinette around the hand. f. Proximal mold.

- Bivalve at 90 degrees from flexion creases (at antecubital fossa and volar wrist) to avoid risk to soft tissues. Do not cut in concave parts of the cast.
- Repeat neurovascular exam after casting and after bivalve.

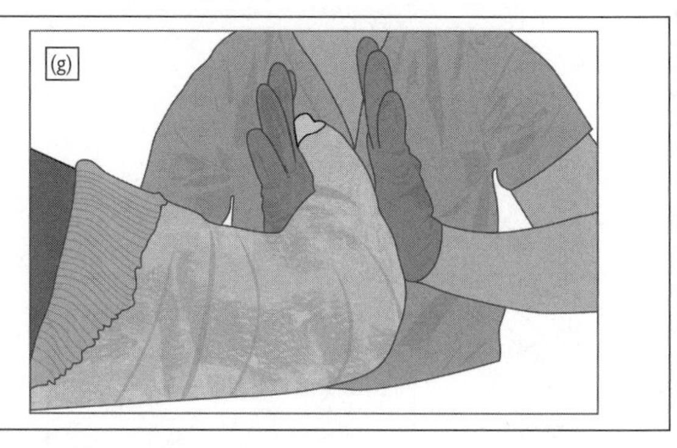

Figure 2.15 (Continued) g. Distal interosseous mold.

Pitfalls

- Failing to have enough layers of Sof-Roll at the beginning and end of splint
- Adjusting angle/position of elbow or wrist after starting to wrap will create unwanted folds
- Not adequately padding the olecranon
- Placing adults acutely in a circumferential long arm cast
- Allowing the fiberglass to touch patient's skin. The first webspace (between the thumb and index finger) is a notoriously challenging region to cast, and fiberglass will often be touching the skin. Always check every cast to make sure the skin is protected.
- Applying fiberglass distal to the metacarpophalangeal joints. The patient should be able to grip when cast is completed. Check the X-ray after casting to make sure there is adequate padding and acceptable positioning.

Splinting and Casting the Lower Extremity

Long Leg Posterior Slab Splint

Gear List

- Six 5 inch rolls of plaster
- Four 6 inch Sof-Rolls
- Four 6 inch Ace wraps

Patient Positioning (see Figure 2.16)

- Patient supine in Trendelenberg
- Bump (folded blanket or sheet) under buttock on affected side
- Suspend foot and/or leg with Kerlix from IV pole(s); IV pole should be at lowest height to allow for adjustment.

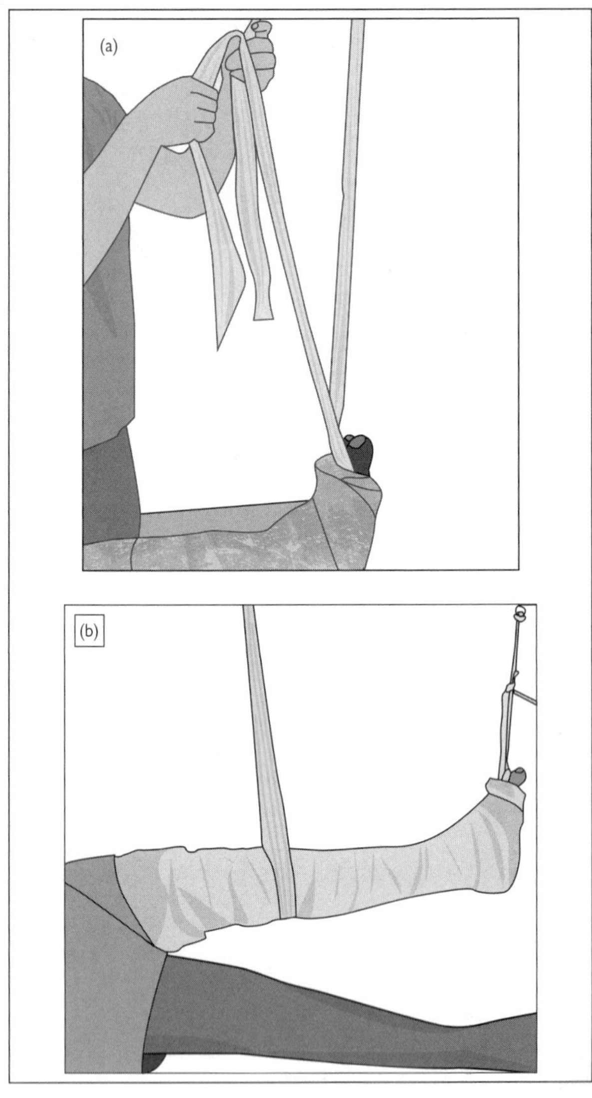

Figure 2.16 Long Leg Bulky. a. Hanging the leg from the toe. b. Hanging the leg by dropping bed from Trendelenburg position.

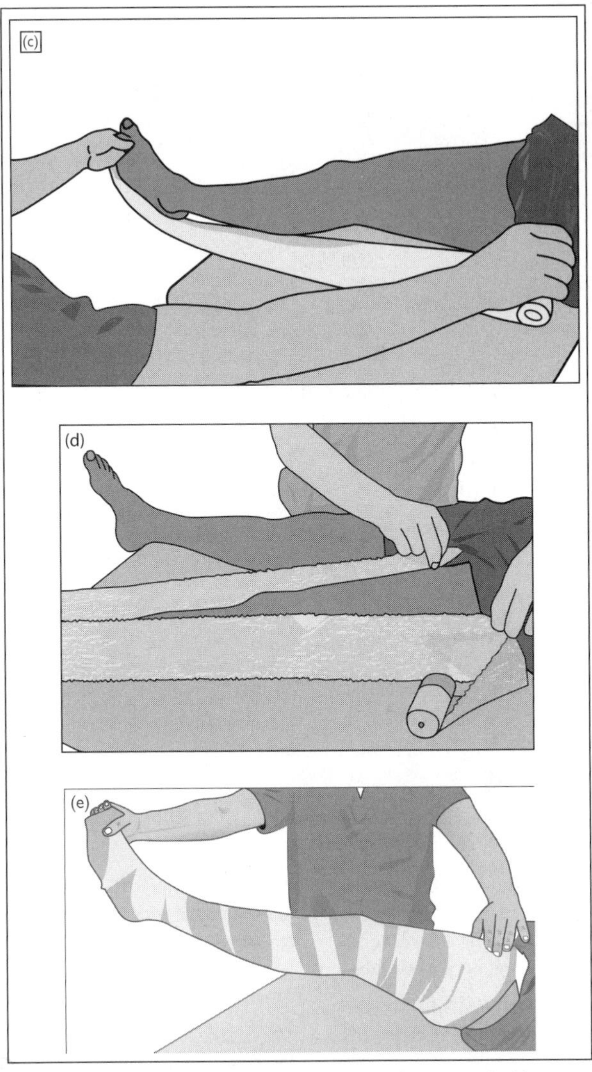

Figure 2.16 (Continued) c. Measuring posterior slab. d. Measuring side slab. e. Wrapping Sof-Roll.

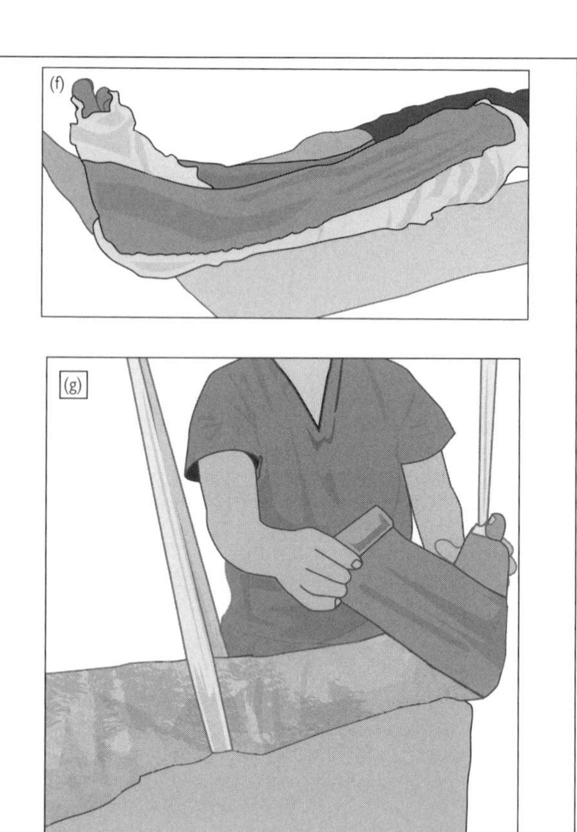

Figure 2.16 (Continued) f. Applying the posterior slab. g. Applying the Ace wrap to the long leg bulky.

- An alternative to suspending the foot with kerlix is to have a qualified assistant hold the leg and support the foot.
- Once the leg is secured to the pole, slowly take bed out of Trendelenberg to elevate the leg.

Preparing the Splint

- Measure posterior slab of plaster from distal foot to proximal thigh (crease between buttocks and thigh) (10–12 layers).
- Measure U component of plaster from medial thigh to lateral thigh (greater trochanter) (10–12 layers).

Technique

1. Apply one layer of Sof-Roll from end of foot to proximal thigh.

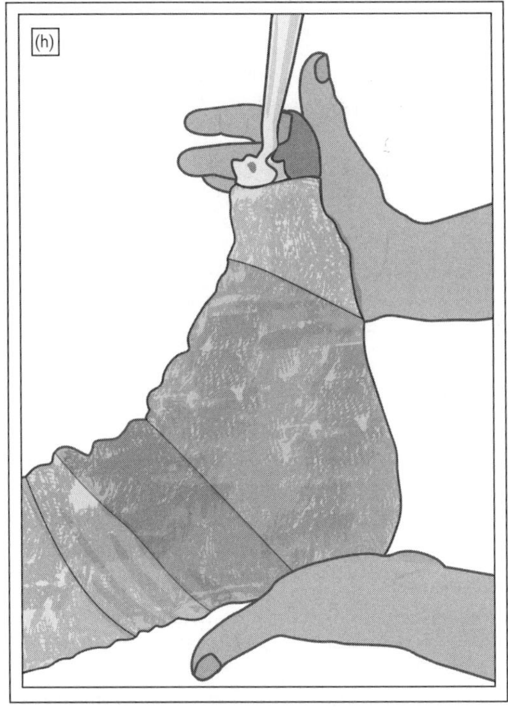

Figure 2.16 (Continued) h. Dorsiflexion of the ankle.

2. Apply bulky Jones cotton from foot to proximal thigh.
3. Apply overwrap of Sof-Roll to the bulky Jones cotton. Use minimal amount of Sof-Roll.
4. Dunk plaster in cool water.
5. Squeeze out excess water.
6. Stretch and smooth plaster for laminating.
7. Apply posterior slab onto leg from foot up to gluteus maximus; assistant can hold posterior slab while "U" is applied. Apply the "U" from the greater trochanter along the lateral margin of the leg around the bottom of the foot and back up the medial aspect of the leg, analagous to a big stirrup.
8. Position ankle in neutral dorsiflexion and hold at 90 degrees.
9. Apply outer layer of Sof-Roll.
10. Secure with Ace wraps.
11. Make sure the foot is held in neutral and knee in extension. Avoid equinus positioning.

Pearls and Pitfalls

Pearls

- Need to extend splint all the way to gluteus maximus
- Ensure adequate padding at proximal plaster.

Pitfalls

- Avoid equinus position (plantar flexion).
- Attempting this splint alone especially without much experience is very difficult.

Bulky Jones Splint for the Ankle

Gear list

- Three or four 5 inch rolls of plaster
- Three 6 inch Sof-Rolls
- Two 4–5 inch Ace wraps

Patient Positioning (see Figure 2.17)

- Prone with knee flexed if the fracture is stable and does not require active maintenance of the reduction
- Patient sitting upright with knee flexed or in figure 4 position *if the fracture requires reduction*

Preparing the Splint

- Measure posterior slab of plaster from distal foot to proximal leg. This should be just below the popliteal fossa when the knee is flexed (10–12 layers).
- Measure U component of plaster from medial to lateral knee (10–12 layers).

Technique

1. Apply one layer of Sof-Roll from end of foot to proximal leg.
2. Apply bulky Jones cotton from foot to proximal thigh.
3. Apply overwrap of Sof-Roll to the bulky Jones cotton. Use minimal amount of Sof-Roll.
4. Dunk plaster in cool water.
5. Squeeze out excess water.
6. Stretch and smooth plaster for laminating.
7. Plaster onto leg; assistant can hold posterior slab while U (stirrup) is applied.
8. Position ankle in neutral dorsiflexion and hold.
9. Apply outer layer of Sof-Roll.
10. Secure with Ace wraps.
11. Hold mold until plaster is hard. There are multiple techniques for molding the cast in 90 degrees of dorsiflexion.

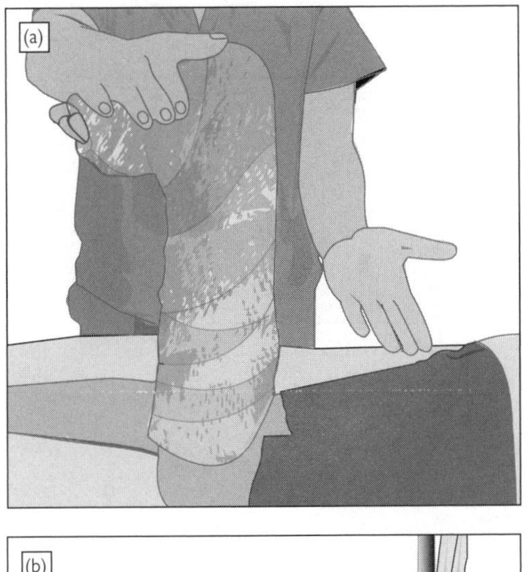

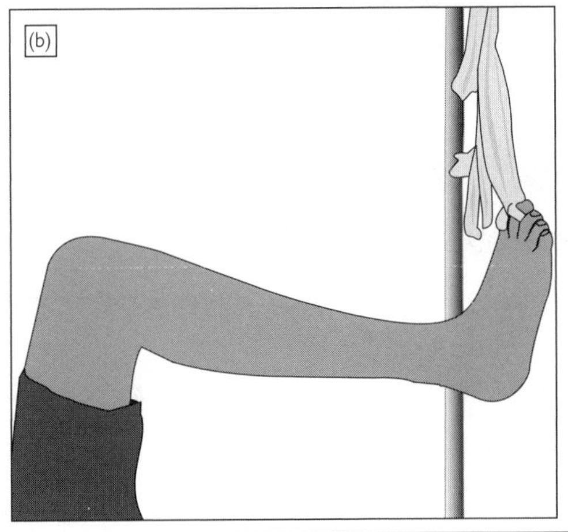

Figure 2.17 Bulky Jones posterior slab splint for the ankle. a. Prone positioning. b. Figure of 4 positioning. The IV pole should be positioned by the patient's contralateral shoulder, so that the patient's legs look like a "4" when viewed from above.

Pearls and Pitfalls

Pearls

- Pad heel, malleoli, and Achilles as they are often the points of skin breakdown.
- Always check splint in full knee flexion to avoid impingement at posterior knee.

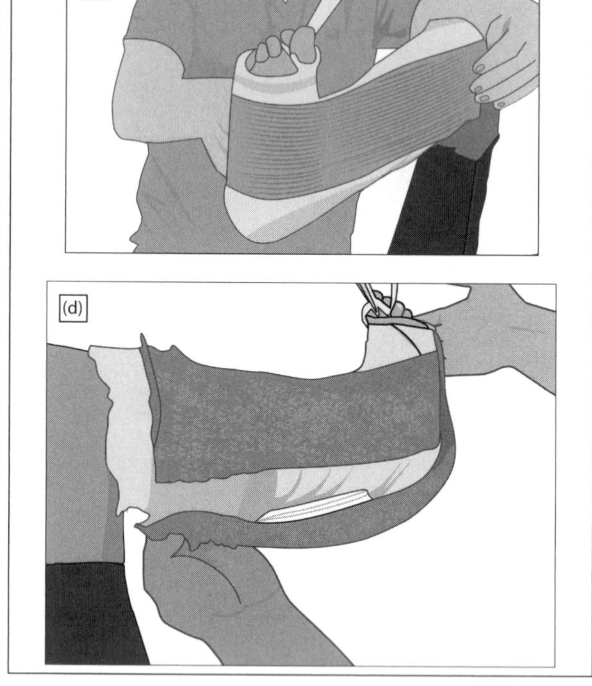

Figure 2.17 (Continued) c. Measuring side component. d. Application of plaster.

Pitfalls

- Avoid equinus position (plantar flexion). Equinus can only be used if it is needed to maintain the talus in the ankle joint. Fractures that require equinus to maintain reduction are by definition unstable fractures that will require surgery. An equinus position is never acceptable in a fracture that will be treated nonoperatively. Check the X-ray after casting to make sure there is adequate padding and acceptable positioning.

Long Leg Cast

Gear List

- Appropriately sized stockinette to patient's leg
- Four small, three medium, and three large Sof-Rolls (for pediatrics, 2, 3, and 4 inch. For adult sized legs, 3, 5, and 6 inch)
- Four small, three medium, and three large fiberglass (see above)
- Two or three large elastic wraps (if bivalving)

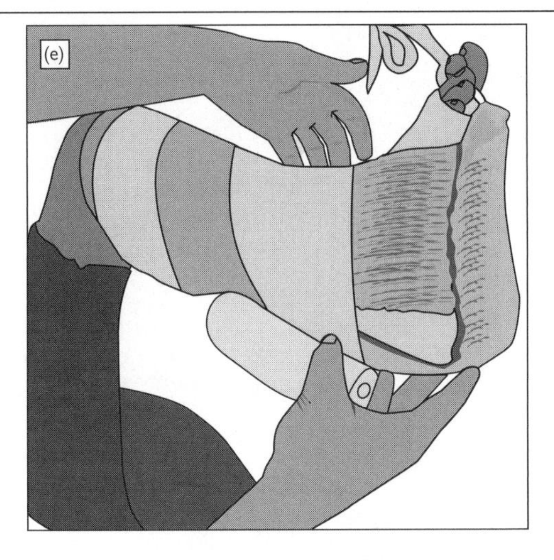

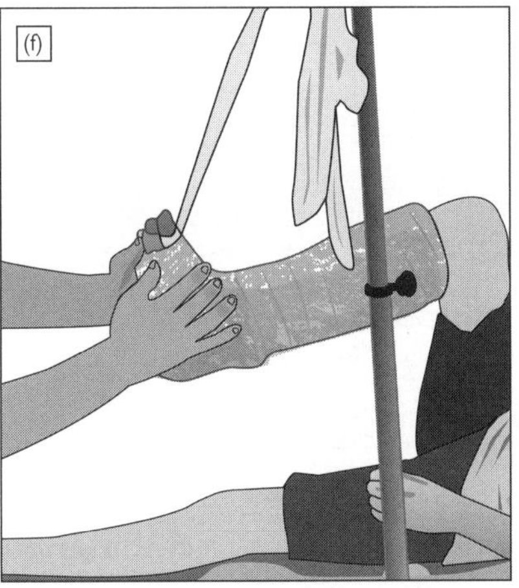

Figure 2.17 (Continued) e. Wrapping Sof-Roll over plaster. f. Technique #1 for dorsiflexion at 90.

Patient Positioning

- Supine with a bump under the affected side hip.
- You will need a qualified assistant. S/he should put on appropriate gloves to protect from fiberglass before the leg is ever elevated as s/he will need to hold the position the entire time.

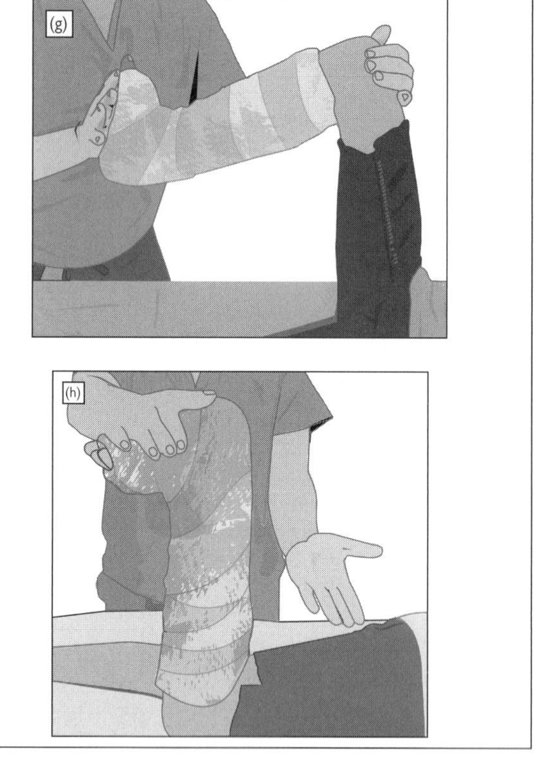

Figure 2.17 (Continued) g. Technique #2 for dorsiflexion at 90. h. Technique #3 for dorsiflexion at 90.

- Some casts require full extension for the injury (see Chapter 4, "Lower Extremity Injuries," under specific fracture injury section); if they do not, the following guidelines can be used:
  - Walking children: Knee at 45, ankle at 90 so toes clear floor when child is using crutches
  - Crawling children: Knee 70–90, ankle 90 with an extended toe box so child can crawl in cast
  - Adults: Knee at 0, ankle at 90. Bending the knee can prevent noncompliant patients from ambulating on the cast, the trade-off for doing this may be a flexion contracture.

Preparing the Cast

- Obtain all materials prior to applying any layers.
- Ensure materials are all at arms reach from the extremity.

- Position and reduce the injury prior to beginning and then do not move from that position until the rigid material has hardened.

Technique (see Figure 2.18)

1. Measure the stockinette and place over extremity. Stockinette should start in groin and end just beyond toes. If the proximal thigh and leg are drastically different in size, two sizes of stockinette can be used and overlapped at the knee. Ensure there are no folds in the popliteal fossa or anterior ankle.
2. Starting with the small size of padding, wrap from the toes proximally in a 50–50 technique. (Each Sof-Roll should cover 50% of previous wrap.)
3. Increase the size of the roll of padding as you move proximally.
4. Ensure a thicker cuff of padding at the very distal and proximal end of the cast.
5. The anterior knee and heel often become thin in padding; they may require that an extra pad be placed directly over these prominences.
6. Dunk the fiberglass in cool water starting with the small size and roll from the toes proximally.
7. Always leave at least 1 cm extra padding visible at the proximal and distal ends of the cast.
8. Use larger rolls as you progress proximally.
9. Use a 50–50 technique with the fiberglass (each fiberglass wrap should cover 50% of previous wrap).
10. Places enough layers from toes proximally to ensure adequate rigidity.
11. Fold ends of stockinette over fiberglass to give padded end and cover with last layer to secure.
12. Mold if necessary. Hold the foot in a dorsiflexed position as the cast hardens.
13. Bilvalve only when completely dry if necessary (in most cases) and cover with elastic wrap.

Pearls and Pitfalls

Pearls

- Casts placed for knee extensor mechanism injuries or tibial spine avulsion should be at full extension.
- If reliable help is not available or if the fracture is very unstable and maintenance of position during the entire duration of cast placement is impossible, the provider can start by making a short leg cast, allowing it to dry, and then extending above the knee into a long leg cast.
- Performing a supracondylar medial and lateral mold will minimize your cast's likelihood of slipping down or off the leg.

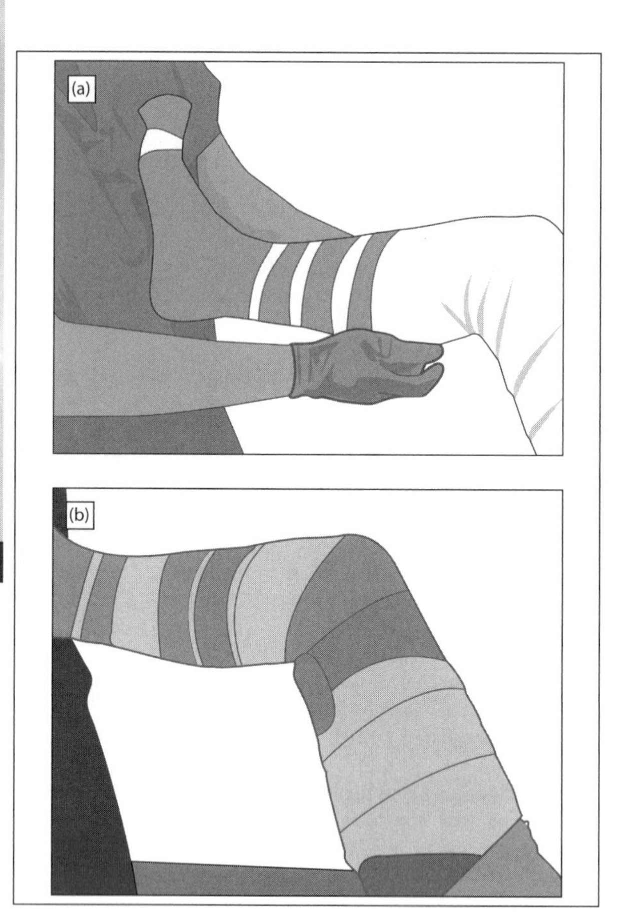

Figure 2.18 Long leg cast. a. Application of fiberglass distally. b. Application of fiberglass about knee.

Create the mold by pressing in on the medial and lateral aspect of the cast with your palms just proximal to the metaphyseal flare.

Pitfalls

- Because this takes time and a lot of effort to complete, the assistant often will have difficulty not moving the extremity through the entire procedure. It is your job to continually check positioning and remind them that any movement is potentially a problem.

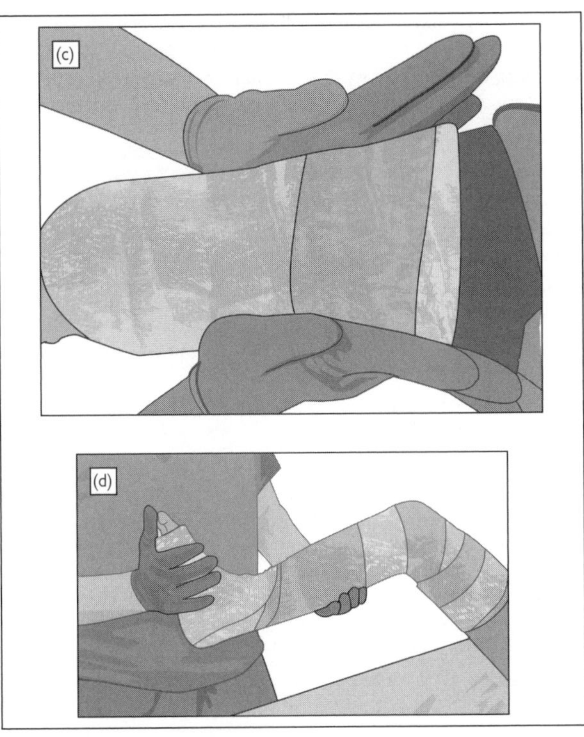

Figure 2.18 (Continued) c. Proximal supracondylar mold. d. Dorsiflexed position.

- The knee and heel are difficult to pad; check these areas before placing fiberglass.
- Check the X-ray after casting to make sure there is adequate padding and acceptable positioning.

Short Leg Cast

Gear List

- Appropriately sized stockinette to patients leg
- Two small and two medium Sof-Rolls (for pediatrics, 2 and 3 inch. For adult-sized legs, 3 and 5 inch)
- Three small and two medium fiberglass (see above)
- One large elastic wrap (if bivalving)

Patient Positioning

- For unstable fractures supine is usually preferred. Techniques to aid in maintaining reduction of unstable fractures are: hang leg in figure 4 position, dangle leg off bed with knee bent, or have qualified assistant hold reduction with knee bent.

- For stable fractures, any of the above techniques work in addition to placing the patient prone with the knee bent to 90 degrees.

Preparing the Splint

- Obtain all materials prior to applying any layers.
- Ensure they are all at arm's reach from the extremity.
- Position and reduce the injury prior to beginning and then do not move from that position until the rigid material has hardened.

Technique (see Figure 2.19)

1. Measure the stockinette and place over extremity. Stockinette should start at inferior pole of the patella and end just beyond toes. Ensure there are no folds at the anterior ankle.
2. Starting with the small size of padding, wrap from the toes proximally in a 50–50 technique.
3. Increase the size of the roll of padding as you move proximally.
4. Ensure a thicker cuff of padding at the very distal and proximal ends of the cast.
5. The heel, and occasionally the malleoli, often becomes thin in padding; may require that an extra pad be placed directly over these prominences
6. Dunk the fiberglass in cool water starting with the small size and roll from the toes proximally.
7. Always leave 1 cm extra padding visible at the proximal and distal ends of the cast.
8. Use larger rolls as you progress proximally.

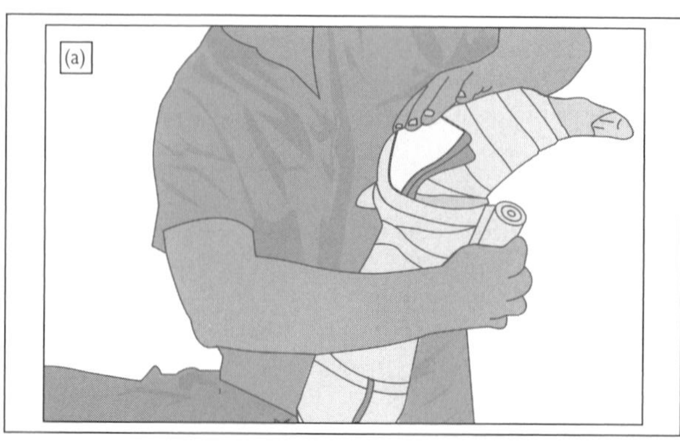

Figure 2.19 Short leg cast. a. Padding the heel with layers of Sof-Roll.

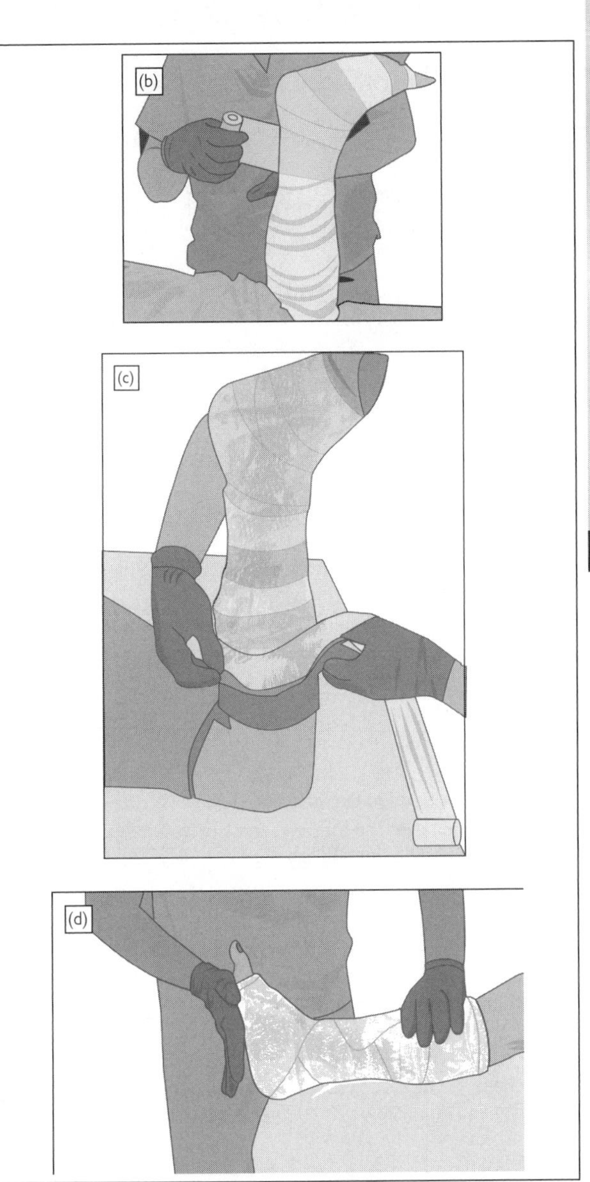

Figure 2.19 (Continued) b. Application of fiberglass about the ankle. c. Folding down the stockinette proximally. d. Dorsiflexed position 1.

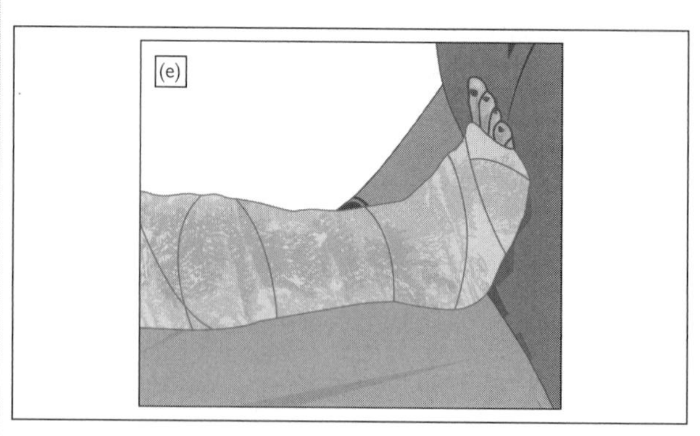

Figure 2.19 (Continued) e. Dorsiflexed position 2.

9. Use a 50–50 technique with the fiberglass (each fiberglass wrap should cover 50% of previous wrap).
10. Place enough layers from toes proximally to ensure adequate rigidity.
11. Fold ends of stockinette over fiberglass to give padded end and cover with last layer of fiberglass to secure.
12. Mold if necessary. Hold heel in dorsiflexion.
13. Bilvalve only when completely dry if necessary (in most situations) and cover with elastic wrap.

Pearls and Pitfalls

Pearls

- The positioning of the joints matters in this cast. Ankles should almost always be at 90 degrees. Avoid equinus.
- If the injury involves the toes, the cast can be extended beyond the distal phalanx to immobilize the toes.

Pitfalls

- Because this takes time and a lot of effort to complete, the assistant often will have difficulty not moving the extremity through the entire procedure. Continually check positioning and remind the assistant that any movement is potentially a problem.
- The knee and heel are difficult to pad; check these areas before placing fiberglass.
- Check the X-ray after casting to make sure there is adequate padding and acceptable positioning.

Chapter 3

Upper Extremity Injuries

Physical Examination of the Upper Extremity

Surface Anatomy

- See Figure 3.1

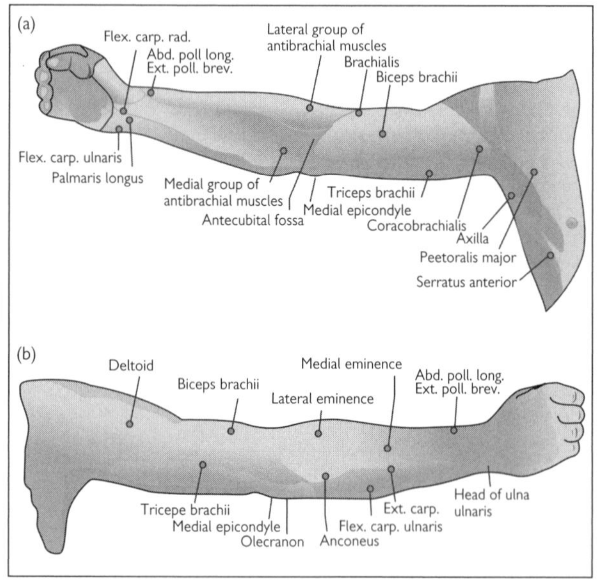

Figure 3.1 Line drawing showing the surface anatomy of the upper extremity, from clavicle to hand. a. Anterior. b. Posterior.

Neurovascular Exam

- Document patient's cutaneous sensation and any deficits. Sensation should be documented by peripheral nerves and any deficits should be noted (see Figure 3.2)
- Vascular exam (use of Doppler for difficult to palpate pulses); capillary refill; always document the radial and ulnar pulses.

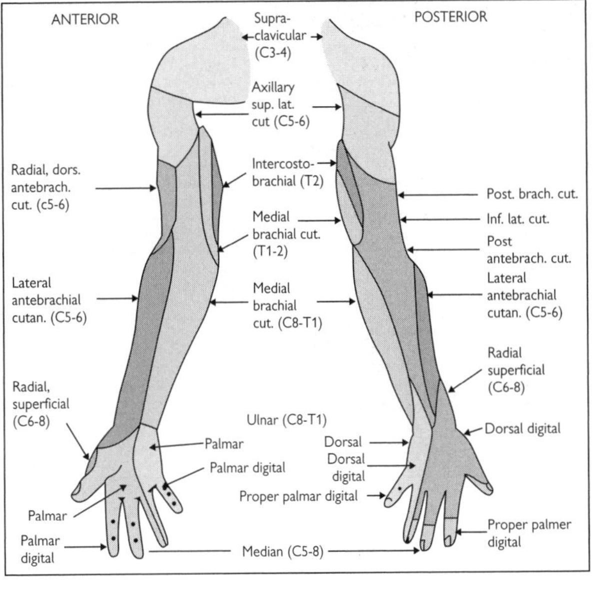

Figure 3.2 Cutaneous innervations of the upper extremity.

Motor Exam

- Motor strength is graded from 0 to 5:
  - Grade 0: No movement
  - Grade 1: Flicker of movement only
  - Grade 2: Movement with gravity eliminated
  - Grade 3: Movement against gravity
  - Grade 4: Movement against resistance
  - Grade 5: Normal power

While motor grading is clearly subjective, common sense should be employed. For example, no one with a humerus fracture should be graded as 5/5 with elbow flexion, as their power will be limited by pain.

Motion	Muscle	Nerve	Root
Shoulder abduction	Deltoid	Axillary	C5
Elbow flexion	Biceps, Brachialis	Musculocutaneous	C5–6
Elbow extension	Triceps	Radial	C7
Wrist extension	ECRL, ECRB, ECU	Radial (PIN)	C6
Wrist flexion	FCR, FCU	Median, Ulnar	C7–8
Finger extension	ED	Radial (PIN)	C7
Thumb extension	EPL	Radial (PIN)	C8
Grip/Finger flexion	FDS, FDP	Median (AIN)	C8
Finger abduction	Interosseous	Ulnar	T1
Thumb abduction	APB	Median	C8

AIN: anterior interosseous nerve, APB: abductor pollicis brevis, ECRL: extensor carpi radialis longus, ECRB: extensor carpi radialis brevis, FCR: flexor carpi radialis, FCU: flexor carpi ulnaris, FDS: flexor digitorum superficialis, FDP: flexor digitorum profundus, PIN: posterior interosseous nerve.

Documenting the Physical Exam

- Skin
  - Example: Skin is intact without breaks, edema or ecchymosis.
- Palpation/Deformity
  - Example: Nontender to palpation of bony prominences. No bony deformity appreciated.
- Pulses
  - Example: 2+ radial and ulnar pulse
- Sensation
  - Example: Sensation intact to light touch throughout the distribution of the axillary, musculocutaneous, median, radial, and ulnar nerves
- Motor
  - Example: Strength is 5/5 for deltoid, biceps, triceps, elbow flexion/extension, wrist flexion/extension, grip, and finger abduction.
- Compartments (there are two compartments in the arm and three in the forearm)
  - Example: Compartments soft and compressible in the arm and forearm

Disposition Disclaimer

All orthopaedic injuries require follow-up. Some injuries require closer and more timely follow-up than others.

Shoulder and Humerus Injuries

Clavicle Fractures

Symptoms and Findings

- History of trauma: A fall onto the involved shoulder, a direct blow, or falling onto an outstretched hand
- Complains of pain in the clavicle and/or shoulder, will not raise arm secondary to pain
- Tenderness over the clavicle and/or fracture site
- There may be a palpable bony deformity. If the deformity is great enough, the skin may be tented or opened (open fracture).

Imaging (See Figure 3.3)

- X-rays: AP and 15 degree cephalad-oblique radiograph
- CT scan is not necessary

Classification

- Fractures are generally classified according to location within the clavicle—middle third (80% of all fractures), distal third (15%), or proximal third (5%)

Primary Stabilization and Management

- Reduction maneuvers are generally not utilized

Admit and Discharge Guidelines

Most clavicle fractures do not require an emergent orthopaedic surgery consult.

- Request an orthopaedic consult in the emergency department for:

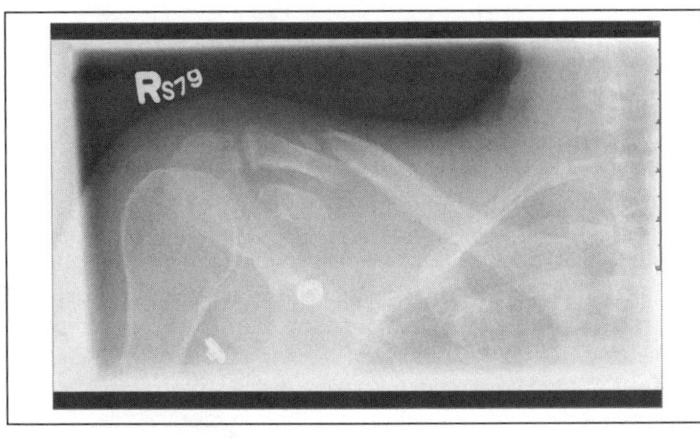

Figure 3.3 Clavicle fracture X-ray.

  - Fractures with significant tenting of the skin, which may cause skin breakdown
  - Open fractures, and fractures associated with neurovascular compromise
- Refer to an orthopaedic surgeon for outpatient follow-up for:
  - Displaced middle-third fractures
  - Displaced lateral-fifth fractures
  - Painful nonunions

Definitive Treatment

- Clavicle fractures are generally treated with a sling for 2–4 weeks with active range of motion exercises of the hand, wrist, and elbow. Extended use of a sling should be discouraged due to the risk of shoulder stiffness.
- Surgery is rarely indicated or necessary. Indications for surgery include open fractures, some displaced and/or angulated fractures, and painful nonunions. ORIF may be performed with plate and screws or tension wire, and intramedullary fixation has also been described.

Acromioclavicular Separations

Symptoms and Findings

- History of a fall onto the acromion or shoulder, typically with the arm adducted and often during contact sports such as hockey, wrestling, or football
- Injury may also occur with a fall onto an outstretched arm.
- Complain of pain in the shoulder or clavicle; pain ranges from minimal to severe
- On exam, patient typically has pain to palpation of the acromioclavicular joint.
- In the more severe injuries, the patient may have a gross deformity of the shoulder with a downward sag of the shoulder and arm.
- While rare, in the most severe AC injuries, patients may also have a pneumothorax, so a thorough primary survey is mandatory.

Imaging (See Figure 3.4)

- X-rays: AP chest, scapular-Y, and axillary views
- A stress view may be required to differentiate between different types of AC injuries, but these are not routinely required in the emergency setting. A stress view is performed by strapping 10–15 lb weights to both wrists and then taking an AP radiograph so that the injured shoulder can be compared to the contralateral side.
- CT scan is generally not necessary

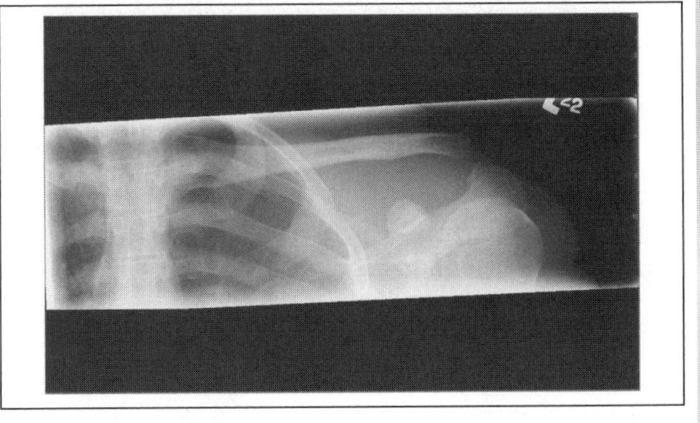

Figure 3.4 AC separation X-ray.

Classification

- Fractures are classified based on the degree of ligamentous injury, and the displacement of the clavicle and treatment is determined by the degree of injury.

Primary Stabilization and Management

Grade	AC ligament	CC ligament	Clavicle displacement	Treatment
I	Sprained but intact	No injury	No displacement	Sling for 1–2 weeks
II	Torn	Sprained but intact	No displacement	Sling for 1–2 weeks
III	Torn	Torn	Dislocated	Sling for 1–2 weeks, ORIF for athletes
IV	Torn	Torn	Dislocated posteriorly through/into trapezius	ORIF
V	Torn	Torn	Clavicle 100% displaced superiorly	ORIF
VI	Torn	Torn	Dislocated inferiorly	ORIF

Note: Grades IV, V, and VI are rare.

Admit and Discharge Guidelines

- Many AC injuries do not require orthopaedic consultation, as uncomplicated AC injuries (I–III) can be treated with a sling with orthopaedic follow-up. Closed reduction is generally not performed, but open reduction is performed for Grade IV–VI injuries.

- Request a general surgery/trauma consult in the emergency department for possible chest tube placement if the patient has a pneumothorax (most common in Grade VI injuries).
- Request an orthopaedic consult in the emergency department for:
  - Injuries with significant tenting of the skin, which may cause skin breakdown
  - Injuries with associated open fractures
  - Injuries associated with neurovascular compromise
  - Injuries that require surgical treatment (Grades IV–VI)
- Refer to an orthopaedic surgeon for close outpatient follow-up for:
  - Painful osteolysis (often seen in power lifters)
  - Development of osteoarthritis

Definitive Treatment

- Uncomplicated AC injuries can be treated with a sling.
- Surgical options include distal clavicle excision, ligamentous reconstruction, or ORIF using metal hardware whereby the clavicle is fixed to the acromion.

Sternoclavicular (SC) Separations

Symptoms and Findings

- Injury may be caused by a direct force on the clavicle most commonly causing a posterior dislocation or an indirect force to the joint when the shoulder is compressed with a lateral force
- Commonly caused by sporting injuries (i.e., football) or motor vehicle collision (MVC)
- Patient presents with pain in the clavicle, may be supporting the affected side with the contralateral arm.
- There may be swelling and tenderness over the SC joint, and pain with shoulder ROM (range of motion).
- The medial clavicle lies over the brachial plexus and major vascular structures, so a thorough neurologic and vascular assessment is required

Imaging

- X-rays: Serendipity view—40 degrees, cephalic tilt AP

Primary Stabilization and Management

- Sprain without dislocation: Treat with ice.
- Anterior dislocation: Closed versus open reduction; may use a figure-of-8 brace if stable once relocated; may use a sling for 1 week if unstable once relocated; ice.

- Posterior dislocation: CT scans should be performed on all posterior dislocations given the proximity of the clavicle to major neurovascular structures. In some cases, open reduction (surgery) will be required.

Admit and Discharge Guidelines

- Request an orthopaedic consult in the emergency department for all anterior and posterior dislocations.
- Posterior dislocations are associated with a high complication rate, including pneumothorax, superior vena cava (SVC) injury, esophageal rupture, thoracic outlet syndrome, and compression of vasculature. If such an injury is found, the general surgery/trauma team should be consulted in the emergency department in addition to orthopaedics.
- As mentioned above, CT scans should be performed on all posterior dislocations given the proximity of the clavicle to major neurovascular structures. Prior to open reduction, an angiogram may be needed for surgical planning, and a cardiothoracic surgeon must be available throughout the surgery.
- Patients with an SC sprain may follow up with orthopaedics on an as-needed basis

Definitive Treatment

- Simple sprains and reduced anterior dislocations are treated with ice and a sling if needed.
- Posterior dislocations may be treated with closed reduction; however, they may require surgery for reduction. If open reduction is needed, a reconstructive procedure is usually performed to ensure the clavicle is held to the sternum.

Scapula Fractures

Symptoms and Findings

- Scapula fractures are relatively uncommon. They are almost always the result of high-velocity accidents such as motorcycle or motor vehicle crashes. As such, patients should be carefully screened for other injuries.
- Patient presents with pain in the scapula or back, may be supporting the affected side with the contralateral arm in an adducted position.
- Often have swelling and tenderness to palpation over the scapula
- A thorough neurovascular examination is needed as concomitant injury to the brachial plexus and vascular structures may occur as a result of the trauma.
- While uncommon, compartment syndrome (see page 292) may occur and should be considered in all scapula fractures.

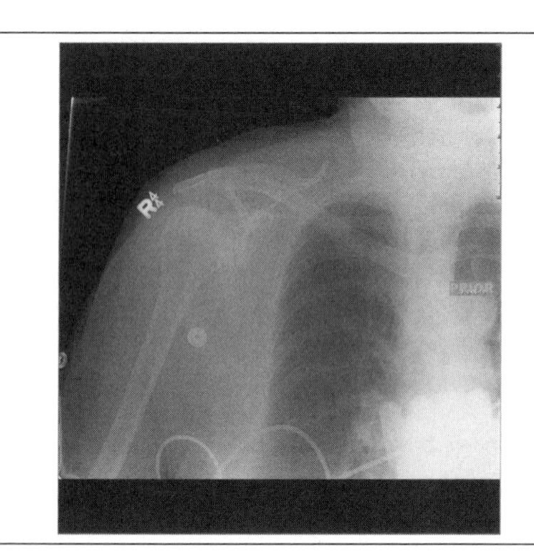

Figure 3.5 Scapula fracture X-ray.

Imaging (See Figure 3.5)

- Often found on a screening chest X-ray as part of the trauma workup
- X-rays: Standard shoulder trauma series (AP, scapular-Y, and axillary views)
- Best imaged by CT scan

Primary Stabilization and Management

- Obtain a CT scan of all scapula fractures as the management of different scapula fractures (scapular body vs. scapular neck vs. glenoid fractures) may vary significantly depending on the extent of injury.

Scapular Body Fractures

- Isolated scapula body fractures do not require immobilization.
- Sling for comfort and early shoulder range of motion exercises to prevent stiffening

Scapular Neck Fractures

- Most scapular neck fractures do not require immobilization.
- Sling for comfort and early shoulder ROM exercises to prevent stiffening

Glenoid Fractures

- Most glenoid fractures do not require immobilization.
- Sling for comfort and early shoulder ROM exercises to prevent stiffening

Admit and Discharge Guidelines

- Scapula fractures are often associated with serious thoracic injury, such as pneumothorax, pulmonary contusion, injury to the great vessels, or spine fractures. Management of these injuries by the trauma team or general surgery team should take priority over any orthopaedic consultation
- Request an orthopaedic consult in the emergency department for: All scapula fractures.
- Refer to an orthopaedic surgeon for outpatient follow-up for: Painful nonunion, chronic scapulothoracic pain, and continued shoulder weakness.

Scapular Body Fractures

- Isolated scapula body fractures do not require treatment beyond a sling for comfort; patients should be advised to perform early range of motion exercises.

Scapular Neck Fractures

- Scapula neck fractures with limited displacement or angulation do not require treatment beyond a sling for comfort; patients should be advised to perform early range of motion exercises.
- Scapula neck fractures with greater than 40 degrees of angulation or 1 cm medial translation may require open reduction and internal fixation.
- Comminuted scapula neck fractures or fractures with concomitant displaced clavicle fractures may require open reduction and internal fixation.

Glenoid Fractures

- Glenoid fractures that are nondisplaced and involve less than 25% of the articular surface do not require surgical treatment; patients should be given a sling for comfort and advised to perform early range of motion exercises.
- Open reduction and internal fixation of glenoid fractures is required if greater than 25% of the articular surface is involved or if there is intra-articular step-off greater than 5 mm.

Definitive Treatment

- Scapula fractures are usually treated with a sling for comfort and early mobilization. Pendulum exercises are performed by having the patient bend at the waist and extend the elbow, allowing the arm to hang perpendicular to the ground. The patient then gently swings the arm in a clockwise and counterclockwise motion, passively moving the shoulder joint.
- Surgical treatment is by open reduction and internal fixation using metal hardware.

Anterior/Posterior/Inferior Shoulder Dislocations

Symptoms and Findings

- Most commonly caused by indirect injuries. Anterior dislocations occur with the shoulder in abduction, extension, and external rotation; posterior dislocations occur with the shoulder in adduction, flexion, and internal rotation
- Injury may be caused by a direct force on the shoulder, and the direction of the force determines the type of dislocation (posteriorly applied force causes an anterior dislocation while an anteriorly applied force causes a posterior dislocation).
- Dislocations are common following seizures or electrical shock. These dislocations are most commonly posterior.
- Inferior dislocations are very uncommon; they occur with a hyperabduction of the humerus, which causes the humeral head to become jammed below the glenoid. Patients usually present with pain in the shoulder and tenderness to palpation of the joint.
- Patients with anterior dislocations will often hold the arm slightly abducted and externally rotated.
- Patients with posterior dislocations often will not have a striking deformity—they often hold the arm in the "sling" position of slight internal rotation and adduction.
- Patients with an inferior dislocation have the most dramatic clinical presentation—they present with the arm pointing upward.
- A thorough neurovascular exam is paramount. The circumflex axillary nerve runs around the head of the humerus and may be injured with dislocations, resulting in partial or complete paralysis of the deltoid.
- It is important to obtain a history of prior dislocations, as some patients may be voluntarily dislocating their shoulder. Additionally, a history of chronic dislocation and/or ligamentous instability may necessitate outpatient follow-up with a shoulder specialist.

Imaging (See Figure 3.6)

- X-rays: Standard shoulder trauma series (AP, Grashey, and axillary views)
- CT scan is not generally needed for diagnostic purposes but should be obtained if there is suspicion of an associated fracture of the scapula, glenoid, or humerus.
- Postreduction X-rays to verify relocation and reevaluate for fractures are mandatory

Primary Stabilization and Management

- Most shoulder dislocations can be treated with closed reduction.
- Prior to attempting closed reduction, it is important to document a complete physical exam, especially if there is

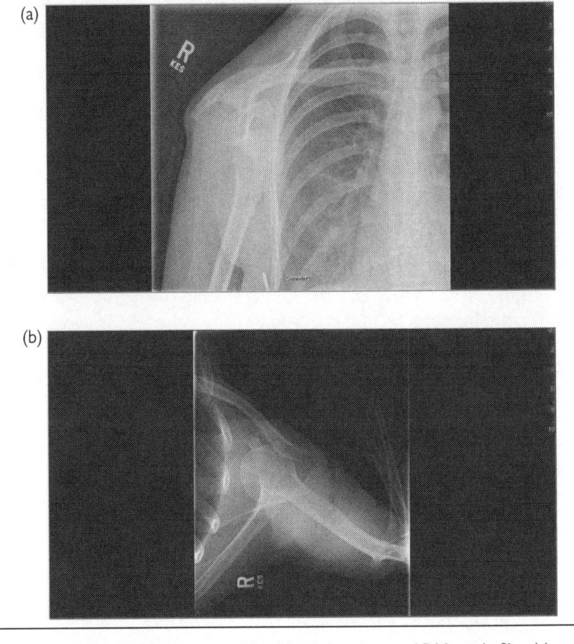

Figure 3.6 Shoulder dislocation a. Shoulder dislocation on AP X-ray. b. Shoulder dislocation on axillary X-ray.

evidence of neurologic damage. The exam should also be repeated and documented following any and every reduction maneuver.

- Radiographs should also be evaluated before reduction to determine if there are other osseous lesions, such as a Hill-Sachs lesion (cortical depression of the humerus), glenoid lip fracture (known as a "bony Bankart lesion"), greater tuberosity fracture, acromion fracture, or coracoid fracture.
- Open reduction may be necessary if soft tissue is interposed within the dislocation. For example, in anterior dislocations the humeral head may "buttonhole" through the subscapularis, prohibiting closed reduction.

How to Perform Closed Reduction of an Anterior Shoulder Dislocation

- Multiple methods exist for reducing an anterior shoulder dislocation. All the methods share two primary features:

mobilization of the humeral head from its perch inferior to the coracoid and reduction of the humeral head into the glenoid.
- The traction–countertraction method requires two people. One person holds a sheet that wraps around the patient's axilla and chest while the other person holds the affected shoulder. Gentle traction and countertraction are applied in opposing axial directions. Once the humeral head begins to move, gentle internal and external rotation combined with abduction will generally reduce the shoulder.
- The Stimson technique is a gentle, one-person reduction maneuver. The patient lies prone on the bed with weight hung from the wrist to mobilize the humeral head through distraction. After the humeral head has begun to move, the arm is gently rotated or rocked to help with the reduction while an anterior force is placed on the scapula. The anteriorly directed force on the scapula anteverts the glenoid, allowing reduction.
- The Milch method may be performed by a single provider. With the patient lying supine on a bed, the provider places one hand on the humeral head and one hand on the patient's forearm. The arm is abducted and externally rotated with one arm until the humeral head is mobile. Then, the humeral head is reduced into the glenoid with the contralateral hand.

Admit and Discharge Guidelines

- Request an orthopaedic consult in the emergency department for:
  - Any dislocation with an associated fracture or neurologic injury
  - An open dislocation, and irreducible dislocations
- Refer to an orthopaedic surgeon for outpatient follow-up for:
- Almost all patients with a shoulder dislocation, who should be seen in follow-up by an orthopaedic surgeon for reevaluation and mobilization with physical therapy.
- Patients who need particularly close follow-up include those with:
  - Rotator cuff tears (a particular concern in patients who cannot actively abduct the shoulder)
  - Minimally displaced greater or lesser tuberosity fracture (may only be seen on postreduction X-rays)
  - Recurrent dislocators

Definitive Treatment

- Closed reduction is the definitive treatment for shoulder dislocations, but staged surgical reconstruction procedures may be required to regain shoulder stability long-term.

Proximal Humerus Fractures

Symptoms and Findings

- Injury is most frequently caused by a fall on an outstretched hand. Other mechanisms of injury include high-energy trauma such as an MVC, direct trauma, electric shock, and seizure.
- Patient presents with pain in the proximal humerus and may be supporting the affected side with the contralateral arm.
- Pain with palpation and shoulder ROM. There may be swelling over the proximal humerus.
- The axillary nerve is frequently injured in a fracture dislocation of the proximal humerus, and a thorough neurovascular exam is mandatory. The usual sensory testing of the axillary nerve (skin sensation just above the deltoid insertion) is unreliable. Nerve injury may present as weakness or numbness but also may present late as delayed recovery of active shoulder motion after dislocation.

Imaging (See Figure 3.7a)

- X-rays: Standard shoulder trauma series (AP, Grashey, and axillary views)

Classification (See Figure 3.7b)

- Neer classification: Classification system based on the number of fracture fragments
- Four possible fracture fragments or "parts": Anatomical neck, surgical neck, greater tuberosity, and lesser tuberosity
- Fractures are considered to be displaced (or in "parts") if the fragments are greater than 10 mm apart or if there is 45 degrees of angulation.
- Classification system guides treatment

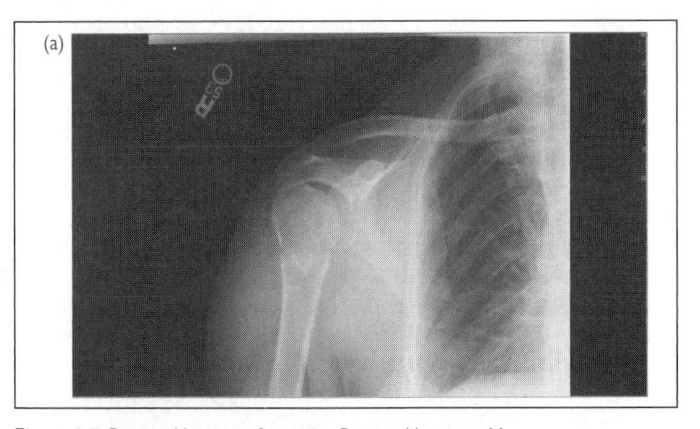

Figure 3.7 Proximal humerus fracture a. Proximal humerus X-ray.

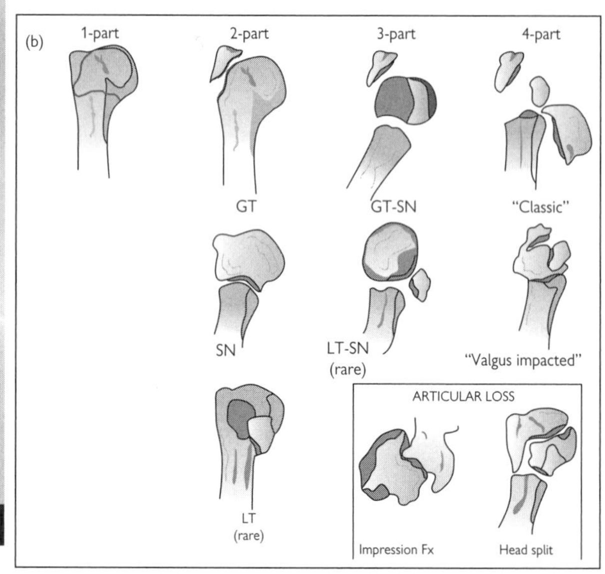

Figure 3.7 (Continued) b. Neer classification of proximal humerus fractures. Source: Neer. C.S. Displaced proximal humerus fractures: I. Classification and evaluation. *J Bone Joint Surg Am.* 1970; 52:1077–1089.

Primary Stabilization and Management

- Minimally displaced or nondisplaced fractures are treated with sling for comfort and early pendulum exercises.
- Surgery may be required if the proximal humerus is broken in more than one place (three or more parts), if there is significant displacement (>10 mm), or if the proximal humerus is considered unstable. Greater tuberosity fractures are generally displaced due to the attachment of the rotator cuff, and these fractures usually require ORIF.

Admit and Discharge Guidelines

- Request an orthopaedic consult in the emergency department for: Injuries with neurologic compromise, multipart or comminuted fractures, fractures with significant displacement of fragments, and fractures that appear to be unstable. Surgical intervention will range from percutaneous pinning to hemiarthroplasty.
- Refer to an orthopaedic surgeon for outpatient follow-up for: A minimally or nondisplaced fracture of the proximal humerus and fractures of the greater or lesser tuberosity

Definitive Treatment

- Nonoperative treatment is with a sling and early mobilization.
- Surgical management depends upon fracture type and surgeon preference. Greater tuberosity fractures require ORIF. Multipart fractures may be treated with ORIF, intramedullary nail, or hemiarthroplasty.

Physeal Injuries of the Proximal Humerus

Symptoms and Findings

- Frequently due to a fall backward on an outstretched hand. Injury can also be due to direct trauma.
- Birth trauma (hyperextension or rotation of the arm during birth) can cause proximal humerus injuries in newborns.
- Patients present with pain in the proximal humerus and often hold the arm in internal rotation.
- Pain with palpation and shoulder ROM; there may be swelling or ecchymosis over the proximal humerus.

Imaging (See Figure 3.8)

- X-rays: Standard shoulder trauma series (AP, Grashey, and axillary views)
- Ultrasound may be needed in newborns.

Classification

- Neer-Horowitz classification is an alternative to Salter-Harris.
  - Grade I: Less than 5 mm displacement
  - Grade II: Displacement less than 1/3 the width of the shaft
  - Grade III: Displacement less than 2/3 the width of the shaft
  - Grade IV: Displacement greater than 2/3 the width of the shaft

Primary Stabilization and Management

- Grade I and II displaced fractures should be reduced with gentle manipulation then placed in a sling. If a single attempt to properly

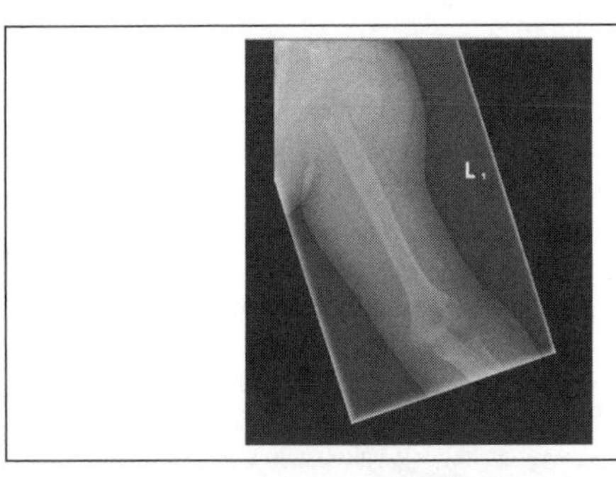

Figure 3.8 Physeal injury of the proximal humerus.

reduce the fracture is unsuccessful, orthopaedic consultation is recommended. If the fracture does not easily reduce, an orthopaedic surgeon should be consulted as there may be soft tissue interposition (biceps tendon) preventing the reduction.

- An orthopaedic surgeon should be consulted prior to any attempted reduction of grade III and IV displaced fractures, as some surgeons will prefer to reduce the fracture in the operating room and then place percutaneous pins.

Admit and Discharge Guidelines

- Request an orthopaedic consult in the emergency department for: Suspected birth trauma, injuries with neurologic compromise, irreducible Grade I or II injuries, Grade III or IV injuries, open fractures, and fractures that appear to be unstable.
- Refer to an orthopaedic surgeon for outpatient follow-up for: Any fracture requiring a reduction maneuver, limb length inequality, loss of motion, and growth arrest.

Definitive Treatment

- Treatment varies depending on the age of the patient and the fracture pattern. Surgery is rarely indicated in children under 12 as extensive remodeling is possible. Surgery is more common in adolescents.
- Stable proximal humerus fractures will generally be treated with a sling.
- Surgery should be considered for fractures with less than 50% opposition or greater than 45 degrees angulation (percutaneous pinning or ORIF), though there is evidence to support nonoperative management even in cases of severe displacement and angulation.

Humeral Shaft Fractures

Symptoms and Findings

- Most commonly caused by a direct impact to the arm
- Less frequently, injured through indirect trauma, such as a fall onto an outstretched arm in an elderly person or a throwing injury with extreme muscular contraction in a younger person
- Present with pain, swelling, and/or deformity of the arm
- There may be pronounced deformity and instability or crepitus on exam.
- A careful neurovascular exam, particularly of the radial nerve, is mandatory. Radial sensory deficit will manifest as diminished light touch sensation over the dorsum of the first web space. Radial motor deficit will result in weakness or absence of thumb and finger extension as well as likely weakness in wrist extension.

Imaging (See Figure 3.9)

- X-ray: AP and lateral of the humerus; shoulder and elbow series
- CT is generally not required.

Classification

- Classification is descriptive.
- Location on the humerus: Proximal third, middle third, distal third
- Displacement or comminution. The displacement of these fractures is determined by the fracture's location relative to the deltoid and pectoralis major insertion.
- Pattern: Transverse, oblique, spiral, segmental, comminuted

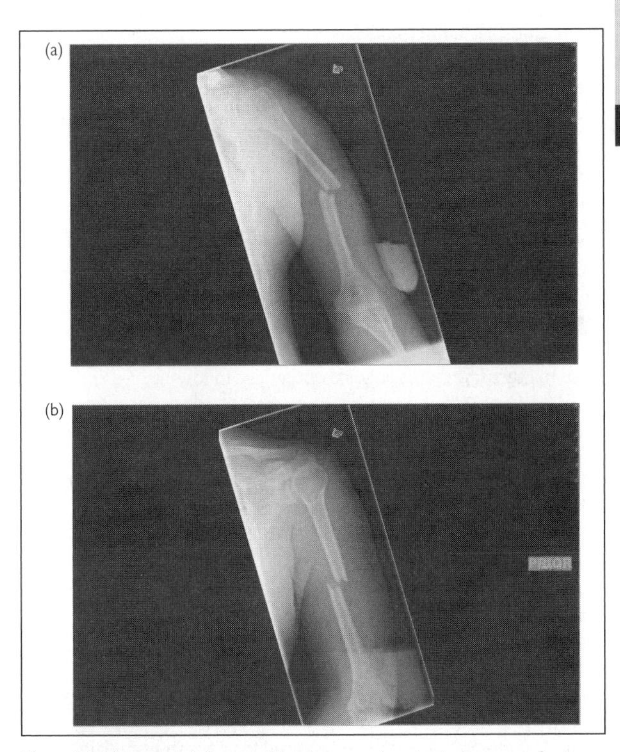

Figure 3.9 Humeral shaft fracture: Note how even in a simple, transverse fracture, understanding of malrotation is better understood with two views. a. Humeral shaft fracture seen on AP X-ray. b. Humeral shaft fracture seen on lateral X-ray.

Primary Stabilization and Management

- A coaptation splint should be applied in the emergency setting for all humerus fractures.
- Acceptable reduction parameters
  - Three cm shortening
  - Thirty degrees of varus angulation
  - Twenty degrees of anterior angulation
  - Fifteen degrees of malrotation
- Inpatient admission and surgery are usually required for open fractures, fractures with associated vascular injury, and "floating elbow" (injury involving fracture of the humerus and fractures of the radius and ulna). A coaptation splint should be applied in the emergency department to provide temporary pain relief and immobilization.

Admit and Discharge Guidelines

- Request an orthopaedic consult in the emergency department for:
  - Radial nerve palsies, especially those that develop after fracture manipulation
  - Open fractures
  - Fractures that cannot be reduced to within acceptable levels of tolerance
  - Suspected pathologic fracture
  - Multi-trauma patients
  - Severely comminuted fractures
  - Bilateral fractures
- All humerus fractures require outpatient follow-up.

Definitive Treatment

- The majority of humerus fractures do not require surgical management. Treatment is with a coaptation splint followed by functional bracing (anterior and posterior molded plastic components with velcro straps to maintain adequate compression) after 1–2 weeks.
- Surgery may be required for: Pathologic fractures, fractures with intra-articular extension, and fractures where the acceptable reduction parameters cannot be achieved or maintained in the coaptation splint. Surgical options include ORIF with plate and screws or intramedullary fixation.
- Surgical intervention may be required for persistent radial nerve palsy.

Elbow and Forearm Injuries

Distal Humerus Fractures in Adults

Symptoms and Findings

- Low-energy injuries are more common in middle-aged and older individuals, and fractures are the result of a fall onto the elbow or the outstretched arm.
- Higher-energy injuries are more common in younger patients, and fractures are often due to sports or motor vehicle accidents.
- Patient presents with pain about the elbow and refusal to use the affected arm.
- Tenderness with palpation of the distal humerus; significant swelling about the elbow may be present.
- Passive ROM (particularly flexion) of the elbow may be limited by displaced fracture fragments, and any locking or catching should be documented.
- A careful initial neurovascular exam is important as there may be injury to the brachial artery, median nerve, or radial nerve. Median nerve injury is characterized by numbness or paresthesias in the thumb, index finger, and/or the middle finger. The radial nerve is most frequently involved, and this may result in numbness or paresthesias over the dorsum of the first web space as well as possible weakness in extension of the thumb and fingers and dorsiflexion of the wrist.

Imaging (See Figure 3.10)

- X-ray: AP and lateral views of the elbow; traction views may be helpful to delineate the fracture pattern but should only be performed by experienced providers as neurovascular injury may occur.
- Arteriogram should be performed for any pulseless extremity.
- CT scan may be utilized to fully delineate the fracture fragments for preoperative planning; however, it is not mandatory.

Classification (See Figure 3.11)

- Descriptive classification: Based on the columnar/condylar concept and typically subclassified by the amount of angulation or displacement; could consider line drawing with transcondylar, supracondylar, intercondylar
- Supracondylar fracture: Horizontal or oblique fracture above the condyles; often subclassified based on location and mechanism

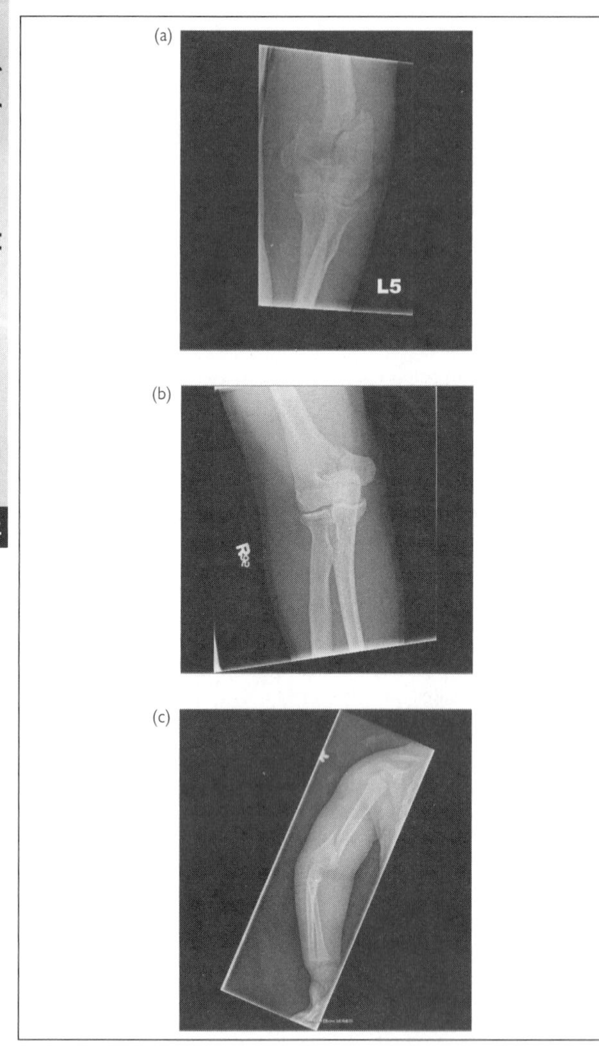

Figure 3.10 Distal humerus fracture X-ray a. Adult distal humerus fracture X-ray—Y- or T-type secondary to a gunshot wound. b. Adult distal humerus fracture X-ray—medial condyle fracture. c. Pediatric distal humerus fracture X-ray.

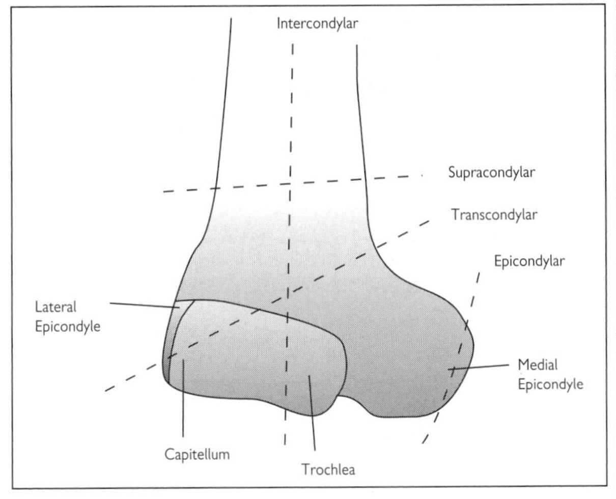

Figure 3.11 The types of distal humerus fractures, including supracondylar, transcondylar, intercondylar, condylar, capitellum, trochlear, and epicondylar.

- Transcondylar fracture: Horizontal or oblique fracture at the level of the condyles
- Intercondylar Fracture: T-shaped fracture between the condyles; often subclassified based on displacement, rotation and comminution; most common fracture pattern and nearly always requires surgery
- Condylar fracture: Fracture that includes the epicondyle and part of the trochlea (medial condyle fractures) or the capitellum (lateral condyle fractures); subclassified based on involvement of the lateral trochlear ridge
- Capitellum fracture: Fracture of the capitellum in the coronal plane; subclassified based on size and comminution
- Trochlea fracture: Shear fracture of the trochlea; almost always due to an elbow dislocation
- Epicondylar fracture: Fracture of the medial or lateral epicondyle

Primary Stabilization and Management

- Regardless of the need for surgical intervention, primary treatment should consist of attempted anatomic reduction of the fracture, with the goal of restoring normal joint articulation, followed by splinting.
- Malalignment of the joint surface or displaced fracture fragments may block elbow joint motion, especially flexion. If the elbow is

blocked or catches on passive range of motion, the fracture will likely require operative management.
- There is a risk of significant swelling with resulting vascular compromise and/or ischemic contracture (Volkmann's contracture) from these injuries. Any patient with significant swelling should be admitted for serial examination and careful monitoring for signs of compartment syndrome.
- A posterior long arm splint is appropriate for almost any fracture about the elbow in an adult. It is important that the splint be well padded and noncircumferential given the risk of swelling.

Admit and Discharge Guidelines

- Admit any patient with significant swelling for serial compartment checks.
- If an injury to the brachial artery is suspected, a vascular surgery consult should be emergently requested, and preparations should be made for urgent surgery.
- Request an orthopaedic consult in the emergency department for:
  - Open fractures
  - Unstable fractures
  - Displaced fractures, comminuted fractures, limited elbow ROM
- All distal humerus fractures require outpatient follow-up.

Definitive Management

- Most distal humerus fractures are treated surgically with ORIF. Nondisplaced fractures may be treated conservatively; however, restoration of the joint surface is mandatory for optimum elbow function.

Supracondylar Elbow Fractures in Children

Symptoms and Findings

- Most fractures are the result of hyperextension onto an outstretched hand +/− a varus or valgus force.
- Rarely, fractures are due to direct trauma on a flexed elbow.
- Patient presents with pain and swelling about the elbow and refusal to use the affected arm.
- Tenderness with palpation of the distal humerus; significant swelling about the elbow may be present.
- A careful initial neurovascular exam is important as there may be injury to the brachial artery, median nerve, or radial nerve.
- Significant deformity about the elbow may be present with obvious fracture angulation and skin tenting or puckering.

Imaging (See Figure 3.12a–c)

- X-ray: AP and lateral views of the elbow
- Nondisplaced fractures may be difficult to visualize, and one should always determine if a fat pad sign is present.
- Arteriogram should be performed for any pulseless extremity.

Classification

- Classified by mechanism and displacement. The classification determines treatment.
- Extension type
  - Type I: Nondisplaced
  - Type II: Displaced fracture with an intact posterior cortex
  - Type III: Completely displaced fracture with posterior displacement of the distal fragment
- Flexion type
  - Type I: Nondisplaced
  - Type II: Displaced fracture with an intact anterior cortex

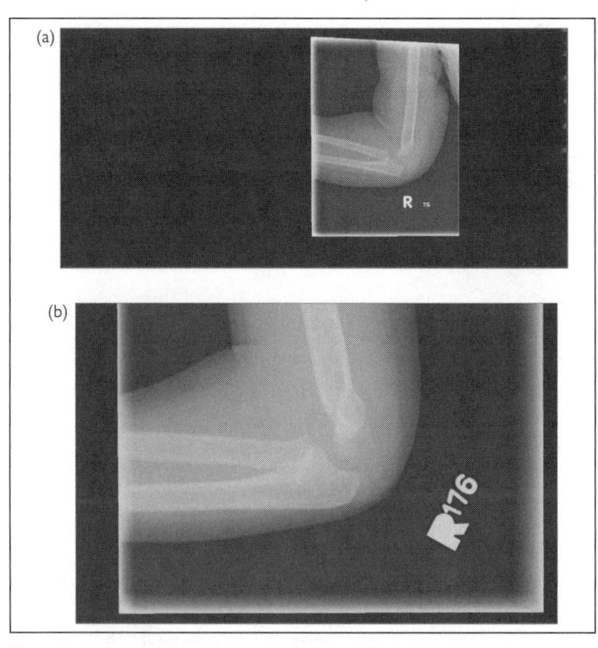

Figure 3.12 Supracondylar humerus fracture in children. a. Type I extension-type pediatric supracondylar fracture on a lateral X-ray. b. Type II extension-type pediatric supracondylar fracture on lateral X-ray.

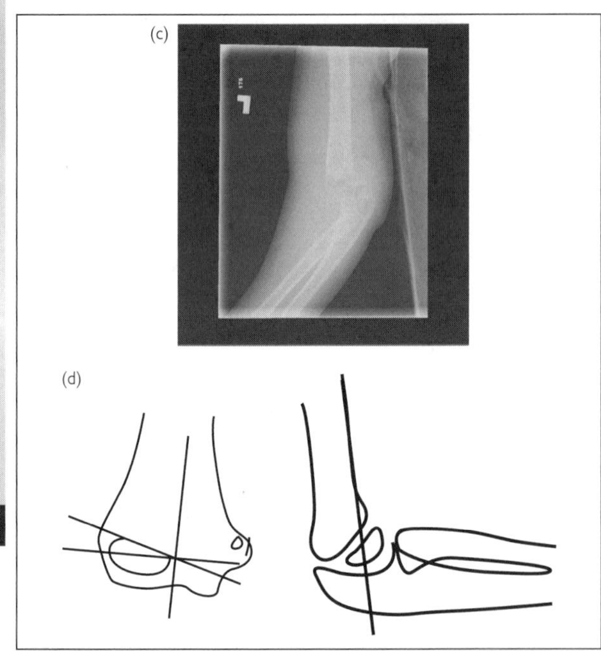

Figure 3.12 (Continued) c. Type III extension-type pediatric supracondylar fracture on lateral X-ray. d. Line drawings demonstrating Baumann's angle and anterior humeral line.

- Type III: Completely displaced fracture with anterior displacement of the distal fragment

Primary Stabilization and Management

- These fractures are associated with significant morbidity and prompt treatment is imperative.
- An orthopaedic surgeon should be consulted for any pediatric supracondylar fracture.
- Prior to calling a consult, pulses should be assessed as a pulseless or unperfused extremity will require urgent surgery for exploration and open reduction. Neurologic status should also be assessed, but unlike a vascular insult, a nerve injury alone does not necessitate surgery.
- There is a risk of significant swelling with resulting vascular compromise and/or ischemic contracture (Volkmann's contracture) from these injuries. Any patient with significant swelling should be admitted for serial

examination and careful monitoring for signs of compartment syndrome.
- No immobilization is required while awaiting orthopaedic consult. Comfort care may consist of pain medication and a pillow to rest the arm on. A well-padded resting elbow splint may help limit the patient's pain if transport is needed.

Admit and Discharge Guidelines

- Request an orthopaedic consult in the emergency department for: All supracondylar fractures in children.

Definitive Treatment

- Definitive treatment is dictated by classification. The key to the treatment of supracondylar humerus fractures is the restoration of Baumann's angle. The anterior humeral line can also be used to judge adequacy of reduction (see Figure 3.12d).
- Extension type
  - Type I: Immobilization for 3 weeks in a bivalved long arm cast; cast in flexion
  - Type II: Reduction and percutaneous pinning
  - Type III: Reduction and percutaneous pinning versus ORIF
- Flexion type
  - Type I: Immobilization for 3 weeks in a bivalved long arm cast; cast in near full extension
  - Type II: Reduction and percutaneous pinning
  - Type III: Reduction and percutaneous pinning versus ORIF

Medial Condyle Fractures in Children

Symptoms and Findings

- May be caused by direct trauma (such as a fall onto the elbow) or indirect trauma, such as a fall onto an outstretched arm
- Medial condyle fractures are associated with elbow dislocations, so after reducing an elbow fracture, the postreduction radiographs should carefully be reviewed for evidence of a medial condyle fracture.
- Tenderness with palpation of the medial condyle, painful range of motion and swelling about the elbow
- A careful initial neurovascular exam is mandatory as the ulnar nerve is frequently injured.
- Pain with resisted wrist flexion (the common flexor tendon originates on the medial condyle).

Imaging (See Figure 3.13)

- X-rays: AP, lateral, and oblique views of the elbow
- Traction views can help differentiate between condylar and epicondylar fractures.

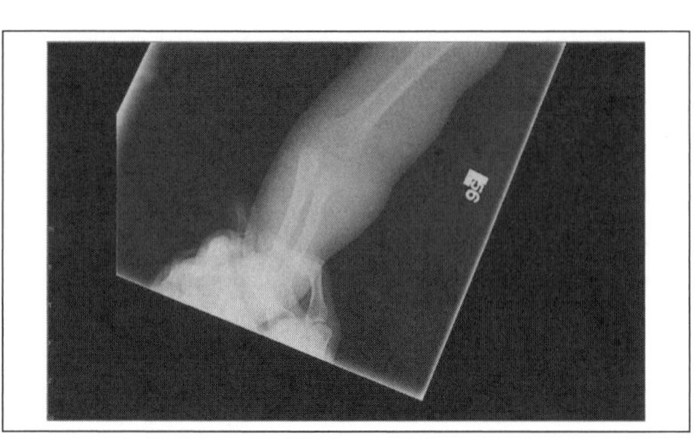

Figure 3.13 Medial condyle fractures in a child on AP X-ray

Classification

- Milch classification: Classified according to whether the fracture line terminates in the trochlear notch (Type I) or in the capitellotrochlear groove (Type II)
- Kilfoyle classification: Classified by displacement; more useful classification as displacement guides treatment
  - Stage I: Nondisplaced
  - Stage II: Less than 1 cm of displacement
  - Stage III: More than 1 cm of displacement

Primary Stabilization and Management

- In general, an orthopaedic surgeon should be consulted for any pediatric fracture about the elbow.
- No immobilization is required while awaiting orthopaedic consult. Comfort care may consist of pain medication and a pillow to rest the arm on. A well-padded resting elbow splint may help limit the patient's pain if transport is needed.

Admit and Discharge Guidelines

- Request an orthopaedic consult in the emergency department for all medial condyle fractures in children.

Definitive Treatment

- Definitive treatment is dictated by classification.
- Stage I: Nondisplaced (and some minimally displaced Stage II) fractures are treated with long arm casting. In the emergency setting, the cast must be bivalved. If there is significant swelling, a posterior slab with buttress splint may be used. The elbow will be held in 90 degrees of flexion with the forearm in neutral.
- Stage II and III: Reduction and percutaneous pinning versus ORIF

Lateral Condyle Fractures in Children

Symptoms and Findings

- Occur by one of two possible mechanisms: Either by "pull-off" or "push-off"
- Pull off: Avulsion injury of the lateral condyle by the common extensor origin; due to a varus stress applied to an extended elbow with a supinated forearm; most common mechanism
- Push off: A fall onto an extended hand causes the radial head to hit the lateral condyle.
- Tenderness with palpation of the lateral condyle, painful range of motion, and swelling about the elbow
- Pain with resisted wrist extension (the common extensor tendon originates on the lateral condyle).

Imaging (See Figure 3.14)

- X-rays: AP, lateral, and oblique views of the elbow

Classification

- Milch classification
  - Type I: Fracture traverses the secondary ossification center of the capitellum lateral to the trochlea; Salter-Harris Type IV fracture; stable fracture pattern; less common than Milch Type II.
  - Type II: Fracture crosses the physis and enters the joint with the fracture line extending into the apex of the trochlea; Salter-Harris Type II fracture; unstable fracture because the trochlea is disrupted.
- Jakob classification: Classified by displacement
  - Stage I: Nondisplaced fracture preserving the articular surface

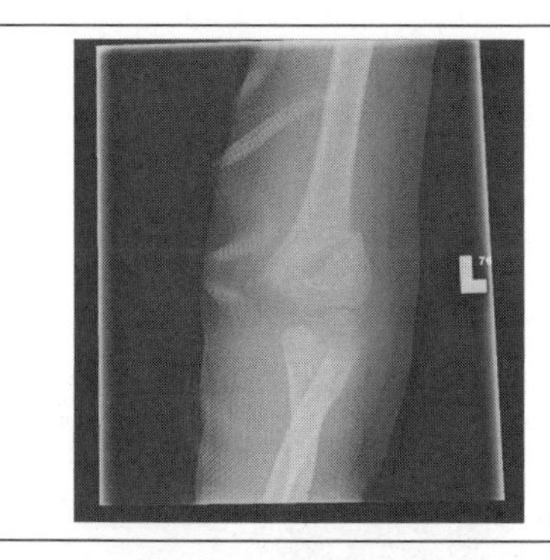

Figure 3.14 Lateral condyle fracture in a child on AP X-ray.

- Stage II: Moderate displacement
- Stage III: Completely displaced fracture and elbow instability

Primary Stabilization and Management

- In general, an orthopaedic surgeon should be consulted for any pediatric fracture about the elbow.
- No immobilization is required while awaiting orthopaedic consult. Comfort care may consist of pain medication and a pillow to rest the arm on. A well-padded resting elbow splint may help limit patient's pain if transport is needed.

Admit and Discharge Guidelines

- Request an orthopaedic consult in the emergency department for: All lateral condyle fractures in children.

Definitive Treatment

- Definitive treatment is dictated by fracture displacement and stability.
- Nondisplaced fractures and fractures with less than 2 mm of displacement are treated with long arm casting. In the emergency setting, the cast will be bivalved. If there is significant swelling, a posterior slab with buttress splint may be used. The elbow will be held in 90 degrees of flexion with the forearm in neutral (thumbs-up position).
- Fractures with greater than 2 mm of displacement will be treated with ORIF.

Medial Epicondylar Apophyseal Fractures in Children

Symptoms and Findings

- Apophyseal injuries of the medial epicondyle are generally avulsion injuries due to indirect trauma. Rarely, direct trauma to the medial epicondyle causes fracture.
- Indirect trauma may be acute or chronic, but both have the same mechanism: avulsion of the flexor muscles. Acute injuries of the medial epicondylar apophysis are usually secondary to elbow dislocation or a fall onto an outstretched hand. Chronic injury is due to overuse from repetitive throwing, which is why the chronic injuries are called "Little League elbow."
- Tenderness with palpation of the medial condyle, painful range of motion, and swelling about the elbow. If the epicondylar fragment becomes trapped within the joint, there may be a mechanical block to elbow range of motion.
- A careful initial neurovascular exam is mandatory as the ulnar nerve may be injured.
- Pain with resisted wrist flexion (the common flexor tendon originates on the medial condyle).

Imaging

- X-rays: AP, lateral, and oblique views of the elbow
- Traction views can help differentiate between condylar and epicondylar fractures

Primary Stabilization and Management

- Most medial epicondylar fractures can be treated nonoperatively with good results
- Nondisplaced and minimally displaced (< 5 mm displacement) fractures can be treated with a posterior slab with buttress splint. The elbow will be held in 90 degrees of flexion with the forearm in neutral or slight pronation.
- Fractures with greater than 5 mm displacement require orthopaedic consultation; an attempt at reduction is appropriate but avoid more than one attempt before the consultant arrives.

Admit and Discharge Guidelines

- Request an orthopaedic consult in the emergency department for:
  - Fractures with more than 5 mm displacement, fractures with greater than 5 mm displacement after a single attempt at reduction in the emergency room, irreducible fractures (does this suggest they should attempt reduction in the ED?)
  - Fractures with a fragment incarcerated within the elbow joint, ulnar nerve dysfunction, high-demand patients (e.g., elite athletes)
- All medial epicondylar apophyseal fractures require outpatient follow-up with an orthopaedic surgeon.
- Follow-up should occur within 3–4 days to allow early ROM.

Definitive Treatment

- Nonoperative treatment is indicated for nondisplaced and minimally displaced fractures.
- Operative treatment generally consists of ORIF.

Elbow Dislocations

Symptoms and Findings

- Most commonly caused by a fall on an outstretched hand or elbow. Other mechanisms include high-energy injuries, often due to sports or motor vehicle accidents
- Patient presents with pain about the elbow, often guarding the arm.
- The elbow may have significant deformity and swelling.
- A careful initial neurovascular exam is important as there may be injury to the brachial artery, median nerve, and especially the branch known as the anterior interosseus nerve (AIN) or ulnar nerve. AIN injury is more difficult to diagnose as there is

no sensory loss; motor loss involves thumb IP (interphalangeal) flexion (FPL [flexor pollicis longus]) and DIP (distal interphalangeal) flexion of the index and middle fingers (FDP to IF [flexor digitorum profundus to index finger], MF [middle finger]). Neurovascular status should be assessed as quickly as possible as neurologic or vascular compromise of the distal forearm requires immediate reduction.

Imaging (See Figure 3.15)

- X-rays: AP and lateral views of the elbow
- Angiography may be needed to evaluate vascular compromise

Classification

- Classified as simple or complex (associated with fracture) and by the direction of the forearm bones relative to the humerus

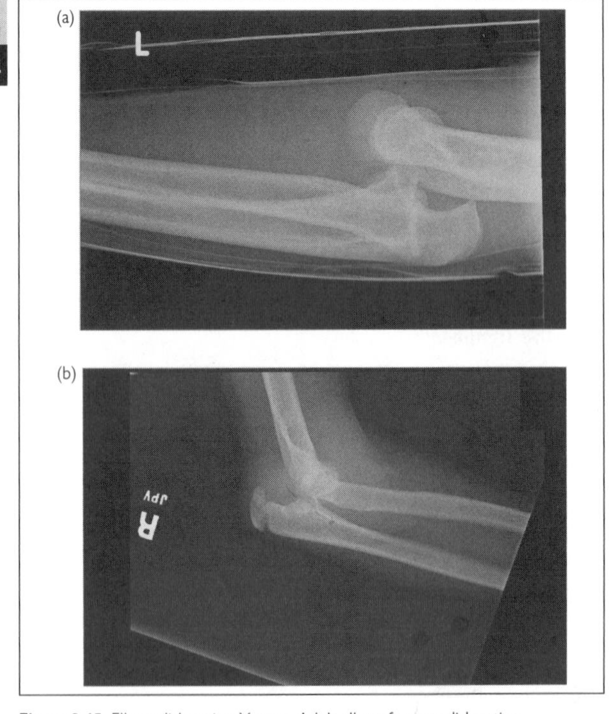

Figure 3.15 Elbow dislocation X-ray a. Adult elbow fracture dislocation. b. Pediatric elbow dislocation.

- Associated fractures include the radial head, medial epicondyle, lateral epicondyle, and the coronoid process.
- Posterolateral dislocations are the most common (80%–90% of all dislocations). Anterior, medial, lateral, and divergent (when the ulna dislocates medially and the radius dislocates laterally) are less common.

Primary Stabilization and Management

- In closed dislocations, acute neurovascular injuries are uncommon but can occur. The ulnar nerve and the AIN are the most commonly injured nerves. Vascular compromise is more common in open dislocations, with the brachial artery at highest risk.
- If a vascular injury exists, the dislocation must be immediately reduced. If arterial flow is not reestablished, the brachial artery will likely require emergent saphenous vein grafting.
- Reduction of a posterior dislocation: Apply traction to the distal forearm with countertraction to the humerus. Correct medial or lateral displacement, then apply distal traction followed by elbow flexion. If only one person is available to perform the reduction maneuver or limited anesthesia/analgesia is available, it is helpful to lay the patient prone on a stretcher (these maneuvers are known as the Parvin maneuver or the Meyn and Quigley maneuver). Gentle traction can then be applied to the distal forearm with the stretcher acting as countertraction.
- Following reduction of a simple dislocation, the elbow should be placed in a posterior slab with buttress splint in 90 degrees of flexion with the forearm in neutral. It is important that the splint be well padded and noncircumferential given the risk of increased swelling.
- Complex dislocations (fracture dislocations) often require surgical treatment. If the radial head is the only fracture, closed treatment may be a reasonable option.
- The "terrible triad of the elbow" injury pattern occurs when the coronoid process and radial head are fractured as a result of an elbow dislocation. These injuries are infamously unstable and require surgery.
- There is a risk of significant swelling with resulting vascular compromise and/or ischemic contracture (Volkmann's contracture) from these injuries. Any patient with significant swelling should be admitted for serial examination and careful monitoring for signs of compartment syndrome.

Admit and Discharge Guidelines

- Admit any patient with significant swelling for serial compartment checks.

- If an injury to the brachial artery is suspected, a vascular surgeon should be emergently requested, and preparations should be made for urgent surgery, including possible saphenous vein grafting of the brachial artery.
- Request an orthopaedic consult in the emergency department for open dislocations, complex dislocations (dislocations with an associated fracture), and dislocations with neurovascular injury or compromise.
- All elbow dislocations require outpatient follow-up

Definitive Treatment

- The choice of definitive treatment depends on stability after reduction. If the elbow joint is stable, a splint may be the definitive treatment. If, however, the joint is not stable (as is the case with most fracture dislocations), surgical treatment is required. Surgical treatment options include ORIF, ligamentous reconstruction, and/or external-fixator immobilization.

Olecranon Fractures

Symptoms and Findings

- May be caused by direct trauma, such as a fall onto the elbow, or indirect trauma, such as a fall onto an outstretched extremity causing a strong triceps contraction
- Patients present with pain about the elbow, often supporting the affected arm with the contralateral extremity, holding the arm in flexion.
- A careful neurovascular exam is mandatory as the ulnar nerve is frequently injured
- Triceps function (active extension of the elbow) should be tested.

Imaging (See Figure 3.16)

- X-rays: AP and lateral views of the elbow

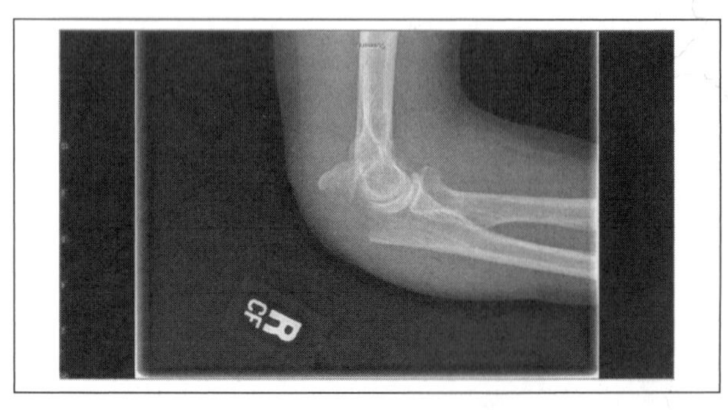

Figure 3.16 Olecranon fracture on a lateral X-ray.

Classification

- Mayo classification describes fractures according to displacement and subluxation/dislocation. Each type is then subclassified according to comminution (A-noncomminuted, B-comminuted).
  - Type I: Nondisplaced or minimally displaced
  - Type II: Displacement of the proximal fragment; elbow is stable.
  - Type III: Displacement of the proximal fragment; elbow joint is unstable (subluxated or dislocated).
- While other classification schemes exist, this classification system is helpful as it guides treatment.

Primary Stabilization and Management

- In the acute setting, a posterior slab with buttress splint should be applied for all olecranon fractures, regardless of possible surgical intervention.
  - Type I fractures are generally treated nonoperatively with immobilization in 30–45 degrees of elbow flexion.
  - Type II and III fractures generally require surgical fixation.
  - Type II and III fractures in poor surgical candidates may be treated nonoperatively.

Admit and Discharge Guidelines

- Request an orthopaedic consult in the emergency department for:
  - Fractures with neurovascular compromise
  - Fracture dislocation
  - Open fractures, and existence of concomitant fractures (e.g., coronoid process or radial head)
- The orthopaedic surgery team should be notified of all Type II and III fractures, so that surgical treatment can be planned.
- Refer to an orthopaedic surgeon for outpatient follow-up: All olecranon fractures.
- Follow-up radiographs should occur within 1 week of injury as Type I fractures may displace (advance to a Type II fracture) and therefore require surgical management.

Definitive Treatment

- Surgical treatment can include intramedullary fixation with a rod or screw, ORIF with a plate and screws, or tension band wiring with Kirschner wires, or surgical excision of the olecranon with triceps tendon advancement.

Radial Head Fractures in Adults

Symptoms and Findings

- Usually due to a fall onto an outstretched hand. Radial head is fractured when it impacts with the capitellum.

- Patients present with pain and limited motion about the elbow. Patients are usually tender to palpation over the radial head.
- A careful neurovascular exam should be performed, particularly in high-energy injuries, which may be associated with concomitant fractures or dislocations.
- Passive range of motion of the elbow should be performed to ensure there is no mechanical block to motion. A hematoma block can be helpful (see Chapter 9 on elbow injections) to reduce pain and permit a complete exam.
- Medial collateral ligament stability should be tested as a valgus deformity may result if it is incompetent.
- If a patient complains of wrist pain, injury to the interosseous ligament and/or distal radioulnar joint should be considered.

Imaging (See Figure 3.17)

- X-rays: AP, lateral, and oblique (Greenspan) views of the elbow
- Nondisplaced fractures may be difficult to visualize, and one should always determine if a fat pad sign is present.
- X-rays of the forearm and wrist should also be obtained, as other injuries are frequently associated with radial head fractures.
- CT scan may be useful in preoperative planning but should not be routinely ordered.

Classification

- Mason classification describes fractures according to displacement, comminution, and dislocation.
  - Type I: Nondisplaced
  - Type II: Displaced single fracture at the margins of the radial head or through the radial neck

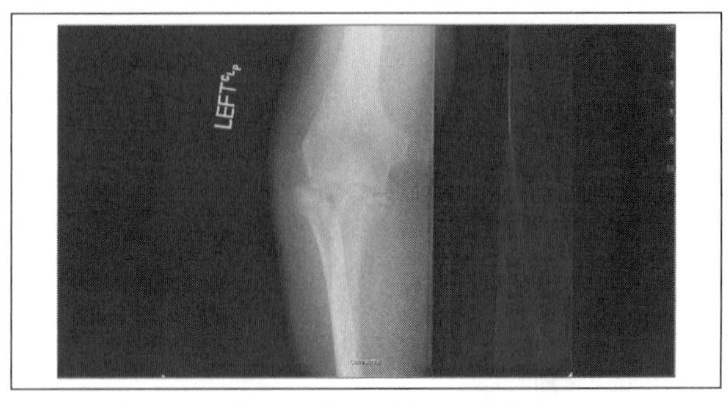

Figure 3.17 Comminuted fracture of the radial head with displacement as seen on an AP elbow X-ray.

- Type III: Comminuted fracture of the entire radial head
- Type IV: Radial head fracture with elbow dislocation

Primary Stabilization and Management

- Treatment of radial head fractures depends on both classification and the existence of associated injuries.
- A posterior slab with buttress splint should be applied to all radial head fractures in the acute setting.
- Concomitant injury of the forearm interosseous ligament should be considered in all radial head fractures. Longitudinal disruption of this ligament is called an Essex-Lopresti lesion, and it is associated with fractures involving the proximal radioulnar joint (usually a Type III or IV radial head fracture) and/or the distal radioulnar joint. This injury is often discovered postoperatively when the radius migrates proximally after radial head excision. A missed diagnosis often necessitates a second surgery.
- The "terrible triad of the elbow" injury pattern occurs when the coronoid process and radial head are fractured as a result of an elbow dislocation. These injuries are infamously unstable and require surgery.

Admit and Discharge Guidelines

- Request an orthopaedic consult in the emergency department for:
  - Fractures with neurovascular compromise
  - Fracture dislocation (Type IV)
  - Open fractures
  - Existence of concomitant fractures (e.g.. coronoid process or olecranon)
- The orthopaedic surgery team should be notified of any fracture that might require surgical treatment.
- Refer to an orthopaedic surgeon for outpatient follow-up for: All radial head fractures.

Definitive Treatment

- Isolated, stable fractures of the radial head (Type I and some Type II) do not generally require surgery. Type II fractures can usually be treated nonoperatively, unless the fracture blocks elbow range of motion or is associated with other injuries making the elbow unstable.
- Nonoperative treatment usually consists of a sling or posterior slab with buttress splint and early range of motion.
- Type III and IV fractures generally require open reduction and internal fixation or radial head replacement.

Radial Head and Neck Fractures in Children

Symptoms and Findings

- Usually due to a fall onto an outstretched hand. Unlike in adults, fracture of the radial head is rare as it is covered in cartilage. The radial neck or physis is more commonly injured.
- Patients present with pain and limited motion about the elbow. Patients are usually tender to palpation over the radial head and have exacerbation of their pain with supination and pronation of the forearm.

Imaging (See Figure 3.18)

- X-rays: AP, lateral, and oblique (Greenspan) views of the elbow

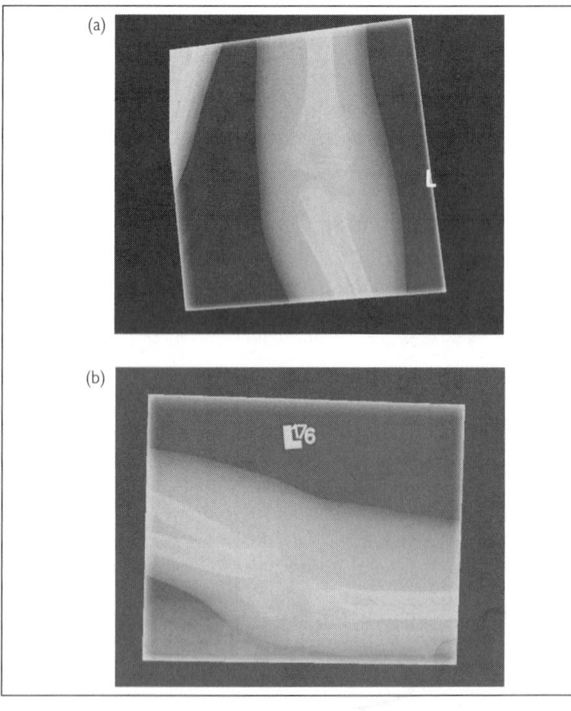

Figure 3.18 Pediatric radial head and neck fracture seen on a. AP X-ray and b. lateral X-ray.

Classification

- O'Brien classification describes fractures according to the degree of angulation:
  - Type I: Less than 30 degrees of angulation
  - Type II: Between 30 and 60 degrees of angulation
  - Type III: More than 60 degrees of angulation

Primary Stabilization and Management

- Treatment of radial head and neck fractures depends on both the classification and the existence of associated injuries.
- Type I fractures can be treated with immobilization in a long arm, bivalved cast with follow-up at 10–14 days.
- Reduction maneuvers are required for Type II fractures. If the fracture can be successfully reduced, surgery may be avoided if the fracture is stable after reduction. After reduction, stable fractures can be placed in a bivalved, long arm cast; unstable fractures will require percutaneous pinning. Reduction maneuvers should only be performed by orthopaedic surgeons, given the possibility of subsequent surgery and the significant complications that may result from this fracture.
- Some Type II fractures and almost all Type III fractures require surgical treatment.

Admit and Discharge Guidelines

- Request an orthopaedic consult in the emergency department for:
  - Any radial head fracture requiring reduction (Type II and III fractures)
  - Open fractures, and existence of concomitant fractures (e.g., coronoid process, olecranon, or medial epicondyle)
- Refer to an orthopaedic surgeon for outpatient follow-up for: All radial head and neck fractures in children.

Definitive Treatment

- Nonoperative treatment consists of a bivalved, long arm cast.
- Surgical treatment requires reduction (closed reduction is often possible, but open reduction may be needed) followed by pinning with Kirschner wires after reduction is achieved. A long arm cast is then applied.

Radial Head Subluxations (Nursemaid's Elbow)

Symptoms and Findings

- Classically, this injury occurs in children younger than 5 who are pulled or swung by their hand or forearm. Parents may report hearing a snap when the injury occurred (see Figure 3.19).
- Children will generally refuse to use their arm and will hold the arm in a flexed and pronated position. Patients will have pain with palpation of the radial head and with supination.

- Injury is due to stretching of the annular ligament resulting in displacement of the radial head.
- A careful neurovascular exam should be performed. If a neurovascular deficit is found, another injury should be suspected, as radial head subluxation alone should not cause a neurovascular deficit.

Imaging

- Radiographs are generally not required to make the diagnosis.
- If there is uncertainty as to the injury, AP and lateral radiographs of the elbow may be obtained. X-rays of the contralateral elbow are often helpful in elucidating the injury.
- This injury is rare in children older than 7 years old. Radiographs should be obtained in any child over 7 years of age.

Primary Stabilization and Management (See Figure 3.20)

- Treat with reduction maneuver.
- Apply gentle traction by holding the patient's forearm in one hand and applying countertraction with the other hand.
- Gradually supinate the forearm and flex the elbow.
- A clunk should be felt as the radial head is reduced. The patient should have pain relief and improved range of motion.
- An alternative reduction maneuver utilizes hyper-pronation (see Figure 3.20b)
- Immobilization is not needed.

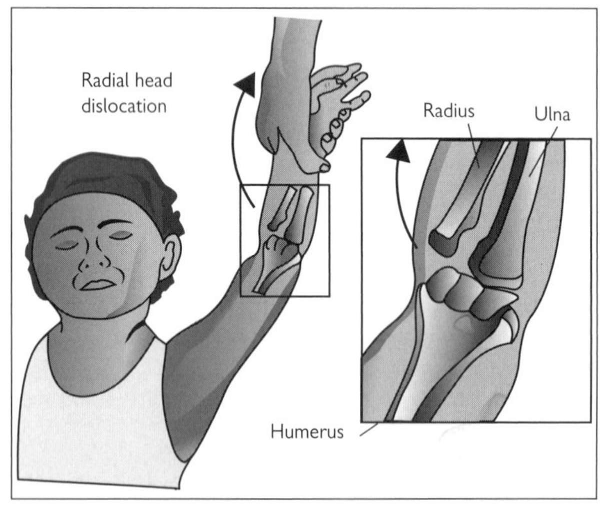

Figure 3.19 Line drawing of a nursemaid's elbow.

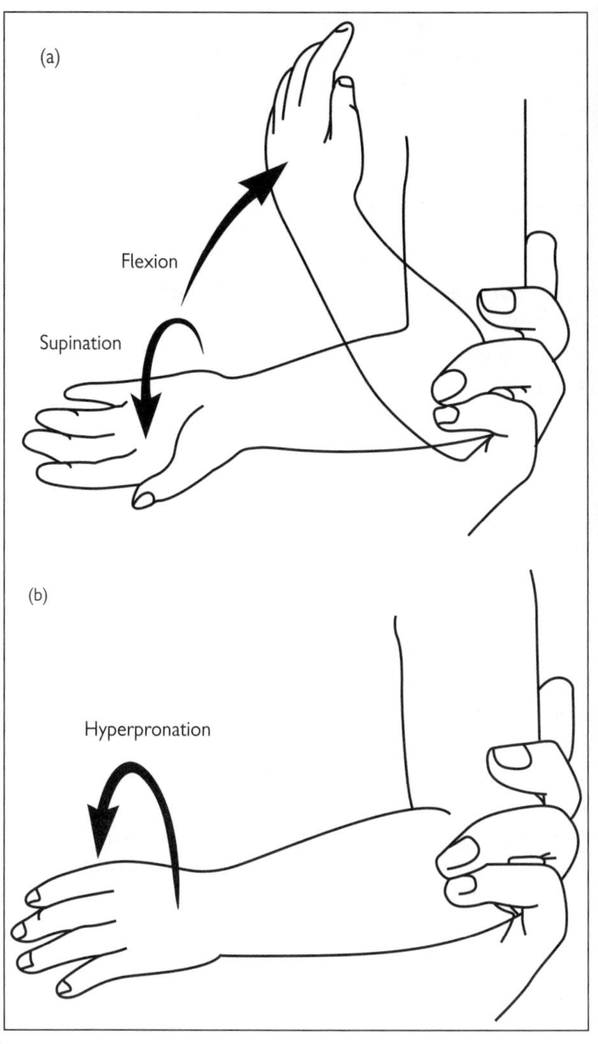

Figure 3.20 Diagram demonstrating reduction technique for a nursemaid's elbow. a. Reduction by supination and flexion. b. Reduction by hyperpronation.

Admit and Discharge Guidelines

- Request an orthopaedic consult in the emergency department for: Irreducible dislocations.
- Refer to an orthopaedic surgeon for outpatient follow-up for: Recurrent subluxations.

Definitive Treatment

- Reduction is the definitive treatment. Open reduction is indicated in rare cases where the annular ligament has become interposed between the radial head and the capitellum.

Proximal Ulna Fracture and Radial Head Dislocation (Monteggia Fractures)

Symptoms and Findings

- May be caused by direct trauma or by fall on an outstretched hand, often with forced pronation of the forearm or forced abduction of the elbow.
- Patients present with significant pain and swelling of the proximal forearm and elbow. They will have pain with elbow ROM, as well as supination and pronation.
- A careful initial neurovascular exam is important, as there may be injury to the radial or posterior interosseous nerve.
- Evaluate for possible compartment syndrome if there is significant swelling about the forearm. A baseline exam must be performed initially, then repeated at intervals with careful documentation given the risk of significant swelling.

Imaging (See Figure 3.21)

- X-rays: AP and lateral views of the elbow, forearm, and wrist
- CT scan is generally not indicated in the emergency setting.

Classification (See Figure 3.22)

- The Bado Classification describes injuries according to mechanism, direction of the dislocation, and the location and angulation of the fracture.

Type	Mechanism	Dislocation	Fracture
I	Forced pronation	Radial head anterior	Proximal 1/3 ulna fracture; apex anterior
II	Axial loading	Radial head posterior	Proximal 1/3 ulna fracture; apex posterior
III	Forced abduction	Radial head lateral	Proximal ulna metaphyseal fracture
IV	Forced pronation	Radial head anterior	Proximal 1/3 of the radius and ulna at the same level

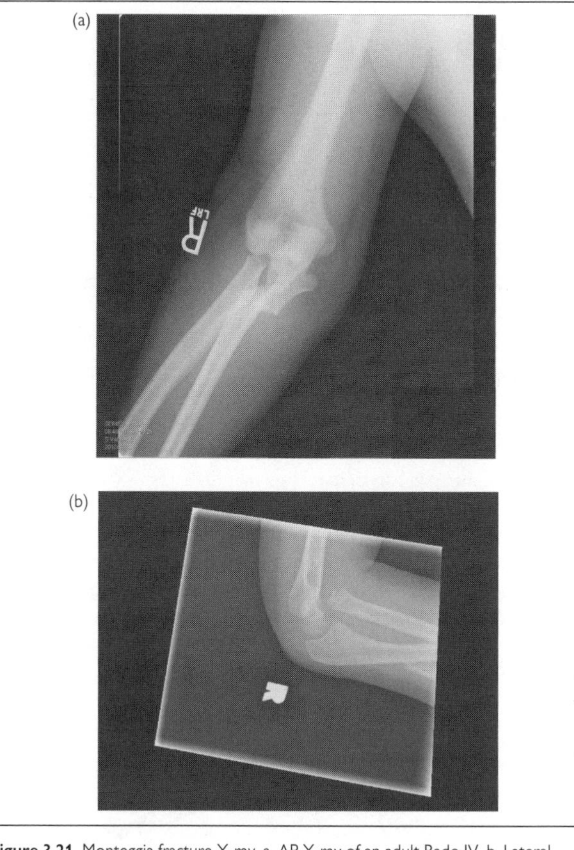

Figure 3.21 Monteggia fracture X-ray. a. AP X-ray of an adult Bado IV. b. Lateral X-ray of a pediatric Bado I.

Primary Stabilization and Management in Adults

- In adults, all Monteggia fractures require open reduction and internal fixation of the ulna fracture and closed versus open reduction of the radial head.
- Orthopaedic surgery should be consulted in the emergency department, and the patient should be admitted for urgent surgery.
- While awaiting surgery, patients should be placed in a well-padded posterior slab splint, and the affected extremity should be elevated to reduce swelling.

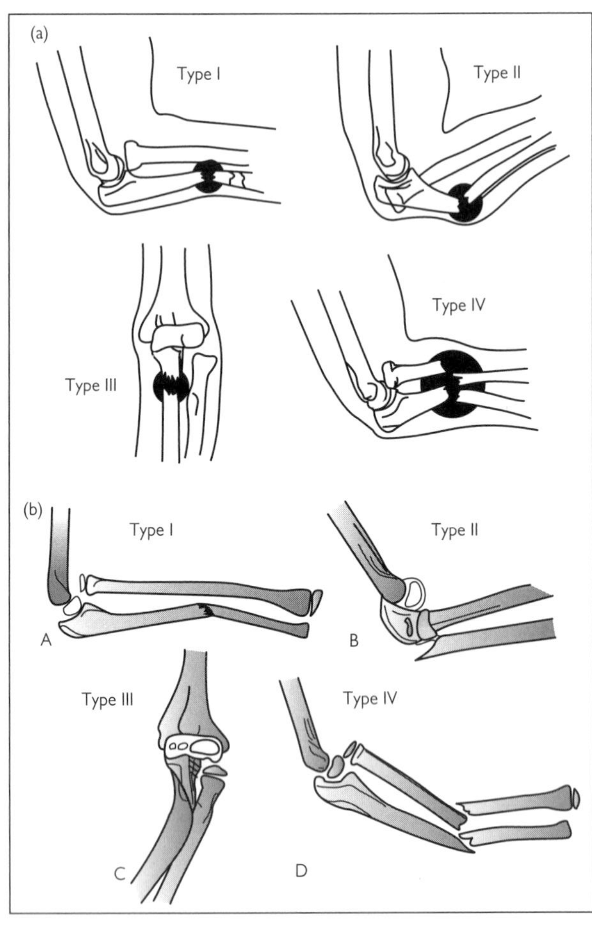

Figure 3.22 Monteggia Fractures. a. Line drawing of the four types of adult Monteggia fractures. b. Line drawing of the four types of pediatric Monteggia fractures.

- There is a risk of significant swelling with resulting vascular compromise and/or ischemic contracture (Volkmann's contracture) from these injuries. Any patient with significant swelling should be admitted for serial examination and careful monitoring for signs of compartment syndrome.

Primary Stabilization and Management in Children

- In children, some Monteggia fractures can be treated with closed reduction and immobilization. ORIF is needed for unstable fractures or irreducible dislocations.
- Orthopaedic surgery should be consulted in the emergency department.
- No immobilization is required while awaiting orthopaedic consult. Comfort care may consist of pain medication and a pillow to rest the arm on. A well-padded resting elbow splint may help limit patient's pain if transport is needed.
- Manipulation of the elbow and/or splint application should be deferred until a decision about surgery is made, as any manipulation increases the risk of significant swelling.
- There is a risk of significant swelling with resulting vascular compromise and/or ischemic contracture (Volkmann's contracture) from these injuries. Any patient with significant swelling should be admitted for serial examination and careful monitoring for signs of compartment syndrome.

Admit and Discharge Guidelines

- Admit any patient with significant swelling for serial compartment checks.
- Request an orthopaedic consult in the emergency department for: All Monteggia fractures.

Definitive Treatment in Adults

- Surgical treatment options include ORIF or external fixation.

Definitive Treatment in Children

- Closed reduction followed by long arm cast application is generally the treatment of choice. The reduction maneuver is performed by applying a stress opposite to the injury mechanism; in other words, Type I injuries result from extension forces and are reduced by traction and flexion.
- Open reduction and internal fixation may be required if reduction cannot be obtained or maintained. Surgery is often required when fractures are complete (through both cortices of the bone).

Shaft Fractures of the Radius and/or Ulna in Adults

Symptoms and Findings

- Fractures of the radial or ulnar shaft are due to high-energy mechanisms in adults. They are often secondary to motor vehicle accidents, direct trauma, falls, and gunshot wounds.
- Isolated fractures of the ulnar shaft are called "nightstick fractures," as they are usually suffered when victims protects their head from being struck with their forearm. The subcutaneous location of

the ulna provides little soft tissue protection from direct trauma, which also increases the possibility of an open fracture.

- Patients present with pain, swelling, and ecchymosis. Significant deformity may be present, and the wrist and elbow should be carefully examined to evaluate for possible concomitant injury.
- A careful neurovascular exam should be performed and documented. The radial and ulnar arteries may be damaged, as can the median, radial, and ulnar nerves.
- An initial compartment examination should be performed and documented. Compartment pressures should be measured in any patient with significant pain, tight compartments, or pain with passive extension of the fingers. Compartment syndrome is a surgical emergency, therefore compartment pressure measurements should only be performed by individuals able to emergently perform fasciotomy (generally orthopaedic and general surgeons). See Chapter 9 for further details on compartment syndrome.

Imaging (See Figure 3.23)

- X-rays: AP and lateral of the forearm; AP, lateral, and oblique of the wrist; AP and lateral of the elbow
- CT scan is generally not required in the emergency setting. CT angiogram may be required if an arterial injury is present.

Classification

- Descriptive classification: Closed versus open, undisplaced versus displaced, comminution; segmental; location; rotation; angulation
- Proximal ulna fracture with radial head dislocation: Monteggia (see page 112)
- Distal-third radius fracture with radioulnar dislocation: Galeazzi or Piedmont (see page 120)

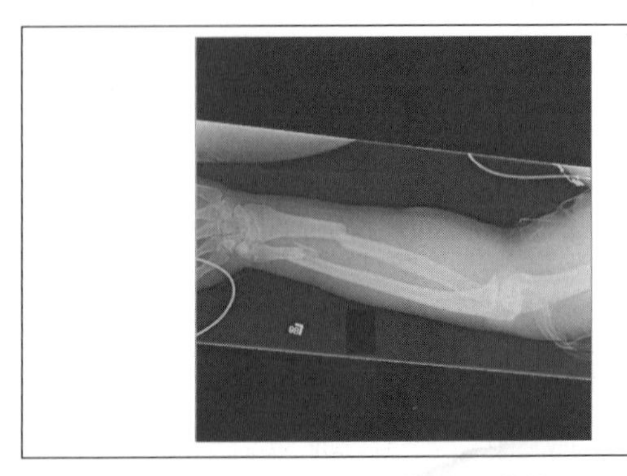

Figure 3.23 Both bone forearm fractures in adult on AP X-ray.

Primary Stabilization and Management

- Most fractures of the radial and ulnar shaft will require definitive surgical fixation.
- A well-padded sugar tong splint should be applied. If the fracture involves the proximal third of the radius or ulna, a posterior slab with buttress in 90 degrees of elbow flexion is preferred.
- The forearm should be elevated and iced to reduce swelling prior to definitive fixation. If patients are to be discharged from the emergency department with plans for outpatient surgery, they and their family should be carefully instructed in the signs and symptoms of compartment syndrome with instructions to return immediately to the hospital if any such symptoms develop.
- There is a risk of significant swelling with resulting vascular compromise and/or ischemic contracture (Volkmann's contracture) from these injuries. Unless the fracture is minimally displaced and due to a low-energy mechanism, these patients should generally be admitted for serial examinations and careful monitoring for compartment syndrome.

Admit and Discharge Guidelines

- Admit any patient with significant swelling for serial compartment checks.
- Request an orthopaedic consult in the emergency department for:
- All shaft fractures of the radius/ulna unless the fracture is closed, minimally displaced, neurovascularly intact with minimal swelling, and no concomitant injuries.
- Refer to an orthopaedic surgeon for outpatient follow-up for: All shaft fractures of the radius and ulna, which should have outpatient follow-up with an orthopaedic surgeon.

Definitive Treatment

- Nondisplaced or minimally displaced fractures may be treated nonoperatively in a long arm cast. Given the probability of swelling, no adult should be placed in a cast in the acute setting. A sugar tong splint or a posterior slab splint should be used until the swelling has subsided (usually 5–10 days).
- Fractures with significant displacement should be reduced prior to splinting, particularly if there is any tenting of the skin. Traction followed by a reduction maneuver should be attempted; however, these fractures are inherently unstable, so improvement of fracture displacement, rather than anatomic realignment, is usually the best that can be achieved.
- Surgical treatment options include ORIF and external fixation.

Shaft Fractures of the Radius and/or Ulna in Children

Symptoms and Findings

- Injuries are usually caused by a fall onto an outstretched hand causing indirect trauma to the radial and ulnar shaft. Fractures may also be the result of direct trauma.
- Patients present with pain and swelling. Significant deformity may be present, and the patient's wrist, elbow, and shoulder should be carefully evaluated for possible concomitant injury.
- A careful neurovascular exam should be performed and documented. The radial and ulnar arteries may be damaged, as can the median, radial, and ulnar nerves.
- Skin integrity should be carefully evaluated and documented as the superficial location of the ulna allows easy penetration of the skin. Additionally, these fractures may be significantly displaced, increasing the risk of skin breakdown.
- An initial compartment examination should be performed and documented. Compartment pressures should be measured in any patient with significant pain, tight compartments, or pain with passive extension of the fingers. Compartment syndrome is a surgical emergency, therefore compartment pressure measurements should only be performed by individuals able to emergently perform fasciotomy (generally orthopaedic and general surgeons). See Chapter 9 for further details on compartment syndrome.

Imaging (See Figure 3.24)

- X-rays: AP and lateral of the forearm; AP, lateral, and oblique of the wrist; AP and lateral of the elbow

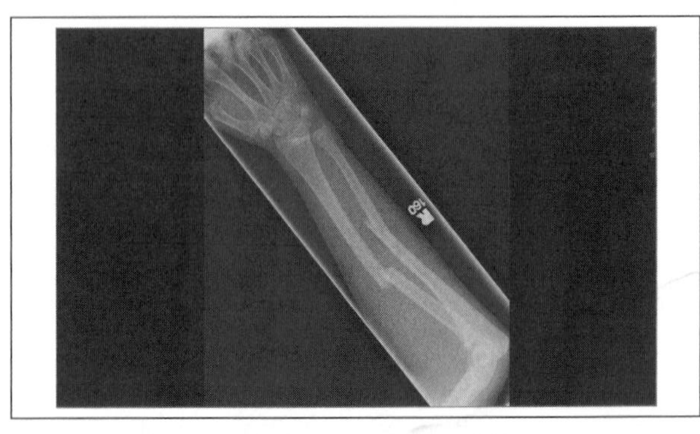

Figure 3.24 Both bone forearm fractures in children on PA X-ray.

- CT scan is generally not required in the emergency setting. CT angiogram may be required if an arterial injury is present.

Classification

- Descriptive classification: Closed versus open, undisplaced versus displaced, location, rotation, angulation, plastic deformation versus greenstick versus buckle versus complete
- Proximal ulna fracture with radial head dislocation: Monteggia (see page 112)
- Distal-third radius fracture with radioulnar dislocation: Galeazzi or Piedmont (see page 120)

Primary Stabilization and Management

- Most fractures can be treated nonoperatively. If a fracture is open, irreducible, unstable, or associated with a compartment syndrome, surgery will be required.
- Children older than 14 years should generally be treated as adults, given their reduced potential for bone remodeling.
- The tolerances for adequate reduction vary with age and fracture location. The younger the child, the greater the potential remodeling of deformities. Angular deformities will generally correct at 10 degrees per year while rotational deformities do not correct.
- Fractures should generally be reduced and splinted under conscious sedation or general anesthesia.
- Ideally, fractures should be reduced using fluoroscopy to optimize reduction. If fluoroscopy is unavailable, serial radiographs may be used.
- If a fracture is nondisplaced and there is minimal swelling, the patient may be placed in a long arm cast acutely. However, if there is any doubt as to the level of swelling, it is always safer to bivalve and overwrap the long arm cast given the risk of compartment syndrome.

Step-by-Step Guide to Reducing and Splinting a Radial Shaft, Ulnar Shaft, or Both Bone Fractures in a Child

1. Prepare materials for a long arm cast
2. Under conscious sedation or general anesthesia and after a full evaluation (including elbow radiographs), hang the patient's arm in traction using an IV pole, finger traps, or Kerlix around the fingers, and weights at the elbow (start with approximately 15 lbs of weight).
3. Obtain traction radiographs and/or use fluoroscopy to determine necessary reduction maneuver.
4. Exaggerate the deformity of the fracture to disengage the fracture fragments.

5. Apply longitudinal traction across the fracture, followed by rotational forces and direct pressure opposite to the deformation.
6. Evaluate provisional reduction by radiographs or fluoroscopy.
7. Apply a long arm cast. The cast should have a 3-point and interosseous mold (see section on long arm cast application in Chapter 2 (Figure 2.14)). The elbow should be positioned in 90 degrees of flexion. Forearm positioning depends on the fracture location:
   a. Proximal-third fractures: Forearm in supination
   b. Middle-third fractures: Forearm in neutral
   c. Distal-third fractures: Forearm in pronation
8. Bivalve and overwrap the cast.
9. Obtain postreduction radiographs, including AP and lateral. Formal radiographs should generally be obtained, unless fluoroscopy images can be saved for future reference.
10. Educate the patients and their family on the signs and symptoms of compartment syndrome. The patients and their family should be instructed to remove the bivalved cast and to return to the hospital should any symptoms develop.

Admit and Discharge Guidelines

- Admit any patient with significant swelling for serial compartment checks.
- Request an orthopaedic consult in the emergency department for: All shaft fractures of the radius/ulna unless the injury is closed and nondisplaced, incomplete (greenstick) or plastic deformation, neurovascularly intact with minimal swelling, and no concomitant injury to the wrist or elbow.
- Refer to an orthopaedic surgeon for outpatient follow-up for: All shaft fractures of the radius and ulna, which should have outpatient follow-up with an orthopaedic surgeon.

Definitive Treatment

- Closed treatment with a long arm cast is the preferred treatment of both bone forearm fractures in children.
- Surgical treatment consists of ORIF or intramedullary fixation.

Wrist and Hand Injuries

Distal-Third Radius Fractures and Radioulnar Dislocation (Galeazzi or Piedmont)

Symptoms and Findings

- Fractures are usually the result of a fall onto an outstretched hand or direct trauma to the forearm or wrist.

- Patients present with pain, swelling, and/or ecchymosis of the distal forearm with tenderness to palpation about the fracture site.
- Significant deformity may be present, and the elbow should be carefully examined to evaluate for possible concomitant injury. Limited elbow range of motion and/or limits with supination/ pronation may indicate a concomitant elbow injury, such as a radial head dislocation.
- A careful neurovascular exam should be performed. The anterior interosseus nerve AIN is at risk with this fracture, and the function of the flexor pollicis longus and flexor digitorum profundus to the index finger should be assessed by having the patient make an "OK" sign with his or her thumb and index finger.
- An initial compartment examination should be performed and documented. Compartment pressures should be measured in any patient with significant pain, tight compartments, or pain with passive extension of the fingers. Compartment syndrome is a surgical emergency, therefore compartment pressure measurements should only be performed by individuals able to emergently perform fasciotomy (generally orthopaedic and general surgeons).
- If a patient complains of elbow pain, injury to the interosseous ligament and/or proximal radioulnar joint should be considered.

Imaging (See Figure 3.25)

- X-rays: PA (posterior-anterior), lateral, and oblique of the wrist and AP and lateral of the elbow
- Instability of the distal radioulnar joint is indicated by ulnar styloid fracture, widening of the distal radioulnar joint on PA view, dislocation on lateral view, and greater than 5 mm radial shortening.

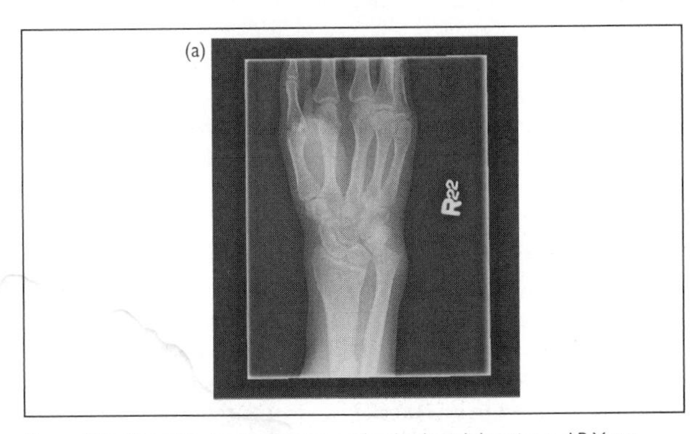

Figure 3.25 Distal-third radius fracture and radioulnar dislocation a. AP X-ray.

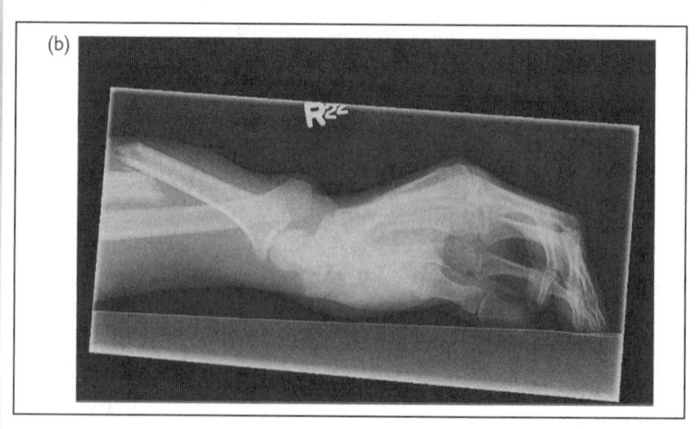

Figure 3.25 (Continued) b. Lateral X-ray.

- X-rays of the contralateral wrist can help in the evaluation of distal radioulnar joint (DRUJ) widening and radial shortening.
- CT-scan may be required to determine the integrity of the DRUJ, but it should not be ordered in the acute setting.

Classification

- In adults, fractures are classified by mechanism.
- Galeazzi fractures result from a fall onto an outstretched hand with the forearm in pronation.
- Reverse Galeazzi (fracture of the ulna and associated disruption of the distal radioulnar joint) fractures result from a fall onto an outstretched hand with the forearm in supination.
- In children, fractures are classified by the direction of displacement of the distal radius fracture or the distal ulnar dislocation.
- Dorsal displacement (apex volar) of the distal radius with the ulna volar to the radius caused by a supination force
- Volar displacement (apex dorsal) of the distal radius with the ulna dorsal to the radius caused by a pronation force
- Galeazzi equivalent: Distal radius fracture with a distal ulnar physeal fracture

Primary Stabilization and Management in Adults

- These fractures almost always require ORIF, which is why they have been referred to as "fractures of necessity."
- In the acute setting, the fracture should be reduced and a molded sugar tong splint should be applied. The splint should be well padded to minimize skin breakdown that might compromise definitive surgical fixation.

Primary Stabilization and Management in Children

- These fractures are generally treated nonoperatively in children 12 and under (children older than 12 should be treated as adults). Surgery is indicated if closed treatments fail to maintain fracture reduction.
- Fractures should generally be reduced and splinted under conscious sedation or general anesthesia.
- Ideally, fractures should be reduced using fluoroscopy to optimize reduction. If fluoroscopy is unavailable, serial radiographs may be used.
- If a fracture is undisplaced and there is minimal swelling, the patient may be placed in a long arm cast acutely. However, if there is any doubt as to the level of swelling, it is always safer to bivalve and overwrap the long arm cast, given the risk of compartment syndrome.

Step-by-Step Guide to Reducing and Splinting a Galeazzi or Galeazzi Equivalent Fracture in a Child

1. Prepare materials for a long arm cast
2. Initiate conscious sedation or general anesthesia.
3. Perform reduction maneuver using fluoroscopy or serial radiographs to:
   - Determine adequacy of the reduction. The fracture should be reduced using forces opposite to those that caused the injury.
   - Reduce radius dorsal/ulna volar fractures with forced pronation and a dorsal-to-volar force on the distal radius.
   - Reduce radius volar/ulna dorsal fractures with traction, forced supination, and a volar-to-dorsal force on the distal radius.
4. Evaluate provisional reduction by radiographs or fluoroscopy.
5. Apply a long arm cast. The cast should have a 3-point mold (see section on splinting). The elbow should be positioned in 90 degrees of flexion. Forearm positioning depends on fracture type.
   - Position radius dorsal/ulna volar fractures in pronation
   - Position radius volar/ulna dorsal fractures in supination
6. Bivalve and overwrap the cast.
7. Obtain postreduction radiographs, including AP and lateral. Formal radiographs should be obtained.
8. Educate the patients and their family on the signs and symptoms of compartment syndrome. The patients and their family should be instructed to remove the bivalved cast and to return to the hospital should any symptoms develop.

Admit and Discharge Guidelines

- Admit any patient with significant swelling for serial compartment checks.

- Request an orthopaedic consult in the emergency department for:
  - Any Galeazzi fracture or Galeazzi equivalent in a child
  - Fractures with neurovascular compromise, open fractures
  - Existence of concomitant fractures
- The orthopaedic surgery team should be notified of any fracture that might require surgical treatment.
- Refer to an orthopaedic surgeon for outpatient follow-up for: All Galeazzi fractures.

Definitive Treatment in Adults

- Nonoperative treatment may be pursued if a fracture is nondisplaced and there is no loss of radial bow. Nonoperative treatment consists of immobilization in a long arm cast with the wrist ulnarly deviated after swelling has subsided (approximately 5–10 days postinjury).
- Definitive surgical fixation usually involves open reduction and plate fixation of the fracture +/− percutaneous pinning of the DRUJ.

Definitive Treatment in Children

- Fractures are generally treated with a long arm cast.
- Surgery is indicated if closed treatments fail to maintain fracture reduction. Surgical management includes percutaneous pinning, intramedullary rods, and open reduction and plate fixation.

Distal Radius Fractures in Adults

Symptoms and Findings

- Most commonly caused by a fall on an outstretched hand
- Patients present with pain, swelling, and bruising about the wrist. Significant deformity may be present.
- The shoulder and elbow should be carefully examined as concomitant injuries are common.
- A careful neurovascular exam is mandatory, particularly as the median nerve may be compromised. Median nerve injury may be secondary to direct compression, traction, or increased pressure within the carpal tunnel.

Imaging (See Figure 3.26)

- X-rays: PA, lateral, and oblique of the wrist and AP and lateral of the elbow
- Traction views should be obtained after the patient has been fully examined and evaluated. If there is an injury around the elbow, traction should be avoided unless it can safely be performed without traction to the elbow.
- CT scan is rarely necessary and should not be ordered in the emergency setting.

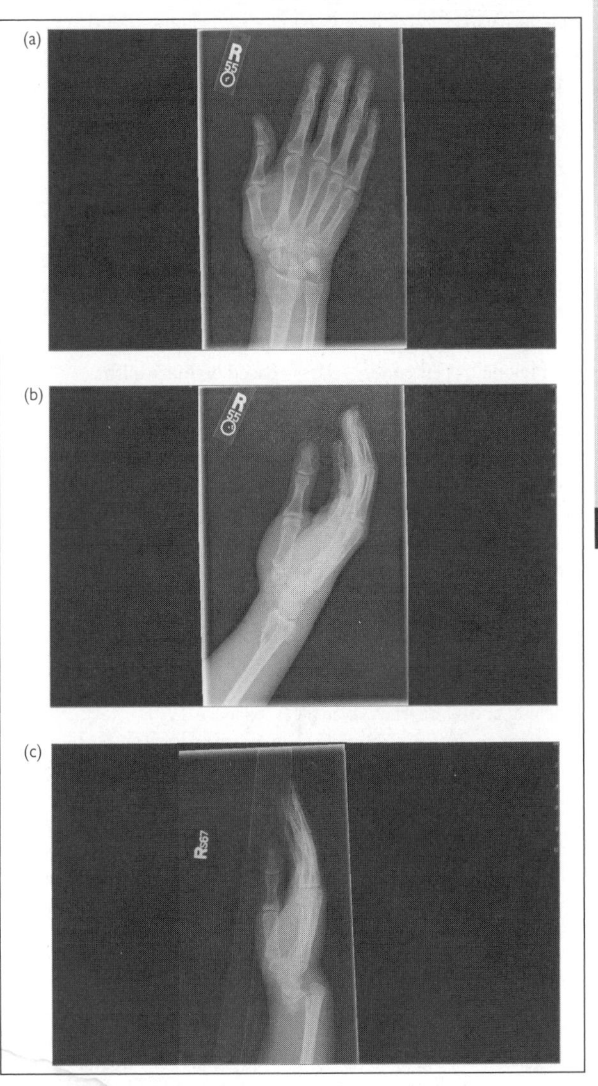

Figure 3.26 Distal radius fractures in adults. a. Colles-type distal radius fracture on PA X-ray. b. Colles-type distal radius fracture on lateral X-ray. c. Smith-type distal radius fracture on lateral X-ray.

Classification

- To improve communication between different services, it is best to describe the fracture rather than use a classification system or an eponym.
  - Open versus closed
  - Intra-articular versus extra-articular
  - Dorsal versus volar angulation
  - Displaced versus nondisplaced
  - Involvement of the radial styloid
  - Involvement of the ulnar styloid/distal ulna
- Multiple classification systems
  - Frykman—classify based on intra-articular involvement.
  - Melone or Fernandez—classify based on mechanism.
- Multiple eponyms
  - Colles—dorsal displacement (apex volar)
  - Smith (a.k.a. reverse Colles)—volar displacement (apex dorsal)
  - Barton—displaced articular lip fracture of the distal radius that may be associated with a carpal subluxation; may be volar or dorsal lip
  - Chaffeur's fracture (a.k.a. Hutchinson)—radial styloid fracture

Primary Stabilization and Management

- Goal of initial management is to restore the normal anatomy of the distal radius. The normal anatomic relationships within the distal radius are 23 degrees of radial inclination, 13 mm radial height, and 11 degrees volar tilt (see Figure 3.27).
- Tolerances for an acceptable reduction are: <5 degrees loss of radial inclination, <5 mm radial shortening, neutral volar/dorsal tilt, and <2 mm articular step-off.
- The need for surgery is determined by fracture stability as well as the quality of the reduction. Some fractures—such as Barton's fractures and fractures involving the radial and/or ulnar styloids—are notoriously unstable and generally require surgery. Even fractures that are perfectly reduced and splinted in the emergency department will often "slide off" and displace after the patient leaves the ER. For this reason, follow-up (within 1 week) and serial radiographs are needed to monitor for displacement, which indicates instability.
- Regardless of the need for surgery, all distal radial fractures, unless nondisplaced, should be closed reduced and placed in a molded sugar tong splint.
- There is a risk of significant swelling with resulting vascular compromise and/or ischemic contracture (Volkmann's contracture) from these injuries. Any patient with significant

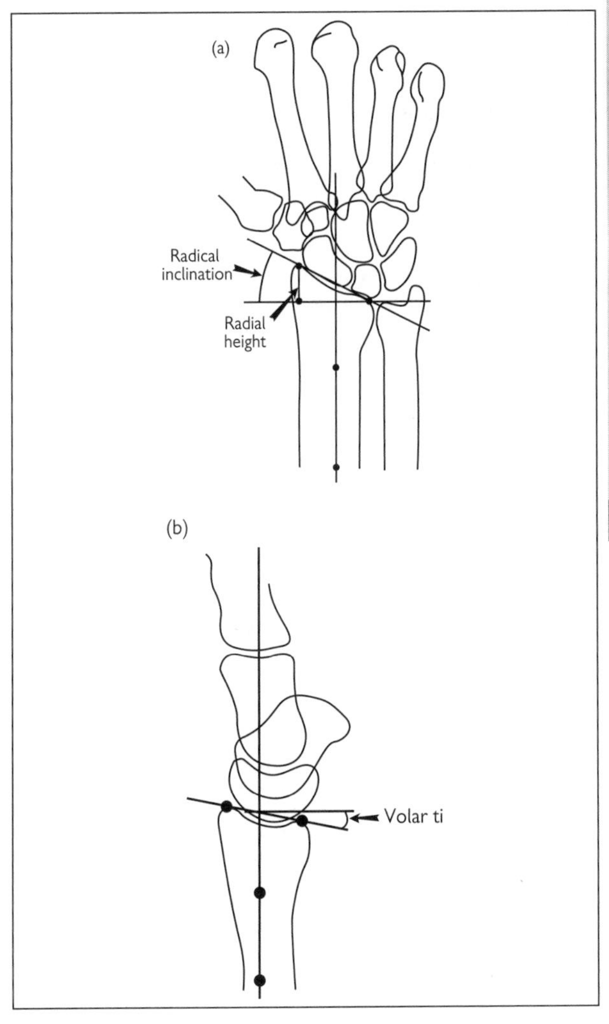

Figure 3.27 Diagram of the distal radius with the normal measurements and tolerances. a. AP schematic showing radial inclination and radial height measurements. b. Lateral schematic showing volar tilt measurement.

swelling should be admitted for serial examination and careful monitoring for signs of compartment syndrome.

Step-by-Step Guide to Reducing and Splinting a Distal Radius Fracture

1. After a full evaluation (including elbow radiographs), hang the patient's arm in traction using an IV pole, finger traps, or Kerlix around the fingers, and weights at the elbow (start with approximately 15 lbs of weights unless the patient is especially thin or muscular).
2. Obtain traction radiographs.
3. Evaluate the radiographs to determine reduction maneuver. Most fractures are dorsally displaced with apex volar, therefore reduction will require a volar force applied from the dorsal side with direct pressure on the apex volar fragment.
4. Measure and prepare splinting materials.
5. Analgesia: Generally, a hematoma block directly into the fracture +/− a bolus of IV pain medication is sufficient analgesia (see Chapter 8, Figure 8.4, page 314). Conscious sedation may be used in certain emergency departments.

6. Perform reduction maneuver. If an assistant is available, the assistant can provide traction while the wrist is flexed and pressure is applied to the displaced fragment. One key point to remember when reducing a distal radius fracture is that the distal radius remains attached to the carpal bones through the radiocarpal ligaments, and these ligaments allow for the transition of traction forces across the hand and into the fracture. In a classic Colles fracture, dorsiflexion of the wrist prior to applying a force will allow the fracture to key in to the fracture plane from its stepped-off position (see Figure 3.28).
7. Apply a well-molded sugar tong splint. REMEMBER: Crooked splints make straight bones! The splint should look like an elongated S if properly molded. See Chapter 3, "Basic Techniques," for more information on splints.
8. Postreduction radiographs should always be obtained to verify fracture reduction and document splint application.

Admit and Discharge Guidelines

- Request an orthopaedic consult in the emergency department for:
  - Open fractures
  - Fractures with associated neurovascular compromise
  - Highly comminuted fractures, tenting of the skin by fracture fragments
  - Irreducible fractures
  - Fractures associated with elbow fractures or dislocations
- The orthopaedic surgery team should be notified of any fracture that might require surgical treatment.

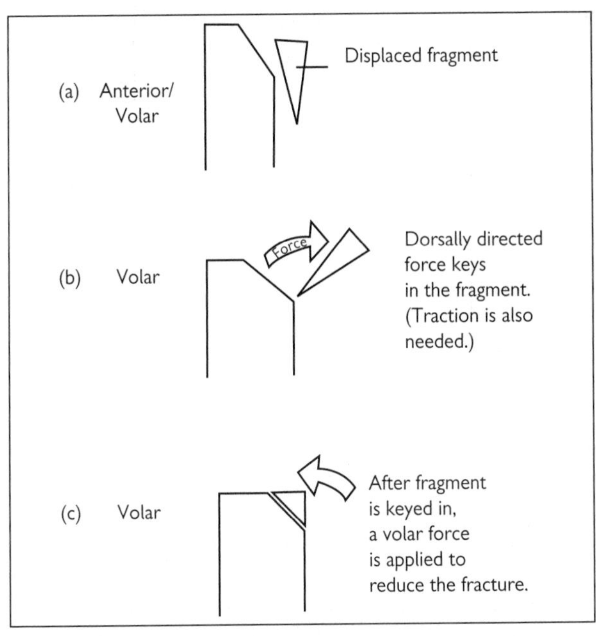

Figure 3.28 Diagram showing how dorsiflexion of the wrist keys in the fragment in a classic Colles fracture.

- Refer to an orthopaedic surgeon for outpatient follow-up for: All distal radius fractures.

Definitive Treatment

- Nonsurgical management is pursued for fractures meeting the tolerances described above. Fractures must be followed closely with serial radiographs so that any displacement is seen as early as possible. A sugar tong splint is initially maintained, followed by short arm cast application.
- Surgical treatment options include ORIF with a plate and screws, closed reduction and percutaneous pinning, and external-fixator placement. Specially designed low-profile external fixators have been designed for distal radius fractures.

Distal Radius Fractures in Children

Symptoms and Findings

- Most commonly caused by a fall on an outstretched hand
- Patients present with pain and swelling about the wrist with tenderness to palpation about the fracture site. The patient may refuse to use the arm.

- The shoulder and elbow should be carefully examined as concomitant injuries are common.
- A careful neurovascular exam is mandatory, particularly as the median nerve may be compromised. Median nerve injury may be secondary to direct compression, traction, or increased pressure within the carpal tunnel. Symptoms of median nerve compression include paresthesias in the thumb, index finger, and/or middle finger as well as a positive Tinel's sign over the median nerve at the wrist.

Imaging (See Figure 3.29)

- X-rays: PA, lateral, and oblique of the wrist and AP and lateral of the elbow

Classification

- Physeal injuries are classified using the Salter-Harris classification.

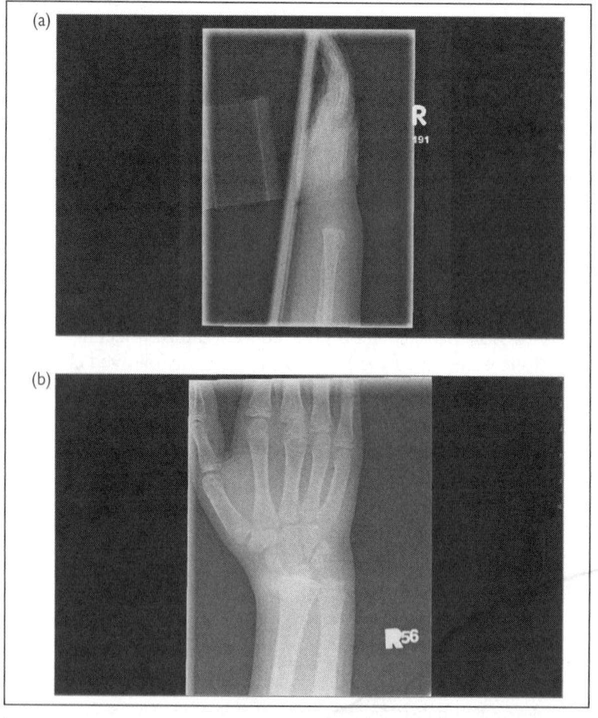

Figure 3.29 Distal radius fracture in children. a. Pediatric buckle distal radius fracture on a lateral X-ray. b. Pediatric distal radius physeal injury on PA X-ray.

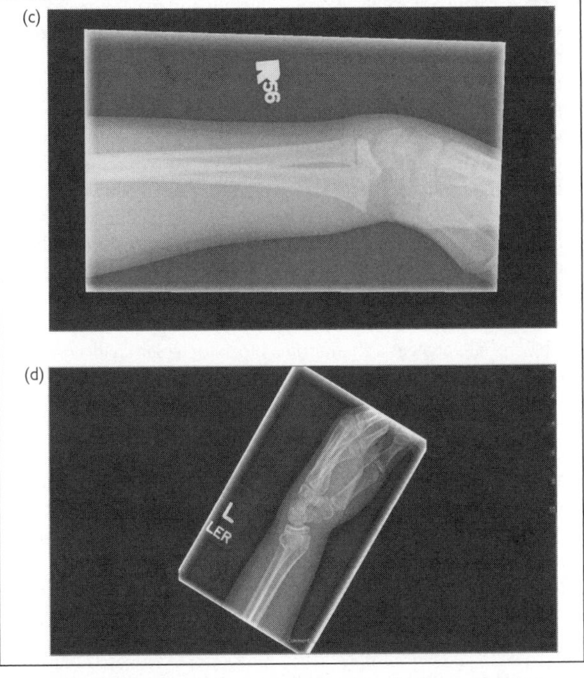

Figure 3.29 (Continued) c. Pediatric distal radius physeal injury on a lateral X-ray. d. Pediatric distal radius metaphyseal injury on a lateral X-ray.

- Metaphyseal Injuries are classified according to the biomechanical pattern:
  - Torus or buckle fracture: The cortex under compression has buckled while the cortex under tension is intact.
  - Incomplete or Greenstick fracture: The cortex under tension has plastic deformation and is disrupted while the cortex under compression is intact.
  - Complete fracture: Both cortices are disrupted.

Primary Stabilization and Management

- Treatment depends upon fracture type.
- Salter-Harris I and II fractures are treated with closed reduction and application of a bivalved, long arm cast with the forearm in pronation. No more than two reduction attempts should be performed for physeal fractures as repeated attempts increase the risk of growth arrest.
- Surgery is generally indicated for irreducible Type I and II fractures and all Type III and IV fractures.

- Torus fractures rarely require reduction and may be treated with a short arm cast. If there is minimal swelling, the cast does not need to be bivalved, but in most cases it should be.
- Incomplete fractures should be treated with closed reduction if there is greater than 10 degrees of angulation. Reduction generally requires completion of the fracture through the compressed cortex, and appropriate analgesia (conscious sedation or general anesthesia) should be given. Following reduction, a bivalved, long arm cast should be applied with the forearm in supination. Reduction should only be performed by an experienced individual.
- Complete fractures can generally be treated with reduction and application of a well-molded, bivalved, long arm cast. Appropriate analgesia (hematoma block for older children versus conscious sedation or general anesthesia) should be given prior to attempting a reduction maneuver. Complete fractures may require surgery if: Swelling prevents application of a well-molded cast, fracture reduction is lost, adequate reduction is not obtained after multiple attempts, or there is the existence of a concomitant elbow injury requiring surgery.
- The benefit of splint immobilization prior to surgical treatment must be balanced against the risk of increased swelling and possible physeal injury. If surgery is planned, the decision about temporary splint application rests with the operating surgeon.

Admit and Discharge Guidelines

- Request an orthopaedic consult in the emergency department for: Any fracture that requires a reduction maneuver or might require surgery, fractures with neurovascular compromise, concomitant fractures of the elbow or hand.
- Refer to an orthopaedic surgeon for outpatient follow-up for: All distal radius fractures.

Definitive Treatment

- Nonsurgical management is performed by application of a well-molded, long arm cast that is initially bivalved then overwrapped.
- Surgical treatment is required for irreducible fractures and Type III and IV injuries. Surgical treatment for both physeal and metaphyseal fractures is typically with closed reduction and percutaneous pinning with the goal of restoring normal anatomic alignment. Open reduction may be necessary for irreducible fractures.

Scaphoid Fractures

Symptoms and Findings

- Most commonly caused by a fall on an outstretched hand
- Patients present with pain made worse with grip and radial wrist swelling

- The classic physical examination finding is tenderness to palpation in the anatomic snuffbox. Even if no fracture is seen on X-ray, a scaphoid fracture should be presumed in any patient with snuffbox tenderness following trauma. The scaphoid tuberosity should also be assessed for tenderness.
- A scaphoid fracture should always be suspected in the presence of a radial styloid or distal radius fracture.
- The wrist, elbow, and shoulder should be carefully examined as concomitant injuries are common.

Imaging (See Figure 3.30)

- X-rays: Scaphoid series that should be performed:
  - PA of the wrist in ulnar deviation
  - PA of the wrist in 45 degrees pronation and ulnar deviation
  - AP of the wrist with 30 degrees supination and ulnar deviation
  - Lateral of the wrist in neutral
- X-rays: PA, lateral, and oblique of the distal radius should be taken if distal radius fracture is suspected and/or cannot be visualized in the scaphoid series.
- CT scan and MRI are frequently used to diagnose occult scaphoid fractures. While CT and MRI provide increased sensitivity in the detection of scaphoid fractures, their utility in the emergency setting is debated. If no fracture is seen on X-ray but there is high clinical suspicion of a fracture, the patient should be treated with thumb-spica immobilization and follow-up visit as an outpatient in 10–14 days for reexamination and repeat imaging as appropriate.

Classification

- Scaphoid fractures may be classified by their displacement, location, or fracture pattern. The anatomic classification system is

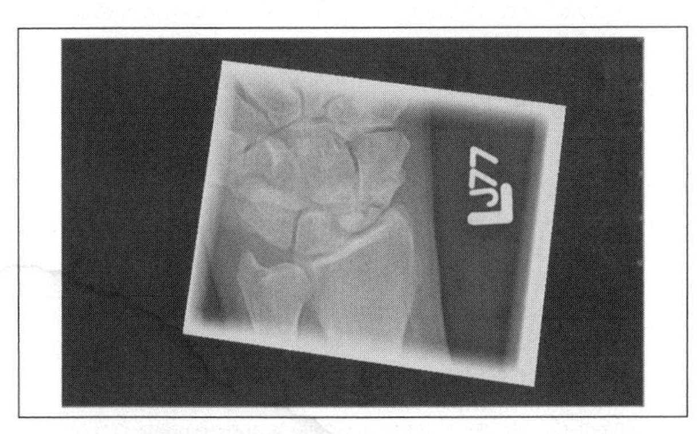

Figure 3.30 Adult scaphoid fracture on PA X-ray.

most useful for predicting outcomes with nonoperative treatment because of the tenuous blood supply to the scaphoid.

- Proximal pole fractures have the poorest blood supply, the highest risk of avascular necrosis (AVN) and nonunion, and the longest time to healing with nonoperative treatment. Conversely, distal pole fractures have the best blood supply, the lowest risk of AVN, and the shortest time to healing with nonoperative treatment. Middle-third or scaphoid waist fractures have intermediate blood supply, risk of AVN, and time to healing with nonoperative treatment.
- Scaphoid fractures associated with dislocations are discussed in the section "Carpal Dislocations and Ligamentous Injuries of the Wrist" (see page 137).

Primary Stabilization and Management

- As a general rule, all displaced scaphoid fractures will require surgery while nondisplaced fractures are initially treated with immobilization. Regardless of surgical planning, all scaphoid fractures should be splinted in the acute setting.
- Isolated distal-third fractures should initially be treated in a radial gutter-thumb-spica splint.
- Isolated middle- and proximal-third fractures should initially be treated in a sugar tong–thumb–spica splint.
- Scaphoid fractures frequently occur with distal radius/ulna fractures. Initial treatment (i.e., reduction and splinting) is the same for these injuries, except that a thumb spica should be added to the sugar tong splint.

Admit and Discharge Guidelines

- Request a hand consult in the emergency department for:
  - Any scaphoid fracture likely to require surgery
  - Scaphoid fractures associated with distal radius/ulna fractures
  - Open fractures
  - Fractures with neurovascular compromise
- Refer to a hand surgeon for outpatient follow-up for: All documented or suspected scaphoid fractures.

Definitive Treatment

- Nonsurgical management consists of thumb spica cast application after initial swelling has subsided. Isolated distal-third fractures are transitioned into a short arm thumb spica cast. Isolated middle- and proximal-third fractures are transitioned from a splint into a long arm thumb spica cast and eventually into a long arm thumb spica cast. Middle- and proximal-third scaphoid fractures must be immobilized longer than distal-third scaphoid fractures as there is an increased risk of avascular necrosis and/or nonunion.

- Surgical treatment of scaphoid fractures consists of open reduction and internal fixation using pins or screws.

Other Carpal Fractures

Symptoms and Findings

- Most carpal fractures result from falls onto an outstretched hand or indirect injury with the wrist in extension. Fractures also occur secondary to direct trauma.
- Hamate fractures are usually due to direct trauma or repetitive trauma (e.g., golf, baseball, racquet sports, and motorcycling). The hamate should be gently tapped to assess for paresthesias indicating involvement of the sensory branch of the ulnar nerve.
- Patients present with pain, swelling, and tenderness with palpation.
- The wrist, elbow, and shoulder should be carefully examined as concomitant injuries are common.

Imaging (See Figure 3.31)

- X-rays: PA, lateral, and oblique of the wrist
- Special radiographic views can help in the visualization of carpal fractures:
  - Carpal tunnel view: Pisiform, trapezium, and/or hamate fractures
  - Scaphoid series: Capitate fractures
  - Lateral or PA oblique views with the wrist pronated: Trapezoid, trapezium, and/or triquetrum fractures
- X-rays of the contralateral wrist can aid in the evaluation of carpal fractures.
- Advanced imaging (CT scan, MRI, and bone scan) may be needed to diagnose difficult-to-visualize fractures, but they should not be routinely ordered in the acute setting.

Classification

- Descriptive classification is preferred:
  - Location: Body, ridge, tubercle, hook
  - Pattern: Oblique, transverse, shear, comminuted
  - Open versus closed
  - Associated fractures and ligamentous injury
- Anatomic relationships dictate injury patterns. Carpal fractures are associated with fractures and dislocations to adjacent structures:
  - Pisiform fractures: Associated with distal radius, hamate, or triquetrum fractures
  - Trapezium fractures: Associated with carpometacarpal dislocations and first metacarpal fractures

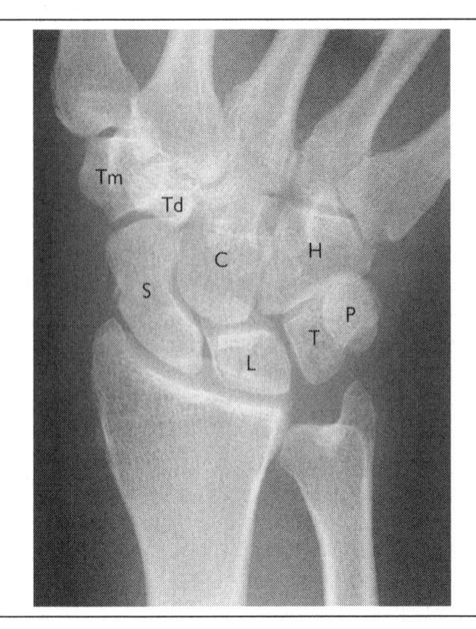

Figure 3.31 X-ray showing the distal radius and ulna with the carpal bones and metacarpal bones. Reprinted with permission from Bucholz, R.W., Heckman, J.D., Court-Brown, C.M., and Tornetta, P. *Rockwood and Green's Fractures in Adults*, 7th ed. Lippincott Williams & Wilkins, 2009.

- Trapezoid fractures: Associated with carpometacarpal dislocations and second metacarpal fractures
- Capitate fractures: Associated with perilunate dislocations, scaphoid fractures, and carpometacarpal fractures and dislocations
- Hamate fractures: Associated with fourth and fifth metacarpal fractures

Primary Stabilization and Management

- As a general rule, displaced carpal fractures will require surgery while nondisplaced fractures can be treated with immobilization. Other indications for surgical intervention are open fractures and nonunion after attempted nonoperative treatment.
- Regardless of surgical planning, all carpal fractures should be splinted in the acute setting.
- Splint choice depends on the location of the fracture as well as associated injuries. The goal of the splint is to immobilize the carpal bone as well as protect it from force transmission, which might cause displacement and nonunion:
  - Radial gutter thumb-spica splints may be used for radial-sided carpal injuries, such as trapezium and trapezoid fractures. Ulnar gutter splints may be used for ulnar-sided carpal injuries, such as pisiform, triquetrum, and hamate fractures.

- A volar slab splint may be used for central carpal injuries, such as lunate or capitate fractures.
- Isolated carpal fractures are difficult to diagnose and are often missed, leading to nonunion and avascular necrosis, requiring surgery. For this reason, every patient who presents with a history and physical exam findings consistent with a carpal fracture should be placed in a splint with instructions to follow-up unless he or she experiences complete resolution of the symptoms.

Admit and Discharge Guidelines

- Request a hand consult in the emergency department for:
  - Any carpal fracture likely to require surgery
  - Carpal fractures associated with fractures or dislocations requiring reduction
  - Open fractures, fractures with neurovascular compromise
- Refer to a hand surgeon for outpatient follow-up for: All documented or suspected carpal fractures.

Definitive Treatment

- Nonsurgical treatment consists of transition into a short arm cast after initial swelling has subsided.
- Surgical management is by ORIF with screws or wires.

Carpal Dislocations and Ligamentous Injuries of the Wrist

Symptoms and Findings

- Injury is usually caused by a fall on an outstretched hand or by direct trauma (often high energy). In perilunate instability injuries, the deforming forces are hyperextension and supination.
- Patients present with pain, ecchymosis, swelling, and/or deformity about the wrist. The dorsum of the wrist is typically more prominent in perilunate dislocations while the palmar surface of the wrist may be more prominent in scapholunate dissociation.
- The patient will have tenderness to palpation of the wrist, and Watson's test is commonly positive:
  - Watson's test: Grasp the patient's wrist placing your thumb over the scaphoid tuberosity at the palmar surface of the wrist. Move the wrist from ulnar to radial deviation. Disruption of the scapholunate ligament should be suspected if there is a more significant "clunk" (representing lunate subluxation), with associated pain, of the injured wrist compared with the contralateral side.
- The wrist, elbow, and shoulder should be carefully examined as concomitant injuries may occur.
- A careful neurovascular exam is important with attention to the median nerve, which may be compressed in the carpal tunnel and require urgent surgical release.

Imaging (See Figure 3.32a–b)

- X-rays: PA and lateral views of the wrist; special radiographic views can help in the visualization of carpal injuries including a clenched fist PA, radial deviation views, and ulnar deviation views.
- X-rays of the contralateral wrist can aid in the evaluation of carpal injuries.
- Ligamentous injuries are difficult to detect on plain radiographs and require knowledge of normal anatomic relationships:
  - Terry Thomas sign: >3 mm gap between the scaphoid and lunate; indicates scapholunate dissociation; best seen on clenched fist PA X-ray
  - "Spilled teacup sign": Palmar tilt of the lunate on lateral radiographs indicating lunotriquetral (LT) ligament (and dorsal radiotriquetral ligament) injury or perilunate dislocation; there will also be loss of colinearity of the capitate, lunate, and radius in perilunate dislocation.
  - Dorsal scapholunate angle > 70 degrees: Dorsal intercalated segmental instability (DISI); dorsal tilt of lunate on lateral radiograph; suggests scapholunate ligament injury
  - Volar scapholunate angle < 35 degrees: Volar intercalated segmental instability (VISI); volar tilt of lunate on lateral radiograph; suggests lunotriquetral (and dorsal radiotriquetral) ligament injury

Classification (See Figure 3.32c)

- Mayfield classification for perilunate instability: The stage of injury is determined by the sequential progression of ligament disruption.

Mayfield stage	Involved joint	Ligament disrupted
I	Scapholunate	Scapholunate
II	Capitolunate	Radioscaphocapitate
III	Triquetrolunate	Distal limb of the radiolunotriquetral
IV	Radiolunate	Dorsal limb of the radiolunotriquetral

- The complete disruption of the radiolunotriquetral ligament leads to dislocation of the lunate.
- Injuries can be purely ligamentous, but transosseous injuries also occur creating a fracture dislocation. These injuries are described by the transosseous injury followed by the ligamentous injury. For example:

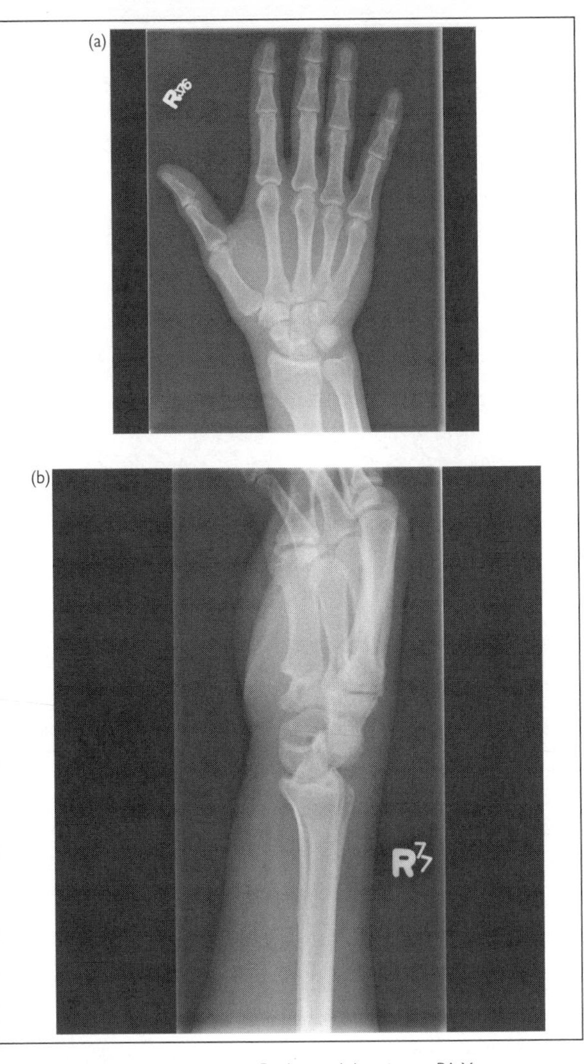

Figure 3.32 Perilunate dislocation. a. Perilunate dislocation on PA X-ray. b. Perilunate dislocation on lateral X-ray.

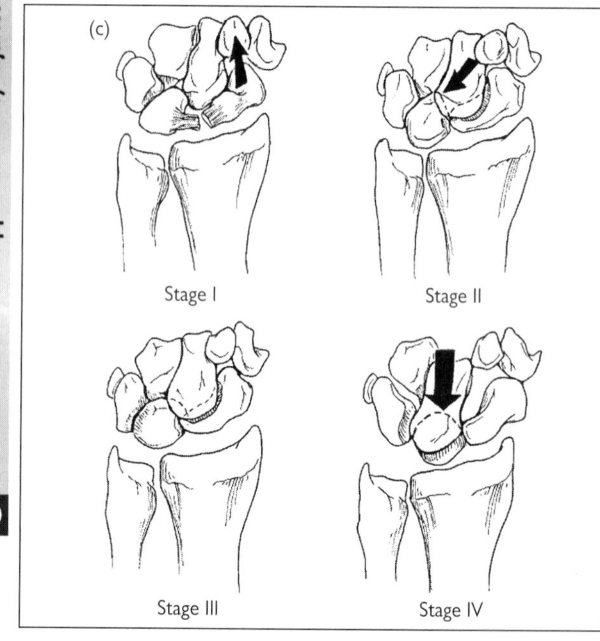

Figure 3.32 (Continued) c. Line drawing of the Mayfield classification. Stages of progressive perilunar instability. Stage I involves disruption of the scapholunate ligamentous complex. In Stage II, the force propagates through the space of Poirier and interrupts the lunocapitate connection. In Stage III, the lunotriquetral connection is violated, and the entire carpus separates from the lunate. In Stage IV, the lunate dislocates from its fossa into the carpal tunnel, the lunate rotates into the carpal tunnel, and the capitate becomes aligned with the radius. ©1998 American Academy of Orthopaedic Surgeons. Reprinted from the *Journal of the American Academy of Orthopaedic Surgeons*, Volume 6(2), pp. 114–120, with permission.

- Transscaphoid perilunate dislocation: Scaphoid fracture + perilunate dislocation
- Transradial styloid perilunate dislocation: Radial styloid fracture + perilunate dislocation

Primary Stabilization and Management

- Prompt reduction of the dislocation and repair of ligamentous injury is imperative, and these injuries are frequently treated with immediate surgery.
- If surgery must be delayed, closed reduction should be performed. While practices vary by hospital, a hand surgeon should generally be notified prior to attempted reduction in case of irreducible dislocation or the development of acute carpal tunnel compression.

- Reduction of carpal dislocations should be performed with longitudinal traction. The patient's arm should be hung in traction using finger traps or Kerlix as would be done for a distal radius fracture (see page 124). Traction alone will sometimes be sufficient to reduce the dislocation. After 5–10 minutes of traction, a reduction maneuver should be performed based on the radiographs. For example, in a volar dislocation of the lunate, reduction requires a dorsal force applied from the volar side to the lunate while a volar force is applied from the dorsal side to the perilunate bones. Sedation analgesia should be given prior to attempting reduction. A volar slab splint should be applied postreduction with modification as needed (i.e., addition of a thumb spica component for a scaphoid fracture). Postreduction radiographs should always be obtained to verify adequate reduction and document splint application.

Admit and Discharge Guidelines

- Request a hand consult in the emergency department for: Open injuries, all dislocations, neurovascular symptoms, especially acute carpal tunnel syndrome, fracture dislocations.
- Refer to a hand surgeon for outpatient follow-up for: All ligamentous injuries or dislocations of the carpal bones.

Definitive Treatment

- Surgical treatment is usually pursued for these injuries. Surgical treatment includes closed versus open reduction with pin or screw fixation and/or ligament repair. External fixation may also be used. Carpal tunnel release is often performed at the same time.

Thumb Fractures and Dislocations

Symptoms and Findings

- Fractures are usually the result of axial loading injuries. The small size of the metacarpal bones means a reduced force is required to fracture these bones compared with most other injuries.
- Dislocations of the first metacarpal are typically caused by an abduction or forced opposition injury, frequently due to falls.
- Patients present with pain and swelling. The skin should be carefully examined for any disruptions, particularly those secondary to animal or human bites.
- The presence and degree of any rotational and/or angular deformity should be noted and documented. The range of motion at the metacarpophalangeal and interphalangeal joint should also be evaluated and documented as compared to the contralateral side.

- Assess thumb MCP joint stability by applying a valgus stress to the thumb with the MCP in flexion and stabilized. An increase in laxity compared to the contralateral side indicates ulnar collateral ligament injury.
- A neurovascular exam—including capillary refill and two-point discrimination—should be performed and documented.
- The wrist and other fingers should be carefully examined to evaluate for other injuries.

Imaging (See Figure 3.33a)

- X-rays: PA, lateral, and oblique of the thumb
- These fractures may also be easily imaged dynamically using a mini C-arm machine

Classification

- Green classification: Classifies by articular involvement and fracture pattern (see Figure 3.33b) .
  - I: Bennett's fracture: Oblique, intra-articular fracture of the first metacarpal base separating a small triangular volar lip fragment from the proximally displaced metacarpal shaft
  - II: Rolando's fracture: Y- or T-shaped intra-articular fracture of the first metacarpal base
  - IIIa: Transverse extra-articular fracture
  - IIIb: Oblique extra-articular fracture
  - IV: Epiphyseal fracture (usually Salter-Harris II)

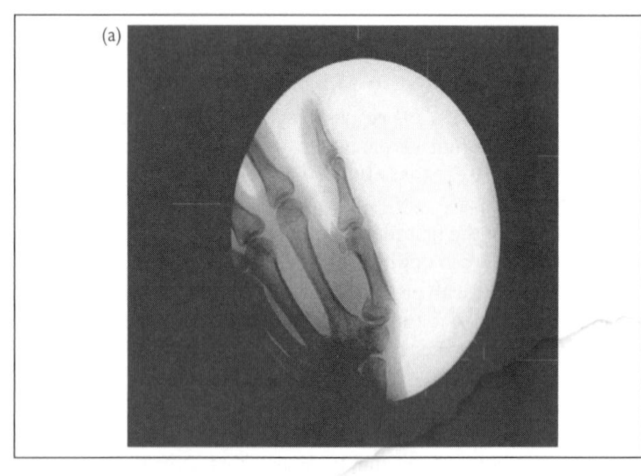

Figure 3.33 Thumb fractures. a. Thumb fracture X-ray.

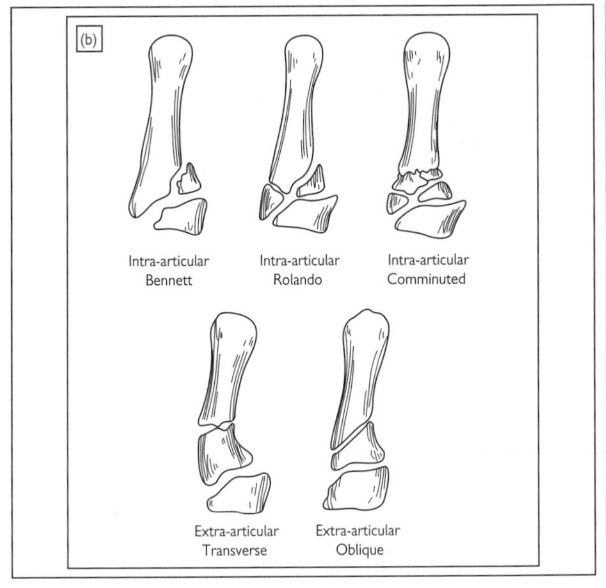

Figure 3.33 (Continued) b. Diagram of Green's classification of thumb fractures

- First metacarpophalangeal joint dislocation: Classified by ligamentous injury (ulnar collateral ligament vs. radial collateral ligament) and bony involvement. Injuries of the ulnar collateral ligament have traditionally been called "gamekeeper's thumb."

Primary Stabilization and Management

- Surgical treatment is required for open fractures, intra-articular fractures, unstable extra-articular fractures, irreducible dislocations, and dislocations with bony involvement (usually the interposed volar plate).
- A hand surgeon should be consulted for all open injuries. Initial treatment in the emergency department should consist of irrigation and antibiotics. A short arm thumb spica splint may be used as a temporizing measure for pain control and soft tissue rest prior to planned operative treatment.
- Reduction should be attempted on all Type III or IV fractures. The goal is anatomic reduction, but up to 30 degrees of displacement may be tolerated. After reduction, a thumb spica splint should be applied.
- Reduction should be attempted on all dislocations without a bony component. A digital or metacarpal block can be

performed prior to attempting reduction (see Chapter 8, Figures 8.1–8.3, pages 312, 313). Open repair is generally required in "complex dislocations" where the volar plate is interposed within the metacarpophalangeal joint. A thumb spica splint should be applied after reduction of the dislocation.

- Regardless of planned surgical treatment, all closed injuries should be placed in a thumb spica splint.

Admit and Discharge Guidelines

- Request a hand consult in the emergency department for:
  - Open fractures
  - Unstable fractures
  - Intra-articular fractures
  - Irreducible dislocations
  - Dislocations with bony involvement
- Refer to a hand surgeon for outpatient follow-up for: All thumb fractures and dislocations.

Definitive Treatment

- Nonoperative management consists of transition to a short arm thumb spica cast once swelling has resolved.
- Surgical treatment of Bennett's fractures, Rolando's fractures, and unstable fractures is by open reduction and internal fixation using screws or closed reduction, percutaneous pinning, and splint application. The decision between open and closed treatment is typically based on whether there is sufficient bone for screw placement (ORIF).

Metacarpal Fractures

Symptoms and Findings

- Fractures are usually the result of axial loading injuries. The small size of the metacarpal bones means a reduced force is required to fracture these bones compared with most other injuries. These injuries are often the result of punching something.
- Patients present with pain and swelling. The skin should be carefully examined for any disruptions, particularly those secondary to animal or human bites.
- "Fight bites" are usually half-moon-shaped lacerations on the dorsum of the hand at the metacarpophalangeal joint. Fight bites always require formal incision and debridement in the operating room.
- The presence and degree of any rotational and/or angular deformity should be noted and documented. The range of motion at the metacarpophalangeal joint should also

be evaluated and documented as compared to the contralateral side.

- Assess rotational deformity by asking the patient to gently flex his or her fingers. All fingers should point toward the scaphoid. Asymmetry suggests a rotational deformity.
- A neurovascular exam—including capillary refill and two-point discrimination—should be performed and documented.
- The wrist, thumb, and other fingers should be carefully examined to evaluate for other injuries.

Imaging (See Figure 3.34)

- X-rays: PA and lateral of the hand; PA, lateral, and oblique of the affected finger
  - Special views: Metacarpal base fractures (particularly the fourth and fifth metacarpal) may be better visualized by taking an AP

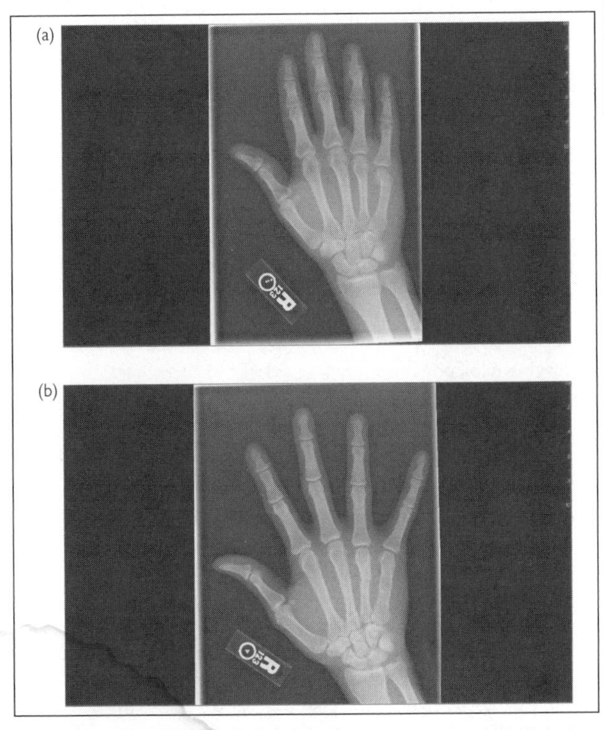

Figure 3.34 Metacarpal fractures. a. 5th Metacarpal head fracture on PA X-ray. b. Transverse 4th metacarpal fracture on PA X-ray.

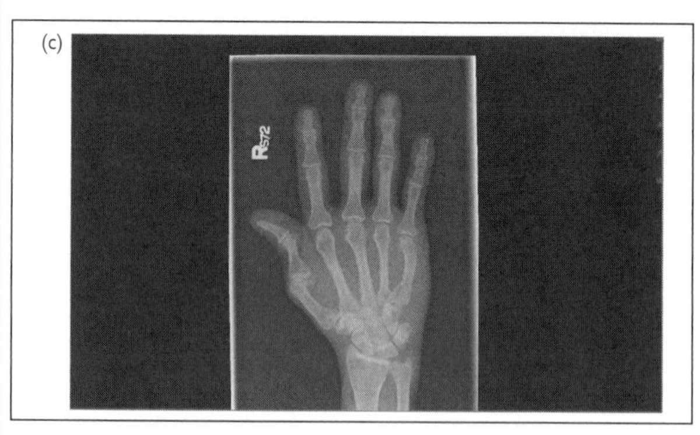

Figure 3.34 (Continued) c. 5th Metacarpal base fracture with intra-articular involvement on PA X-ray.

X-ray with the hand supinated 45–60 degrees from the neutral position.

Classification

- While multiple classification systems and eponyms exist, it is better to describe the fracture pattern than to name or classify the injury. A full description should include:
  - Location: Head, neck, shaft, or base
  - Fracture pattern: Transverse, spiral, oblique, comminuted
  - Open versus closed
  - Intra-articular versus extra-articular
  - Deformity: Angulation, rotation, shortening, translation/displacement
- Unavoidable Eponyms
  - Boxer's fracture: Fracture of the fifth metacarpal neck with volar displacement of the metacarpal head
  - Baby Bennett or reverse Bennett: Fracture/dislocation of the fifth metacarpal base, usually with intra-articular extension

Primary Stabilization and Management

- Metacarpal fractures should be treated with reduction and splinting. There is a significant risk of stiffness of the metacarpophalangeal joints, and the hand should always be immobilized in the "protected position" (also called *intrinsic plus* or *Edinburg position*): wrist in 30 degrees of extension, MCP joints in 90 degrees of flexion, and IP joints in full extension (see Chapter 2, Figure 2.10, page 38). Failure to use the protected position may result in permanent contracture of the intermetacarpal collateral ligaments and permanent loss of hand range of motion.

- Reduction maneuver depends on the deformity. Often, fractures can be reduced as part of splint application. Fractures that require more significant manipulation usually need surgery as a splint is unlikely to hold the reduction.
- Radiographs should always be obtained after splinting to evaluate for adequate reduction and document splint application.
- The need for surgical treatment is determined by fracture stability. In general, open fractures, intra-articular fractures, comminuted fractures, and multiple metacarpal fractures require surgery.

Tolerances for Closed Treatment of Metacarpal Fractures

- No rotational deformity
- Acceptable angulation depends on the metacarpal injured and the location of the injury:
  - Metacarpal head: Requires anatomic reduction; displaced fractures require ORIF versus CRPP.
  - Metacarpal neck: <10 degrees accepted for second and third metacarpal; <30 degrees accepted for fourth metacarpal; <40 degrees accepted for fifth metacarpal (boxer's fracture)
  - Metacarpal shaft: <10 degrees accepted for second and third metacarpal; <30 degrees accepted for fourth and fifth metacarpal
  - Metacarpal base: < 1 mm of articular step-off

Admit and Discharge Guidelines

- Request a hand consult in the emergency department for:
  - Open fractures
  - Unstable fractures
  - Multiple metacarpal fractures
  - Rotational deformity
  - Intra-articular fractures
- Refer to a hand surgeon for outpatient follow-up for: All metacarpal fractures.

Definitive Treatment

- Nonsurgical management generally consists of transition to a short arm cast after swelling has resolved.
- Surgical treatment consists of open reduction and internal fixation using plates and/or screws versus closed reduction and percutaneous pinning. Pins can be directed into the medullary canal or across adjacent metacarpals.

Phalanx Fractures

Symptoms and Findings

- Fractures are usually the result of a direct blow, or twisting or crush injuries.

- Patients present with pain and swelling. The skin should be carefully examined for any disruptions, particularly those secondary to animal or human bites.
- The presence and degree of any rotational and/or angular deformity should be noted and documented. The range of motion at the metacarpophalangeal and interphalangeal joints should also be documented and compared to the contralateral side.
- A neurovascular exam—including capillary refill and two-point discrimination—should be performed and documented.
- The wrist, thumb, and other fingers should be carefully examined to evaluate for other injuries.
- Check rotational deformity

Imaging (See Figure 3.35)

- X-rays: PA and lateral of the hand; PA, lateral, and oblique of the affected finger

Classification

- Fractures should be described rather than classified:
  - Location: Condyle, neck, diaphysis, base
  - Fracture pattern: Transverse, spiral, oblique, comminuted
  - Open versus closed
  - Intra-articular versus extra-articular
  - Deformity: Displacement, angulation, rotation, shortening
  - Associated dislocation and/or tendon injury

Primary Stabilization and Management

- Extra-articular, nondisplaced fractures can generally be treated with dynamic splinting (also known as buddy taping).
- Displaced, extra-articular fractures should be treated with reduction and splinting. Immobilization is required to maintain the reduction, and either a plaster splint or prefabricated aluminum/foam splint should be used. The proximal interphalangeal (PIP) joint should not be immobilized in distal phalanx fractures. Follow-up radiographs are required to ensure the fracture remains reduced.
- Intra-articular fractures of the proximal and middle phalanx require anatomic reduction and splinting. Surgery is required if there is greater than 1 mm of displacement.
- Intra-articular fractures of the distal phalanx are usually due to tendon injury (see Mallet finger and Jersey finger, page 151).
- Distal phalanx fractures are associated with soft tissue damage and nail bed injury (see nail bed injuries in Chapter 9).

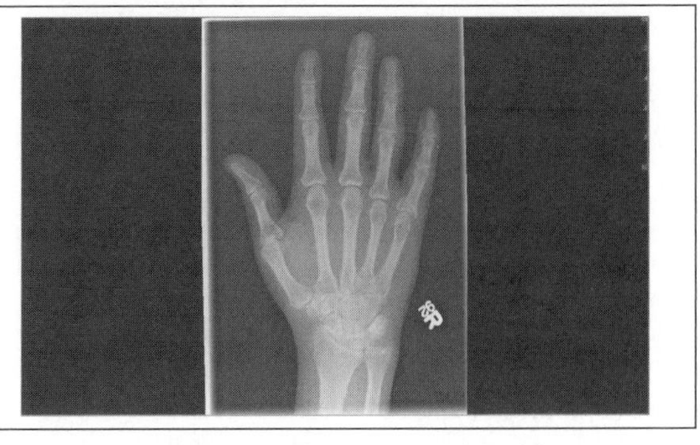

Figure 3.35 Base of the fifth proximal phalanx fracture on PA X-ray.

Admit and Discharge Guidelines

- Request a hand consult in the emergency department for:
  - Open fractures
  - Unstable fractures likely to require surgical fixation
  - Rotational deformity
- Refer to a hand surgeon for outpatient follow-up of all phalangeal fractures and especially for:
  - Intra-articular fractures
  - Fractures that required reduction (i.e., fractures requiring follow-up radiographs)

Definitive Treatment

- Nonsurgical management consists of buddy taping or immobilization using prefabricated or plaster splints.
- Surgical treatment consists of open reduction and internal fixation using plates and/or screws versus closed reduction and percutaneous pinning.

Finger Dislocations

Symptoms and Findings

- Dislocations are usually due to extreme bending or twisting injuries. Dislocations not associated with a fracture are often missed or present late with residual instability.
- Patients present with pain and deformity of the involved joint. The skin should be carefully examined for any lacerations.
- A neurovascular exam—including capillary refill and two-point discrimination—should be performed and documented.
- The wrist and other fingers should be carefully examined to evaluate for other injuries.

Imaging

- X-rays: PA and lateral of the hand is usually sufficient; if concomitant fracture is suspected, PA, lateral, and oblique of the involved digit/joint should also be obtained.

Primary Stabilization and Management

- Open dislocations should be irrigated and debrided prior to reduction. A digital block can be performed for analgesia (see page 311).
- Dislocations should be treated with reduction and radiographs (to verify congruence on the lateral radiograph). The need for and length of immobilization depends upon stability.

Metacarpophalangeal (MCP) Dislocations

- MCP dislocations can usually be reduced with flexion of the MCP joint.
- Dorsal dislocations are most common and, once reduced, are usually stable. These injuries can be treated with dorsal slab (to block full extension) or ulnar gutter splint (with a dorsal block to full extension), followed by buddy taping.
- Volar dislocations are rare and usually require surgical treatment.
- Irreducible or "complex" dislocations require surgical treatment. These injuries are most often due to interposition of the volar plate in the metacarpophalangeal joint.

Proximal Interphalangeal Dislocations (See Figure 3.36a)

- PIP injuries can be reduced with gentle traction and correction of the rotatory deformity.
- Following reduction, the finger can be treated with buddy taping if the joint is congruent on X-ray. Dorsal dislocations are best

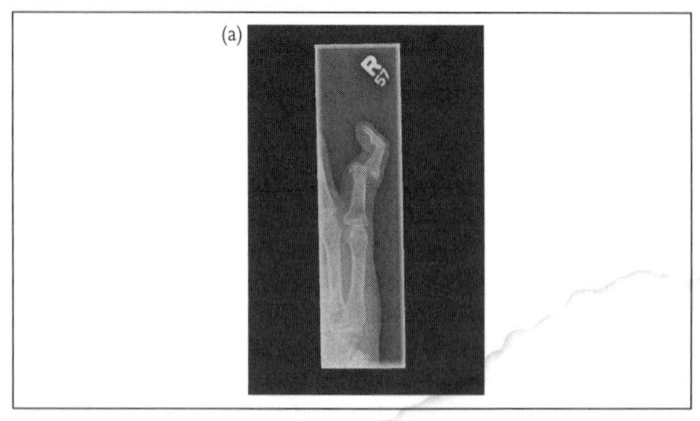

Figure 3.36 Finger dislocation X-ray. a. Dislocation of PIP on PA X-ray.

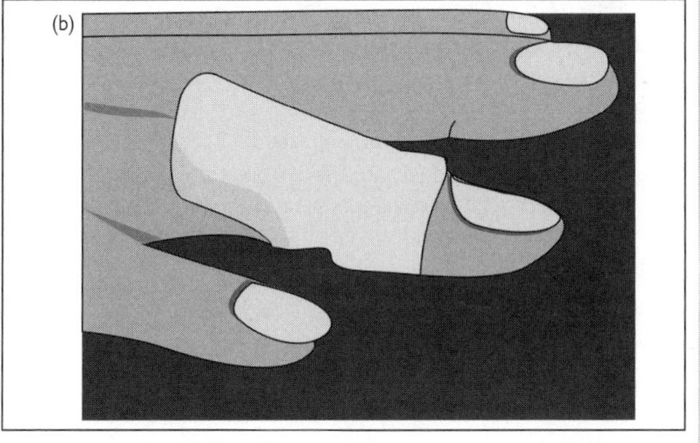

Figure 3.36 (Continued) b. DIP hyperextension splint.

treated with dorsal extension block splint for 1 week followed by buddy taping.

- Dorsal fracture/dislocations of the middle phalangeal base (volar lip is fractured) require reduction and dorsal extension block splinting. Fractures involving greater than 30% of the articular surface are typically unstable and require ORIF.
- Volar fracture/dislocations of the middle phalangeal base (dorsal lip is fractured) require reduction and splinting. Reduction can usually be accomplished through traction and flexion of the PIP joint. Irreducible or incompletely reduced dislocations and/or fractures with >1 mm of displacement require ORIF.

Interphalangeal and Distal Interphalangeal Dislocations

- DIP and IP dislocations are often missed and may present late.
- Acute dislocations can be treated with reduction and splinting.
- Irreducible and chronic dislocations generally require open treatment.

Mallet Finger: Extensor Digitorum Communis (EDC) Avulsion

- Caused by a direct blow that hyperflexes the distal phalanx; often due to jamming a finger when catching a ball
- Patient is unable to actively extend the distal phalanx.
- "Bony mallet": The dorsal lip of the distal phalanx base is fractured by the tendon avulsion injury.
- Treat with hyperextension DIP splint (PIP joint free). Splint must be worn at all times for 6–8 weeks (see Figure 3.36b).
- May also be treated with closed reduction and percutaneous pinning or open reduction and internal fixation for a bony mallet, although most can be treated with splinting

Jersey Finger: Flexor Digitorum Profundus (FDP) Avulsion

- Forceful extension of the distal phalanx causes avulsion of the FDP; the volar lip of the distal phalanx base may be fractured when the tendon avulses.
- Patient is unable to flex the distal phalanx.
- These injuries should be treated with primary repair. The degree of tendon retraction dictates the urgency of surgical repair, and a hand surgeon should be consulted in the emergency setting.

Admit and Discharge Guidelines

- Request a hand consult in the emergency department for:
  - Open fractures
  - Irreducible fractures
  - Acute Jersey finger
- Refer to a hand surgeon for outpatient follow-up for: Chronic dislocations, Mallet finger, chronic Jersey finger

Definitive Treatment

- Nonsurgical management consists of immobilization to allow ligamentous healing.
- Surgical treatment consists of open reduction and internal fixation, closed reduction and percutaneous pinning, or primary repair in certain cases (i.e., FDP avulsion).

Subacute Upper Extremity Injuries

Subacute injuries can often be diagnosed on history and physical examination alone. If there is no history of acute trauma, radiographs do not routinely need to be performed in the emergency setting. With few exceptions, subacute injuries are best treated with conservative management, including rest, anti-inflammatories, and ice. While injections may be performed to diagnose or treat certain pathologies in the outpatient setting, this is not advised in the emergency setting. Injections are not without risks, and much of the diagnostic utility is lost without appropriate follow-up.

Rotator Cuff Tendinitis and Tears

Anatomy

- The rotator cuff is composed of four muscles—supraspinatus, infraspinatus, teres minor, and subscapularis. The rotator cuff muscles have two major functions: dynamic stabilization of the shoulder joint and arm motion.
- Supraspinatus—Initiates abduction of the arm

- Infraspinatus and Teres Minor—Externally rotates the arm
- Subscapularis—Internally rotates and adducts the arm

Symptoms and Findings

- Rotator cuff injuries can be broadly classified into chronic and acute injuries.
- Acute injuries are more common in young (<40 yo) patients and are usually secondary to significant trauma. Rotator cuff tears are associated with dislocations (most commonly the subscapularis muscle is torn) and greater tuberosity fractures, which is the attachment site for the supraspinatus, infraspinatus, and teres minor. Patients with an acute injury will generally have pain about the shoulder and weakness in the affected rotator cuff muscle.
- Chronic injuries are more common in older (>40 yo) individuals. Patients with tendinitis and chronic tears will give a history of pain with shoulder range of motion, particularly with overhead activities. Night pain is a frequent complaint. Impingement testing, such as the Neer test (pain with passive forward elevation of the shoulder) and Hawkins test (pain with passive internal rotation of the shoulder with the arm held in 90 degrees of forward elevation) is frequently positive.

Imaging

- X-rays: Standard shoulder trauma series (AP, Grashey, and axillary views)

Primary Stabilization and Treatment

- Treatment should consist of a sling for comfort and advising the patient to perform elbow and shoulder range of motion exercises to prevent joint stiffness.
- Conservative management, including rest, anti-inflammatories, and ice can significantly help symptoms.

Admit and Discharge Guidelines

- Rotator cuff injuries do not require admission or urgent treatment. Patients should be referred for outpatient follow-up with an orthopaedic surgeon.

Definitive Treatment

- If an acute, complete rotator cuff tear is suspected in a young patient, follow-up with an orthopaedic surgeon should be arranged. Outcomes are improved with timely surgical repair (<3 months).
- Treatment of chronic rotator cuff tears is dependent upon patient factors (such as health status and activity demands), as well as the surgical feasibility of repair. In patients with concomitant arthritis, arthroplasty may be an option.

- Physical therapy is nearly always a component of the treatment plan, regardless of whether the patient undergoes surgery.
- MRI should not be ordered on an acute basis as findings will not alter the initial conservative management of the injury. It is preferable to have the MRI ordered by the treating surgeon.
- CT scan is of little utility in diagnosing rotator cuff tears but may be useful in preoperative planning if a greater tuberosity fracture requires repair. Like MRI, it should only be ordered by the treating surgeon.

Biceps Tendon Rupture

Anatomy

- Biceps means "two heads," and the muscle has both a long head (which originates on the supraglenoid tubercle) and a short head (which originates on the coracoid process). The two heads have a common insertion on the proximal radius. The biceps flexes and supinates the forearm.

Symptoms and Findings

- Patients with an acute biceps tendon rupture will present after the sudden onset of pain, usually after forceful flexion of the elbow against resistance. The patient may report hearing a pop.
- Ruptures of the long head of the biceps classically have a proximal arm bulge, termed a "popeye arm."
- Ruptures of the distal biceps will have elbow swelling, and the loss of muscle substance at the antecubital fossa may be palpable.
- Elbow flexion and supination strength will be decreased compared with the contralateral side, particularly in distal biceps ruptures.

Imaging

- X-rays: Standard shoulder trauma series (AP, Grashey, and axillary views) and elbow (AP and lateral) X-rays

Primary Stabilization and Treatment

- Treatment should consist of a sling for comfort and advising the patient to perform elbow and shoulder range of motion exercises to prevent joint stiffness.
- Conservative management, including rest, anti-inflammatories, and ice can significantly help symptoms.

Admit and Discharge Guidelines

- Biceps tendon ruptures do not require admission or urgent treatment. Patients should be referred for outpatient follow-up with an orthopaedic surgeon.

Definitive Treatment

- Rupture of the long head of the biceps does not result in significant weakness. As such, the injury can generally be treated nonoperatively. Surgical treatment may be performed for cosmesis or in a young athlete or laborer. The short head of the biceps remains intact, and surgical repair only improves flexion strength by approximately 10%.
- Rupture of the distal biceps tendon decreases supination strength by approximately 50%. Nonsurgical treatment is a reasonable option if a patient has low demands on the affected extremity and limited pain from the injury. Surgery should be performed in a timely fashion to prevent scarring of the biceps insertion to surrounding tissue.
- Biceps injuries are diagnosed by physical exam. MRI may be useful to the treating surgeon if the diagnosis is in question or as part of surgical planning. MRI should not be performed on an acute basis.

Biceps Tendinitis

Anatomy

- Biceps means "two heads," and the muscle has both a long head (which originates on the supraglenoid tubercle) and a short head (which originates on the coracoid process). The two heads have a common insertion on the proximal radius. The biceps flexes and supinates the forearm.

Symptoms and Findings

- Patients present with pain in the shoulder and tenderness in the bicipital groove. Provocative maneuvers, such as resisted supination or flexion, may provoke tendinitis symptoms.

Imaging

- X-rays: Standard shoulder trauma series (AP, Grashey, and axillary views)

Primary Stabilization and Treatment

- Treatment should consist of a sling for comfort and advising the patient to perform elbow and shoulder range of motion exercises to prevent joint stiffness
- Conservative management, including rest, anti-inflammatories, and ice can significantly help symptoms

Admit and Discharge Guidelines

- Biceps tendinitis does not require admission or urgent treatment. Patients should be referred for outpatient follow-up with an orthopaedic surgeon.

Definitive Treatment

- Biceps tendinitis is diagnosed clinically. MRI may be performed if concomitant shoulder injury exists or if symptoms are recalcitrant to conservative management.
- Conservative management, including rest, anti-inflammatories, and ice can help symptoms.
- Biceps tendonitis may be treated with a biceps tenodesis.

Subacute Elbow Conditions: Epicondylitis and Bursitis

Anatomy

- The common extensor tendon (extensor carpi radialis brevis, extensor digitorum, extensor digiti minimi, and extensor carpi ulnaris) originates on the lateral epicondyle, and the common flexor tendon (pronator teres, flexor carpi ulnaris, palmaris longus, flexor carpi radialis, flexor digitorum supfercialis) originates on the medial epicondyle.
- The olecranon bursa is a thin sac of tissue that lies between the bony olecranon and the skin of the posterior elbow.

Symptoms and Findings

- Lateral epicondylitis is also called tennis elbow and results from repetitive or overuse injuries to the common extensor origin. Patients complain of chronic pain at the lateral elbow, and symptoms can be reproduced with resisted wrist and finger extension with the forearm in pronation.
- Medial epicondylitis is also called golfer's elbow and results from repetitive or overuse injuries to the common flexor origin. Patients complain of pain at the medial elbow, and symptoms can be reproduced with resisted wrist or finger flexion.
- Olecranon bursitis presents with swelling and pain over the posterior elbow localized to the olecranon. A mass may be palpable, and septic bursitis may be warm to the touch.

Imaging

- X-rays: Standard elbow (AP and lateral) X-rays

Primary Stabilization and Treatment

- Treatment should consist of immobilization for soft tissue rest. Medial or lateral epicondylitis can be managed with acock-up wrist splint and a forearm counterforce strap during activity. Olecranon bursitis should initially be immobilized with an elbow extension splint, avoiding pressure over the olecranon. Conservative management, including rest, anti-inflammatories, and ice can significantly help symptoms

- Olecranon bursitis may be treated with sterile aspiration of the bursa. If there is concern for septic bursitis, the fluid should be sent for culture, gram stain, cell count, and crystal analysis. If gram stain and fluid are consistent with infection, empiric antibiotics may be started and then tailored to the specific pathogen. A splint can help reduce inflammation by allowing for soft tissue rest. The traditional posterior elbow splint can be used, but a volar-based resting splint will be more comfortable as it prevents any additional pressure from being applied to the olecranon. In cases of recurrent bursitis, placement of a drain may be considered with removal after 1–2 days.

Admit and Discharge Guidelines

- Epicondylitis and aseptic olecranon bursitis do not require admission or urgent treatment. Patients should be referred for outpatient follow-up with an orthopaedic surgeon. If bursitis is secondary to a rheumatologic disorder (e.g., gout or rheumatoid arthritis), referral to a rheumatologist is preferred.
- The treatment of patients with septic olecranon bursitis depends upon the degree of infection, concomitant cellulitis, and the patient's overall health. Treatment should be tailored to the patient. Many individuals can be treated with oral antibiotics, resting splint, and close follow-up. Immunocompromised patients may require hospitalization and intravenous antibiotics.

Definitive Treatment

- Conservative management is the first-line treatment.
- Surgical debridement of the bursa is only performed after the failure of maximal conservative management, including antibiotics and repeated aspirations. Elective surgical debridement of the bursa may be performed in patients with multiple recurrences of aseptic bursitis.
- Epicondylitis and bursitis are diagnosed clinically, and advanced imaging modalities are generally not required for diagnosis or treatment.

Flexor Tenosynovitis (Trigger Finger)

Symptoms and Findings (See Figure 3.37)

- Flexor tenosynovitis (stenosing tenosynovitis) or "trigger finger" results when inflammation of the fibrinous flexor sheath causes the flexor tendons to snap when passing through the A1 pulley. A fusiform thickening within the flexor tendon may be palpated, and patients complain of pain as well as locking with finger flexion.

Imaging

- X-rays are not required to make the diagnosis.

Primary Stabilization and Treatment

- A finger splint to block flexion may be applied to prevent the finger from locking.
- Conservative management, including rest, anti-inflammatories, and ice can help symptoms

Admit and Discharge Guidelines

- Trigger finger does not require admission or urgent treatment. Patients should be referred for outpatient follow-up with a hand surgeon.

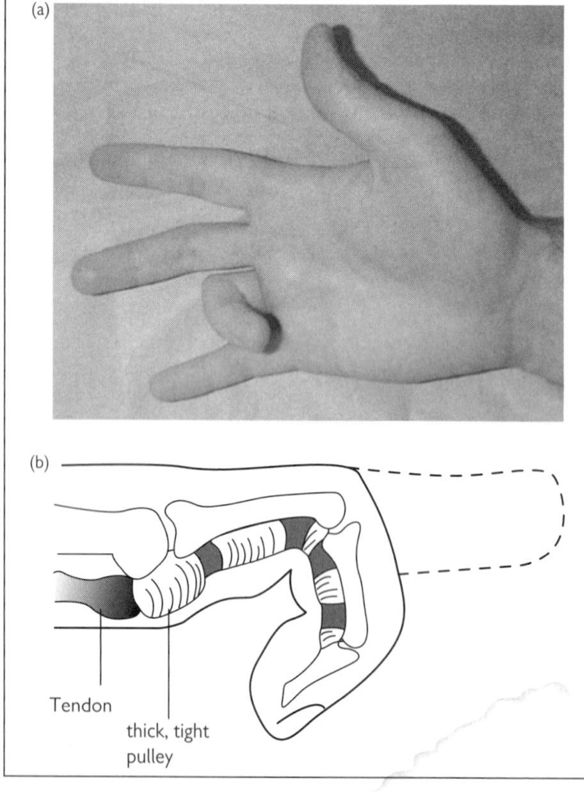

Figure 3.37 Flexor tenosynovitis. a. Examination of trigger finger. b. Diagram of flexor tenosynovitis flexor sheath

Definitive Treatment

- The first-line treatment is injection of steroids into the tendon sheath. If conservative treatment fails, the A1 pulley may be surgically released.

Carpal Tunnel Syndrome

Anatomy

- The carpal tunnel contains nine tendons and one nerve. The flexor digitorum profundus contributes four tendons, the flexor digitorum superficialis contributes four tendons, and the flexor pollicis longus contributes one tendon. The median nerve is also within the tunnel and passes between the tendons of the FDP and FDS.

Symptoms and Findings

- Carpal tunnel syndrome (CTS) is caused by compression of the median nerve within the carpal tunnel. Patients present with paresthesias or numbness in the distribution of the median nerve distal to the carpal tunnel (the tips of the thumb, index, and middle finger). Patients may also have loss of grip strength and wrist pain. CTS may be acute or chronic in nature.
- Acute carpal tunnel syndrome is secondary to trauma, most commonly fractures and dislocations about the wrist. The diagnosis is dependent upon a careful sensory and motor nerve examination and is characterized by acute change in sensation and/or motor in a median nerve distribution.
- Subacute or chronic carpal tunnel syndrome is usually idiopathic, though it has been associated with repetitive motion stress.
- Patients often have worsening of symptoms at nighttime and have increased sensitivity to vibration (such as a steering wheel, hair dryer, or power tool).
- Carpal tunnel syndrome is associated with multiple metabolic disorders, including diabetes, thyroid disease, and alcoholism. It is also associated with pregnancy.
- Tinel's and Phalen's tests are frequently positive. Tinel's test is performed by tapping over the carpal tunnel in an effort to provoke the median nerve; the test is positive if the patient experiences an electric-like sensation (paresthesias) in the median nerve distribution (see Figure 3.38a). Phalen's test is performed by having the patient flex the wrist with the elbow extended for at least 60 seconds; the test is positive if the patient experiences paresthesias in the median nerve distribution (see Figure 3.38b).

Imaging

- X-rays are not required to make the diagnosis.

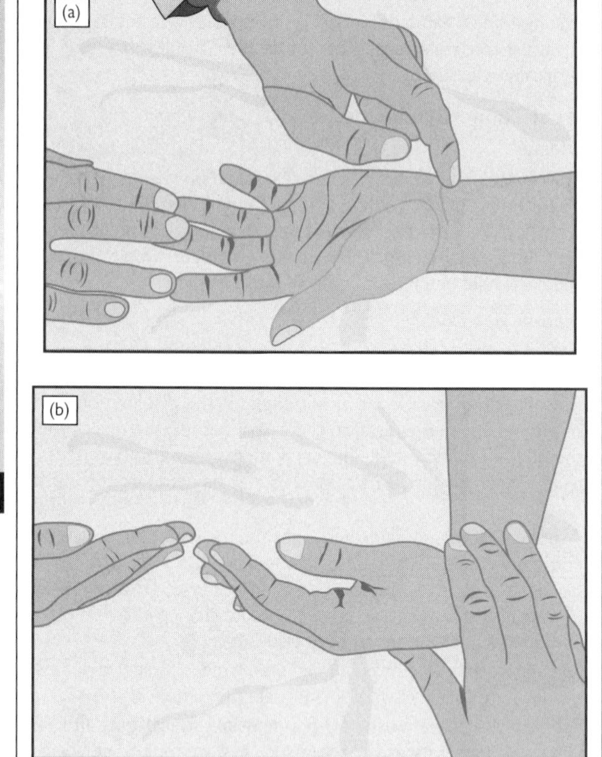

Figure 3.38 Carpal tunnel. a. Tinel's sign. b. Phalen's test.

Primary Stabilization and Treatment

- A volar-based cock-up wrist splint may be applied to allow for rest, but the patient should be instructed to perform finger and hand range of motion exercises. Sleeping with the splint on may reduce symptoms.
- Conservative management, including rest, anti-inflammatories, and ice can significantly help symptoms

Admit and Discharge Guidelines

- Acute carpal tunnel syndrome requires surgical release to prevent permanent injury to the median nerve. A hand surgeon

should be consulted emergently in any patient with acute carpal tunnel syndrome.
- Subacute or chronic carpal tunnel does not require admission or urgent treatment. Patients should be referred for outpatient follow-up with a hand surgeon

Definitive Treatment
- Conservative management, including rest, NSAIDs, ice, and a cock-up wrist splint can help symptoms. The patient may benefit from sleeping in a wrist splint.
- Steroids may be injected directly into the carpal tunnel.
- The carpal tunnel may be released surgically through either an endoscopic or an open approach.

de Quervain's Syndrome

Symptoms and Findings
- de Quervain's syndrome is a tenosynovitis of the first dorsal compartment affecting the two tendons (extensor pollicis brevis and abductor pollicis longus) that run in the compartment.
- Patients complain of pain and swelling over the radial aspect of the wrist and dorsum of the thumb, as well as difficulty with grip.
- Finkelstein's test is used to diagnose de Quervain's syndrome. The test is performed by having the patient grip his or her thumb within a closed fist, followed by ulnar deviation of the fist. The test is positive if it reproduces the patient's pain (see Figure 3.39).

Imaging
- X-rays are not required to make the diagnosis.

Primary Stabilization and Treatment
- A thumb-spica splint may be applied to allow for rest, but the patient should be instructed to perform finger range of motion exercises.
- Conservative management, including rest, anti-inflammatories, and ice can significantly help symptoms

Admit and Discharge Guidelines
- De Quervain's does not require admission or urgent treatment. Patients should be referred for outpatient follow-up with a hand surgeon.

Definitive Treatment
- The first-line treatment is splint immobilization +/− oral antiinflammatory medication, possible therapy with ultrasound

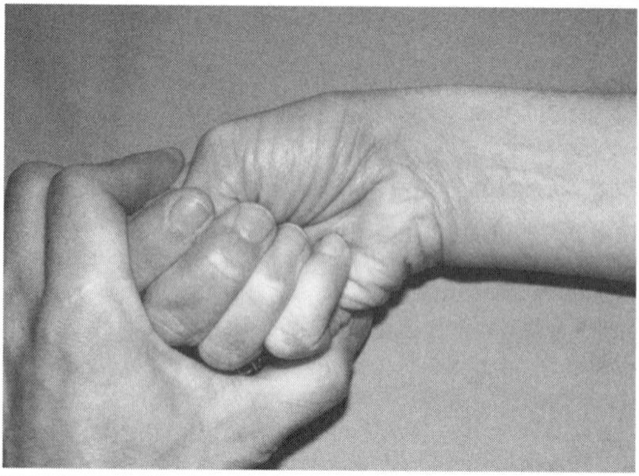

Figure 3.39 De Quervain's syndrome—Finkelstein's test.

or iontophoresis, and possible injection of steroids into the compartment. If conservative treatment fails, the compartment may be surgically released.

Gout

Symptoms and Findings

- Patients generally present with monoarticular joint pain, erythema, and swelling. Joint range of motion is often markedly limited. Tophi may be visible within the skin.
- Patients should be asked about a history of gout.

Imaging

- X-rays of the affected region may demonstrate periarticular erosions and arthritic changes.

Primary Stabilization and Treatment

- Diagnosis requires joint aspiration. The fluid should be sent for culture, gram stain, cell count, and crystal analysis. Lyme studies should always be considered in patients who live in endemic areas. Needle and rod-shaped crystals with negative birefringence should be seen in gram stain. Joint fluid white blood cell count is usually less than 60,000/uL.

Admit and Discharge Guidelines

- The decision to admit a patient with an acute gout attack is a medical decision. It is important to determine whether the joint is superinfected as a joint may have both gout and infection. Patients with a septic joint require prompt surgical debridement.

- Patients should be referred to an internist or rheumatologist for long-term treatment.

Definitive Treatment

- Gout is treated medically with NSAIDs, colchicine, and hypouricemic therapy including allopurinol and probenicid.
- A resting splint and elevation can help improve the patient's symptoms.
- Management of a superinfected gouty joint is as for a septic joint and should be treated accordingly with antibiotics and prompt surgical debridement.

Chapter 4

Lower Extremity Injuries

Physical Exam of the Lower Extremity

Surface Anatomy

- See Figure 4.1

Neurovascular Exam

- Document patient's cutaneous sensation and any deficits. Sensation should be documented BY peripheral nerves and any deficits should be noted (see Figure 5.3).
- Vascular exam (use of Doppler for difficult-to-palpate pulses); capillary refill. The following pulses should be checked routinely: Femoral, DP, PT

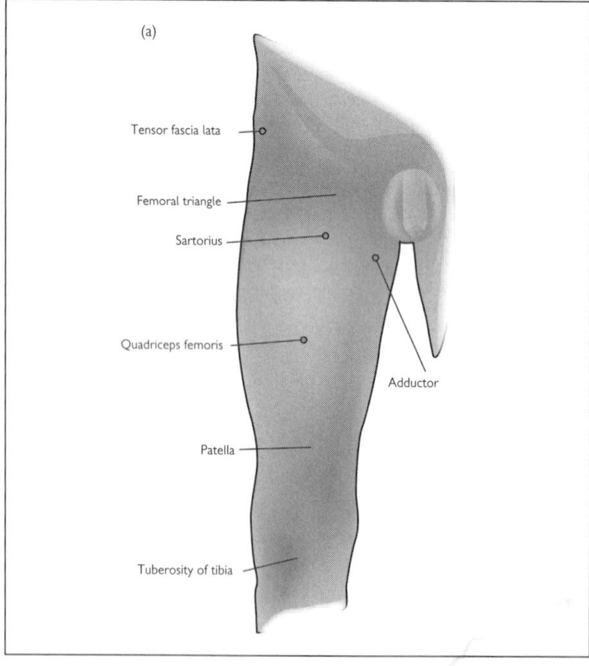

Figure 4.1 a. Anterior lower extremity surface anatomy

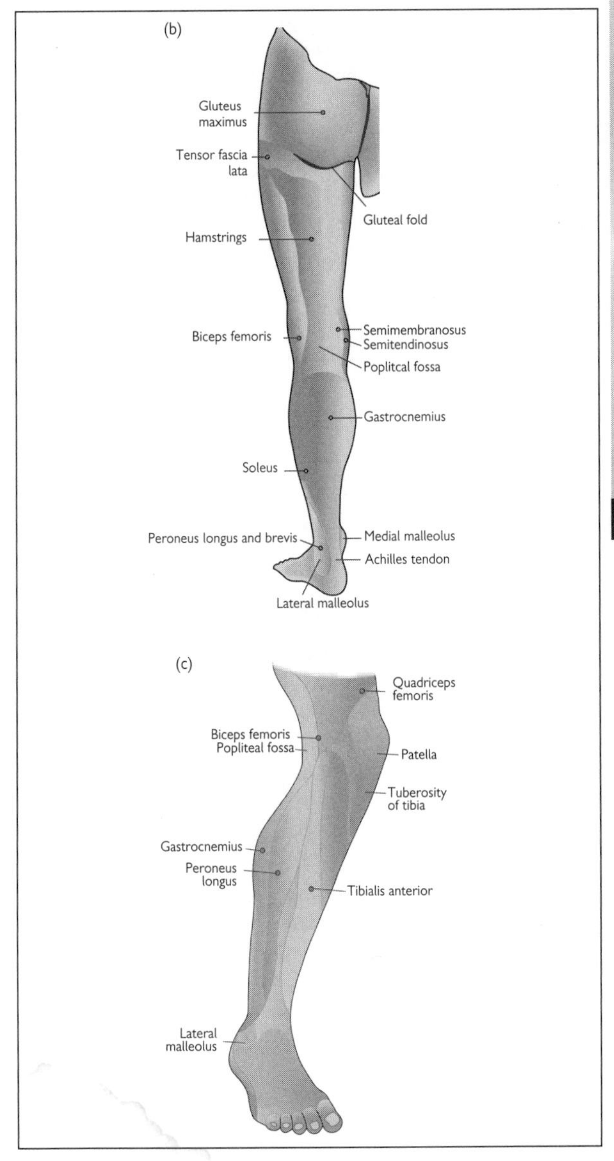

Figure 4.1 (Continued) b. Posterior lower extremity surface anatomy. c. Lateral lower extremity surface anatomy

Motor Exam

- Detailed motor exam of lower extremities important, **especially L5 nerve root**, which travels just anterior to SI joint
- Motor strength is graded from 0 to 5:
  - Grade 0: No movement
  - Grade 1: Flicker of movement only
  - Grade 2: Movement with gravity eliminated
  - Grade 3: Movement against gravity
  - Grade 4: Movement against resistance
  - Grade 5: Normal power

Motion	Muscle	Nerve	Root
Hip flexion	Iliopsoas	Femoral	L2–3
Hip extension	Gluteus	Gluteal	L5-S2
Knee flexion	Hamstrings	Sciatic	L5-S2
Knee extension	Quadriceps	Femoral	L2–4
Ankle dorsiflexion	Tibialis anterior	Deep peroneal	L4
Ankle plantarflexion	Gastrocsoleus	Tibial	S1
Ankle eversion	Peroneus longus	Superficial peroneal	L5-S1
Great toe extension	Extensor hallucis longus	Deep peroneal	L5

Documenting the Physical Exam

- ABCs
  - Especially in younger patients and those that sustain high-energy injury, patients should have formal trauma evaluation.
- Skin
  - Example: Skin is intact without breaks, edema, or ecchymosis.
- Palpation/Deformity
  - Always check the lower extremities for crepitus or bony defect, especially in the obtunded patient. Palpation in the awake patient can focus on areas of tenderness. Deformities should be noted.
- Pulses
  - Example: 2+ dorsalis pedis and posterior tib pulses
- Sensation
  - Example: Sensation intact to light touch in the superficial peroneal, deep peroneal, and tibial nerve distributions
- Motor
  - Example: Strength is 5/5 in the iliopsoas, hamstrings, quadriceps, tibialis anterior, gastrocsoleus, and extensor hallucus longus.

Hip Injuries

Hip injuries are very common. The majority of hip pathology will likely be in elderly patients with poor bone quality involved in low-energy trauma. Hip fractures and dislocations in younger individuals and children are often high-energy injuries, as this portion of the body is accustomed to high physiologic loads at baseline. Hip injuries most commonly present with groin pain; however, the importance of referred pain to other areas such as the knee cannot be excluded. Children with slipped capital femoral ephiphysis (SCFE) can often present with knee pain as the primary complaint.

Anterior and Posterior Hip Dislocations

Hip dislocation is a significant injury with potential for long-term sequelae including posttraumatic arthritis, avascular necrosis (AVN), recurrent dislocations, sciatic nerve palsies, and associated injuries. Unlike the shoulder, the hip is intrinsically stable and takes a large amount of force to dislocate. Related fractures of the femoral head, neck, and acetabulum can occur. These fracture fragments can occasionally be mechanical blocks to reduction. Hip dislocation is one of **the few orthopaedic urgencies.** Damage to **the medial femoral circumflex artery** (MFCA) can predispose to AVN and significant long-term morbidity. Length of time unreduced is directly related to risk of AVN. Best results are reported within the first 6 hours and rates of permanent damage go up after that.

Symptoms and Findings

Anterior: Less common.

- Leg usually held with knee mildly flexed, hip mildly flexed, externally rotated and abducted
- Pain in groin
- Unable to ambulate

Posterior: More common. Associated with femoral head and acetabular fractures. High-energy mechanism in younger patients, often flexed knee, flexed hip, hitting dashboard.

- Hip held flexed, internally rotated, and adducted with a bent knee
- Pain in groin
- Unable to ambulate

Imaging (See Figure 4.2)

- AP and lateral X-rays of the affected side. The lateral is often difficult but is necessary and a good cross-table can suffice. Only the lateral X-ray can confirm anterior versus posterior.
- AP pelvis
- Inlet/outlet or Judet views based on associated pelvic and acetabular injuries

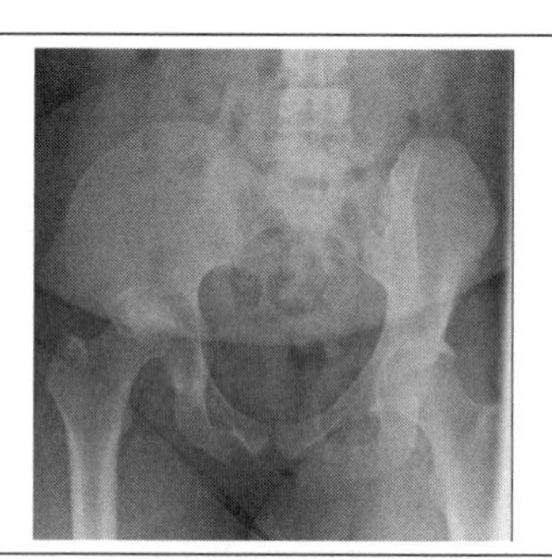

Figure 4.2 Hip dislocation with fracture.

- One should be able to visualize the femoral head and neck adequately to check for associated fractures prior to reduction.
- If there are signs of a femoral head fracture or acetabular fracture, a CT scan of the hip should be performed after the hip is reduced. The CT will need to be repeated if it is done prior to the reduction. The orthopaedic indications for a CT scan prior to reduction are if an adequate lateral plain film cannot be obtained to determine anterior versus posterior dislocation or evaluation of associated fractures cannot be done with plain films.

Classification

- Anterior or posterior
- Posterior is then further subclassified based on associated fractures:
  - I—No fracture or minor fracture
  - II—Large fracture fragment of wall
  - III—Comminuted wall
  - IV—Acetabular floor
  - V—Femoral head fracture

Primary Stabilization and Management

- Native hip dislocations are one of the **orthopaedic urgencies**
- The risk of femoral head AVN doubles from 10% to 20% if a dislocated hip is not reduced within 6 hours from injury. Therefore, these often need to be reduced prior to transfer to a secondary or tertiary care center for management of associated injuries.
- Because of the size and nature of the musculature surrounding the hip joint, conscious sedation with muscle relaxation and sometimes general anesthesia are required to relocate the hip. The authors' preference is fast-acting muscle relaxation in the emergency department, such as etomidate, or general anesthesia

in an operating room to expedite the reduction. One should not wait longer than 6 hours for general anesthesia if conscious sedation is available.

- Posterior dislocations are reduced by traction in the anterior direction (inline traction) on the 60–90 degree flexed hip with gentle internal rotation and adduction.
- Anterior dislocations are reduced with longitudinal (inline) traction, gentle external rotation, and slight abduction.
- For both methods, one should have a significant mechanical advantage such as standing on the stretcher or placing the patient on the floor.
- Slow increases in a constant force are more effective than large "yanks."
- Fluoroscopy can help guide reduction if available.
- Reduction must be confirmed using radiographs and/or CT.

Admit and Discharge Guidelines

- Regardless of final disposition, closed reduction must be performed in the ED or operating room before transfer/admission ideally within 6 hours. An exception is when the dislocated hip is an arthroplasty (i.e., hip replacement), then there is no risk to the blood supply and therefore no urgency.
- Simple first-time hip dislocations that are stable and fully reduced without significant associated injuries (most commonly seen with dislocated total hip prosthesis, which requires much less force to dislocate) can be discharged when cleared by the trauma team or the emergency care team. They can be weight bearing as tolerated with assistive devices. Active and passive range of motion (ROM) can be started and should be limited to flexion of 90 degrees and internal rotation of 10 degrees for 6 weeks. This requires a hip abduction brace and should be in place prior to discharge.
- With many hip dislocations, there are associated nonorthopaedic injuries, and patients are often admitted to a trauma service to manage their multiple injuries. If there are associated injuries such as acetabular fractures, the patient should be admitted for definitive treatment or transferred if the often-complicated repair is beyond the scope of the facility's resources.
- An alternative to abduction bracing is to place the patient in a knee immobilizer at all times, as it is very difficult to get into a position to redislocate with a straight knee.
- Unstable reductions should be reduced and held with skeletal traction in abduction and slight external rotation and admitted or transferred for definitive management.
- Orthopaedic follow-up is required.

Definitive Treatment

- Closed reduction is the treatment for simple dislocations without associated injuries.
- Open reduction with capsulotomy is the treatment if closed treatment fails, often through the anterior approach to spare the blood supply of the femoral head. For closed reduction to be considered failed, it must be attempted under general anesthesia.
- If there are mechanical blocks interfering with the closed reduction, or if there is an unstable reduction due to fracture, or if there is a significant femoral or acetabular fracture, then open reduction with internal fixation and debridement is required.

Femoral Head Fractures

Femoral head fractures are high-energy injuries and almost always occur with femoral head dislocations, 90% posterior and 10% anterior. Because of the mechanism of injury and trauma to the head, they are at risk of AVN. See "Anterior and Posterior Hip Dislocations" for management of the dislocation. Hip dislocations are orthopaedic **urgencies.** Damage to the femoral head can cause AVN, posttraumatic arthritis, changes in symmetric loading of the head, and loss in range of motion.

Symptoms and Findings

- Symptoms and findings will almost all be related to the associated dislocation. See "Anterior and Posterior Hip Dislocations."

Imaging (See Figure 4.3)

- AP pelvis
- AP/lateral hip
- Inlet/outlet or Judet views based on associated pelvic and acetabular injuries

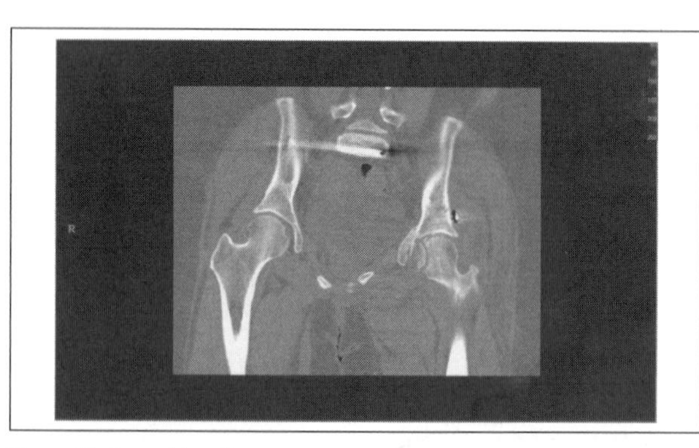

Figure 4.3 Femoral head fracture.

- Thin slice (1–3 mm) CT pelvis/femoral head + neck postreduction

Classification (See Figure 4.4)

- Pipkin (most commonly)
- Brumback (most recently/completely)

Primary Stabilization and Management

- Based on dislocation. Reduce dislocation per "Anterior and Posterior Hip Dislocations" section.
- All nonarthroplasty hip dislocations should be reduced quickly regardless of associated femoral head fractures.
- May need skeletal traction to maintain reduction until definitive management.

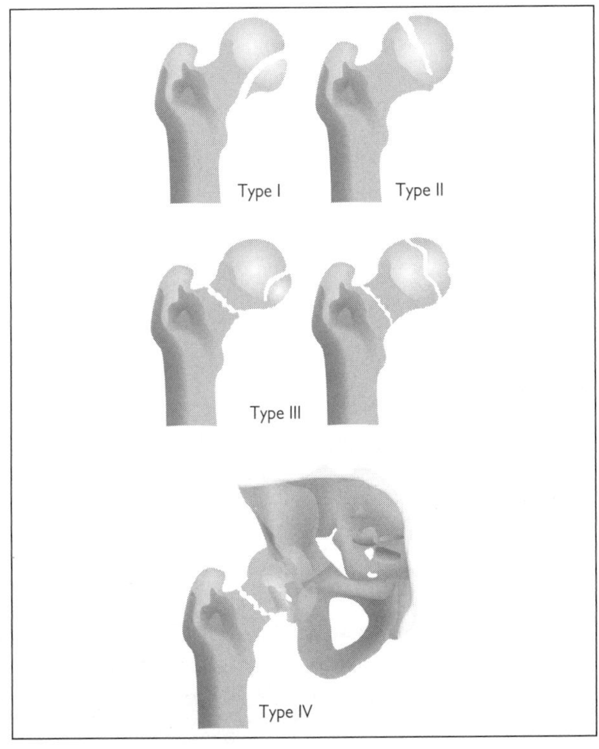

Figure 4.4 Pipkin classification.

Admit and Discharge Guidelines

- Femoral head fractures will almost always need definitive treatment and therefore be admitted or transferred to an appropriate facility after reduction of the hip joint has been completed.

Definitive Treatment

- Definitive treatment is based on the type of femoral head fracture. The range of treatments are from closed reduction and traction for 4–6 weeks, closed reduction and open fragment excision, ORIF, and arthroplasty.

Slipped Capital Femoral Epiphysis (SCFE)

This is most common in overweight children during periods of rapid growth and is likely due to repetitive trauma of the weight on the insufficient physis. If an SCFE is found in a child who does not fit this criteria, a workup for endocrine disorders such as hypothyroidism is indicated.

Symptoms and Findings

- Often has a prodromal chronic slip phase with hip pain and limp
- Acute slips occur in approximately 10% of cases and are associated with trauma.
- Acute pain after a period of the prodromal pain is classified as acute on chronic slip.
- Given an SCFE on one side, 33% will have a bilateral SCFE without prophylactic treatment.
- May have knee pain as only presenting symptom

Imaging (See Figure 4.5)

- AP pelvis
- AP hip
- Frog leg lateral bilateral hips

Classification

Acute versus Chronic

- Acute: May or may not have a prodromal low-grade hip pain and then an inciting incident or trauma and have an acute increase in pain and inability to walk.
- Chronic: Low-grade hip or knee pain. Limp. May progress to acute.

Primary Stabilization and Management

- Chronic slips: Referral to pediatric orthopaedics. Make non-weight-bearing until evaluated by orthopaedist. No reduction or stabilization is indicated.

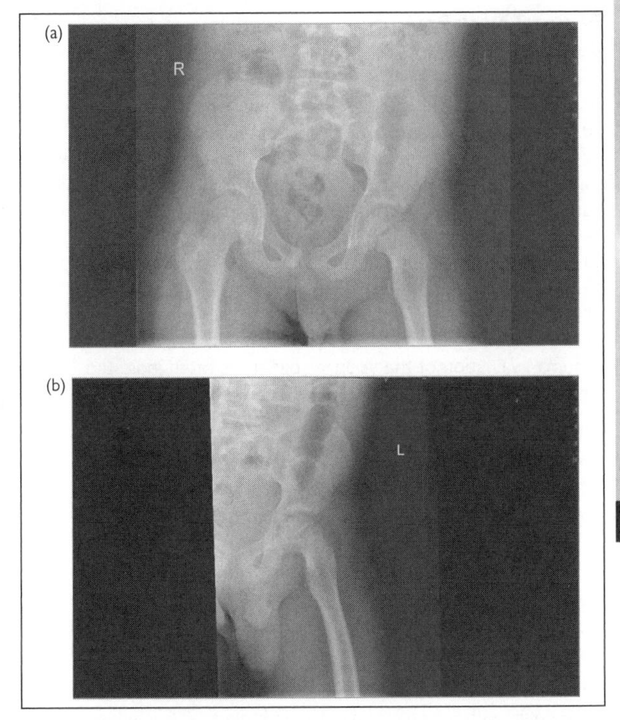

Figure 4.5 SCFE seen on. a. AP pelvis X-ray. b. Lateral X-ray.

- Acute slips: Admission for surgery versus transfer to a facility with pediatric orthopaedic coverage for in situ pinning based on facility and on-call orthopaedist's ability. Do not attempt reduction or stabilization. Bedrest until treatment.

Admit and Discharge Guidelines

- Request an orthopaedic consult in the emergency department for: Acute slip.
- Refer to an orthopaedic surgeon for outpatient follow-up for: Chronic slip.

Definitive Treatment

- In situ pinning
- Some advocate prophylactic pinning of the contralateral hip because of the risk of developing a contralateral SCFE and the low risk of the procedure.

Femoral Neck Fractures

Fractures of the femoral neck occur in two settings. First, and more commonly, are low-energy falls in older patients (>60) with poor bone stock. The second is high-energy injury in patients younger than 50. Because arthroplasty is successful, survival of the femoral head is not important in patients over 60 where the rate of needing revision surgery for an arthroplasty is significantly lower. However, because younger patients (anyone less than 50) have a much higher chance of needing to replace an arthroplasty at some point, attempts are made to preserve the native femoral anatomy. The femoral head is prone to AVN, and femoral neck fractures present a significant risk of this. Therefore, in anyone 50 years old or younger **urgent** closed versus open reduction with internal or percutaneous fixation is important. Trauma evaluation of the younger patient is often indicated for the high-energy mechanism of these injuries.

Symptoms and Findings

- Groin pain
- Pain with ambulation or inability to ambulate
- Affected leg is often shortened and externally rotated if the fracture is displaced.
- Groin or hip pain with log roll or other movements of the leg
- Patients should be evaluated for prefracture hip pain, night pain, and tumor history, and images should be evaluated for possible pathologic fracture.

Imaging (See Figure 4.6)

- AP pelvis with toes touching or taped together (slight internal rotation)
- AP and lateral of affected hip. The lateral will often have to be cross-table.
- CT scan can be done if hip fracture is suspected but not found on XR (X-ray). Occasionally, completely nondisplaced fractures cannot be seen on XR but visualized on CT.
- Stress fractures or nondisplaced fractures may not show up on XR; MRI or bone scan should be obtained as an outpatient. Patients must remain foot flat or non-weight-bearing until imaging can be completed.

Classification

- Historically the Garden system was used; however, this has been shown to not impact treatment options.
- Displaced versus nondisplaced or impacted is a crucial distinction to make.
- Pediatric fractures are based on the Delbert classification that stratifies the risk of AVN. This classification includes fractures from the neck to the intertrochanteric region:

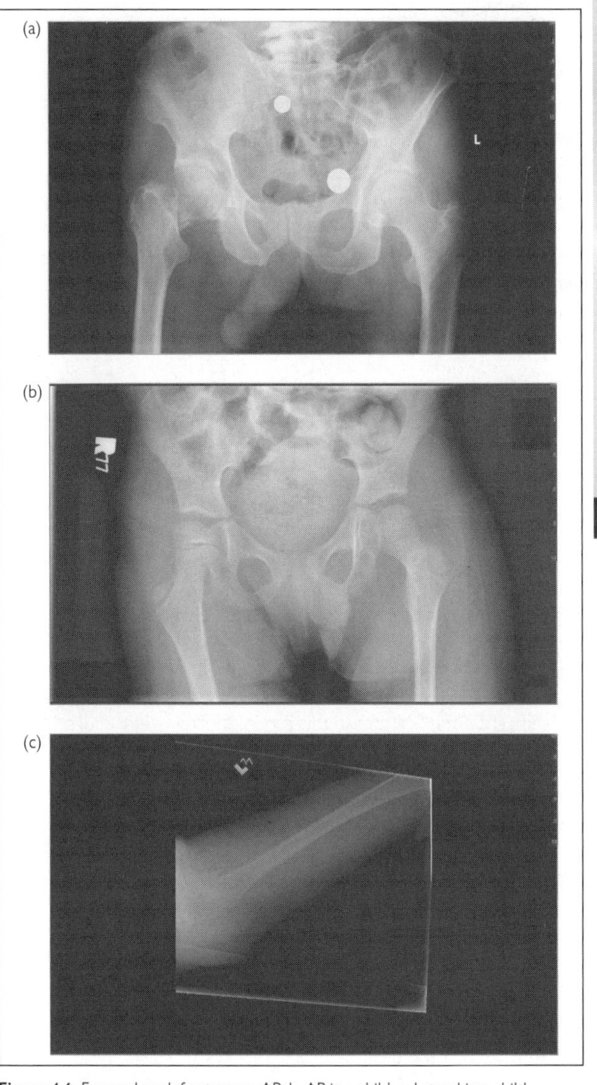

Figure 4.6 Femoral neck fractures a. AP. b. AP in a child. c. Lateral in a child.

- I—Transepiphyseal fracture—AVN almost 100%
- II—Transcervical fracture—AVN 50%–60%
- III—Basicervical fracture—AVN 30%–40%
- IV—Intertrochanteric fracture—Low risk

Primary Stabilization and Management

- There is no primary stabilization for these fractures. Patients should be put in bedrest, with motion limited and a pillow under the knee for slight flexion and external rotation. Be sure to protect the heel from sores.

Admit and Discharge Guidelines

- Request an orthopaedic consult in the emergency department for: All hip fractures.
- Urgent consultation is required for hip fractures in physiologic age <50
- All will need to be admitted or transferred. The admitting service should be decided based on patient's comorbidities. Often admission to a medical service is most appropriate because of existing comorbities and other complications in the geriatric population.
- Preoperative evaluation should be started ASAP as femoral neck fractures will almost always require surgery, and early operation has been shown to decrease complications in the geriatric population.
- Delayed operation secondary to slow medical optimization can decrease the chance of good outcomes. Bedrest can cause significant complications in this group. Mortality in the first month is about 10% and 50% in the first 1–2 years. Delays of >72 hours in getting to surgery significantly increases the first-month mortality rates.
- Young patients and especially those with pediatric femoral neck fractures should be managed urgently (within 6–8 hours). Orthopaedic consultation or transfer should be started immediately.

Definitive Treatment

- There is ongoing debate as to the cost effectiveness and patient satisfaction when comparing total hip arthroplasty, and bipolar or unipolar hemiarthroplasty. Hemiarthroplasty is the treatment of choice in patients >60 with a normal acetabulum. Total hip arthroplasty is the treatment of choice for displaced femoral neck fractures with significant arthritis with pre-existing hip pain or dysfunction.
- Percutaneous fixation can be considered if not displaced or impacted in a good position. This option is especially attractive for young or very ill elderly patients.

- Treatment of the younger patients depends both on their age and the type of fracture. Closed versus open reduction with percutaneous screw fixation and possible capsulotomy to decompress are all considerations.

Intertrochanteric Fractures

Subtrochanteric and intertrochanteric fractures are extracapsular from the hip joint and therefore have significantly lower risk of AVN. Most commonly these fractures are results of low-energy falls in elderly patients. Patients with intertrochanteric when compared to femoral neck fractures tend to be older and more frail. Most are isolated injuries in this age group. If the patient is young, this injury is a high-energy injury and associated injuries should be suspected. Reduction and fixation are not urgent; however, delay to the OR because of medical comorbidities increases morbidity and mortality. Medical evaluation and possible admission should begin immediately in the ED.

Symptoms and Findings

- Groin pain, hip pain
- Inability to ambulate
- Pain with log roll
- Usually neurovascularly unchanged
- As in femoral neck fractures, prefracture pain and tumor history should be evaluated for pathologic fracture.
- Falls are often associated with neurologic or cardiac causes (stroke or myocardial infarction). Complete evaluation of the source of the fall should take precedence, and specific history should be sought regarding the etiology of the fall.

Imaging (See Figure 4.7)

- AP and lateral of hip
- AP pelvis with slight internal rotation (toes taped together)
- AP femur for preoperative planning
- CT is optional and should be obtained if more detail is needed to make the diagnosis.

Classification

- Classification is based on the Evans system and is only important for definitive management.
- Pediatric fractures are based on the Delbert system (see page 178)

Primary Stabilization and Management

- There is no primary stabilization for these fractures. Patients should be on bedrest and motion limited. Be cautious of bedrest complications including ulcers, DVT, and so on.

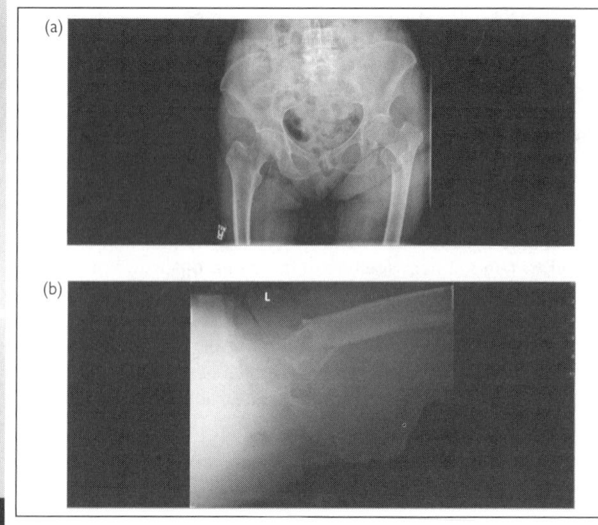

Figure 4.7 a. IT hip fracture AP. b. IT hip fracture lateral.

- Large displacement is rare, but when it occurs, Bucks' traction can be considered for pain relief.

Admit and Discharge Guidelines

- Request an orthopaedic consult in the emergency department for all intertrochanteric fractures.
- UTIs (urinary tract infections) frequently cause the change in exercise tolerance or ability to ambulate in this population. UTIs predispose these patients to this injury. Urine should be checked and treated starting in the ED when appropriate.
- Starting in the ED, the process of preparing for the OR should begin, including preoperative labs, preanesthesia medical evaluation, chest radiograph, EKG, and urinalysis.

Reference to Definitive Treatment

- Surgical fixation is indicated for almost all of these fractures; nonoperative is reserved for bed-bound or severely demented patients.
- Operative fixation includes: Sliding hip screws or helical devices, fixed-angle devices, proximal femoral locking plates, and cephalomedullary devices. Fracture pattern, surgeon preference, and institutional ability will determine which fixation device will be used.

Femur Injuries

Subtrochanteric Fractures

The subtrochanteric region is defined as below the lesser trochanter to 5 cm distal on the femur. There are significant forces acting at this level. The bone is also diaphyseal and metadiaphyseal, making reduction and union more difficult. Reduction is further complicated by some of the largest muscles in the body crossing the fracture site. Proximal fragment is usually flexed, abducted, and externally rotated because of these muscle attachments. These fractures can happen in all age groups and have a bimodal peak in young men (trauma) and elderly women (falls). There is a new subset of subtrochanteric femur fractures that seem to result from the iatrogenic inability to remodel bone from the use of bisphosphonates. Pain in this area with a history of bisphosphonates and negative X-rays should prompt a stress fracture work-up, including an MRI.

Symptoms and Findings

- Groin pain, hip pain
- Inability to ambulate
- Pain with any motion
- Deformity as described above
- As in femoral neck fractures, prefracture pain and tumor history should be evaluated for pathologic fracture in elderly patients.
- Falls are often associated with neurologic or cardiac causes (stroke or MI). Complete evaluation of the source of the fall should take precedence, and specific history should be sought regarding the source of the fall.
- Trauma is often associated with other injuries and the patient should be evaluated fully.

Imaging (See Figure 4.8)

- AP pelvis
- AP/lateral femur
- AP/lateral knee
- AP/lateral tibia if traction to be placed
- Traction radiographs can help show extent of injury and may be needed.
- Postplacement of traction, pin AP/lateral tibia before weight is added, and postweight AP/lateral of the injury with weight are all necessary if traction is used.
- CT is optional for further classification and preoperative planning.

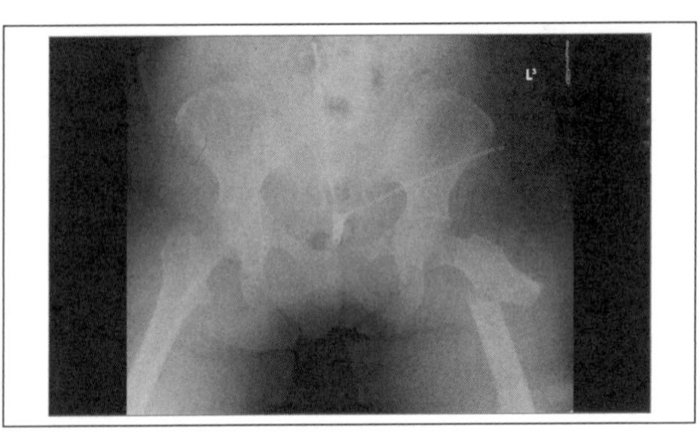

Figure 4.8 Subtrochanteric hip fractures—AP.

Classification (See Figure 4.9)

- Classification is based on the **A**rbeitsgemeinschaft für **O**steosynthesefragen (AO) or Russell-Taylor system, which guides definitive treatment.

Primary Stabilization and Management

- For children who qualify for spica casting, spica casting can be done in the ED or OR by a trained professional. Hip spica casting should not be attempted by a general practitioner.
- All other patients will need admission for treatment. Some advocate bedrest alone while others advocate bedrest with skeletal or skin traction to restore length and for pain control. Skeletal traction should be avoided in young children; skin traction may be used.

Admit and Discharge Guidelines

- Request an orthopaedic consult in the emergency department for: All subtrochanteric femur fractures.
- The process of preparing for the OR should begin, including in the ED, and include preoperative labs, medical evaluation, CXR, EKG, and UA.
- Spica casting may be performed in the ED by a trained specialist with subsequent discharge and follow-up by an orthopaedic surgeon.

Definitive Treatment

- Surgical fixation is indicated for almost all of these fractures that are not amenable to spica casting.
- Two main operations exist: plate fixation and intramedullary/cephalomedullary devices. Fracture pattern, surgeon preference, and institutional ability will determine which fixation device will be used.

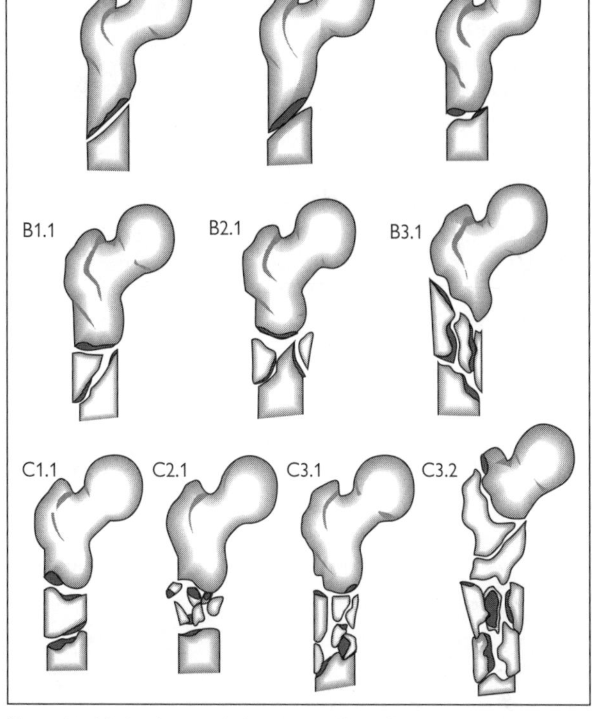

Figure 4.9 AO classification of subtrochanteric femur fractures.

Femoral Shaft Fracture

These fractures can happen at any age. They are a result of trauma, usually high energy. In healthy adults it requires significant force to break the femoral shaft. Small children and elderly may have a lower-energy mechanism. Diaphyseal femoral fractures can have significant blood loss; up to 1 liter of blood can accumulate in the thigh. The proximal fragment tends to lie posterior and abducted. Children <2 years old with this injury have a 50%–80% chance of it being non-accidental. Children with this injury should be evaluated for child abuse.

Symptoms and Findings

- Thigh pain
- Deformity and shortening of the femur

- Swelling and tenderness in the thigh
- Inability to ambulate and reluctance to move hip or knee

Imaging (See Figure 4.10)

- Trauma AP pelvis
- AP/lateral hip
- AP/lateral femur
- AP/lateral knee
- AP/lateral tibia (if using skeletal traction)
- Postcasting AP/lateral femur (spica casting only)
- Posttraction pin AP/lateral tibia
- Posttraction AP/lateral femur. Weight should be added until at length confirmed by serial AP/lateral femur X-rays.
- If severely comminuted or difficult-to-distinguish fracture pattern, CT is optionally used.

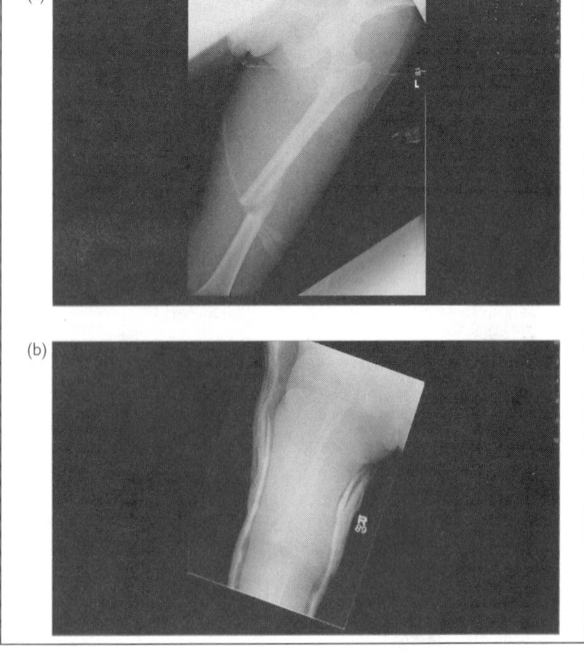

Figure 4.10 Femoral shaft fractures in an: a. Adult on AP X-ray. b. Child on AP X-ray.

Classification

- AO or Winquist-Hansen classification. Both emphasize comminution as the principal factor in classifying these fractures.

Primary Stabilization and Management

- Neonates should be treated with Pavlik harness.
- Children with a femoral canal diameter of 5 mm or less should receive spica casting in the ED/OR. Contraindications to spica casting are abdominal or thoracic trauma, open fractures, and obese or large children.
- All others will likely need an operation, and preoperative work up based on age should be started immediately in the ED.
- Children who do not fit the above criteria and are not skeletally mature: Admission, bedrest, and skin traction.
- Skeletally mature individuals: If the fracture is distal enough to stabilize with a splint and not shortened, a long leg bulky splint should be used with extra care given to the proximal portions of the splint extending as high as possible. The lateral portion of the stirrup can be brought up beyond the hip if well padded.
- Most skeletally mature individuals will require skeletal traction with significant weight (10–25 lbs) to return the femur to length and bedrest in anticipation of treatment in traction versus operative treatment.

Admit and Discharge Guidelines

- Request an orthopaedic consult in the emergency department for: All femur fractures.
- Refer to an orthopaedic surgeon for outpatient follow-up for: Femur fractures treated in spica casting or pavlic harness. See spica casting section for tolerable deformity.
- All others should be admitted with either splint stabilization or traction stabilization and bedrest. Immediate initiation of the preoperative workup is indicated.

Definitive Treatment

- In children, definitive treatment depends upon their size, age, and stability of the fracture pattern. Treatment options include spica casting, flexible intramedullary (IM) rods, rigid IM nailing, submuscular plating, direct plating, and external fixation.
- The current preferred treatment in skeletally mature individuals is reamed antegrade rigid IM nailing. Other options include retrograde nails, submuscular plating, direct plating, and external fixation.

Distal Femoral Shaft (Supracondylar) Fractures

Fractures of approximately the distal 1/3 of the femoral shaft (approximately 8–15 cm from the distal aspect of the femur down to and including the condyles) are considered supracondylar. These are separated from frank shaft fractures because they are typically meta-diaphyseal or metaphyseal and have different muscle forces working to displace them. They are very difficult to control with closed methods and provide a significant challenge with open methods as well. Like other femur fractures, the incidence is often bimodal occurring in young men and elderly women. Injury in young men is a fracture due to significant trauma, while elderly women with this type of fracture are usually the result of moderate trauma on a flexed knee. Children have a high incidence of angular deformity and limb shortening with these injuries if growth is disrupted, which is more pronounced the younger the child is. Child abuse should be suspected if this injury is seen in children less than 1 year of age and should be considered if this fracture occurs in children less than 4 years of age.

Symptoms and Findings

- Knee or thigh pain
- Inability to walk
- Shortening, apex posterior deformity
- Associated injury to knee ligaments; tibial fracture common
- Vascular injury can occur and pulses need to be checked. Any abnormality should prompt further workup including ABI (ankle-brachial index), computed tomography angiography, or angiogram.

Imaging (See Figure 4.11)

- AP/lateral hip
- AP/lateral femur
- AP/lateral knee
- AP/lateral tibia if warranted by any pain
- CT knee if intra-articular
- Varus/valugs stress AP knee if concern for Salter-Harris I injury

Classification (See Figure 4.12)

- Muller adaptation of the AO system:
  - A: Extra-articular/supracondylar
    1–3 Based on comminution
  - B: Unicondylar
    1: Lateral condyle
    2: Medial condyle
    3: Coronal fracture of a condyle (Hoffa fracture)

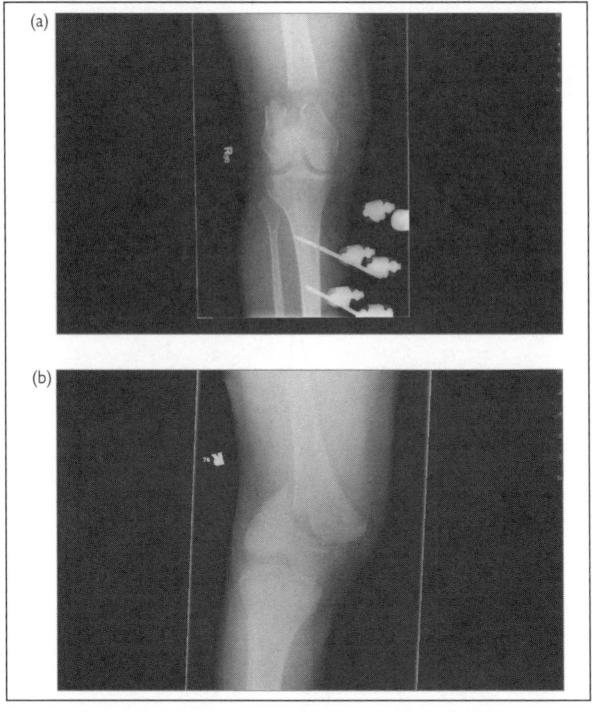

Figure 4.11 Distal femur fracture. a. Adult distal femur fracture. b. Pediatric distal femur fracture.

- C: Bicondylar
  1: No supracondylar comminution (T or Y type)
  2: Supracondylar comminution
  3: Both supracondylar and intracondylar comminution
- Children will often have a Salter-Harris-type fracture through the growth plate.

Primary Stabilization and Management

- Skeletal traction through the tibia will provide the ability to regain length and pain control while awaiting definitive fixation.
- Bucks' traction or skin traction may be used if skeletal traction will not be tolerated (skeletally immature or elderly).
- Reduction and splinting in a long leg splint may provide pain relief but will unlikely be able to maintain reduction. This should be used if neither skeletal nor skin traction is possible. Care should

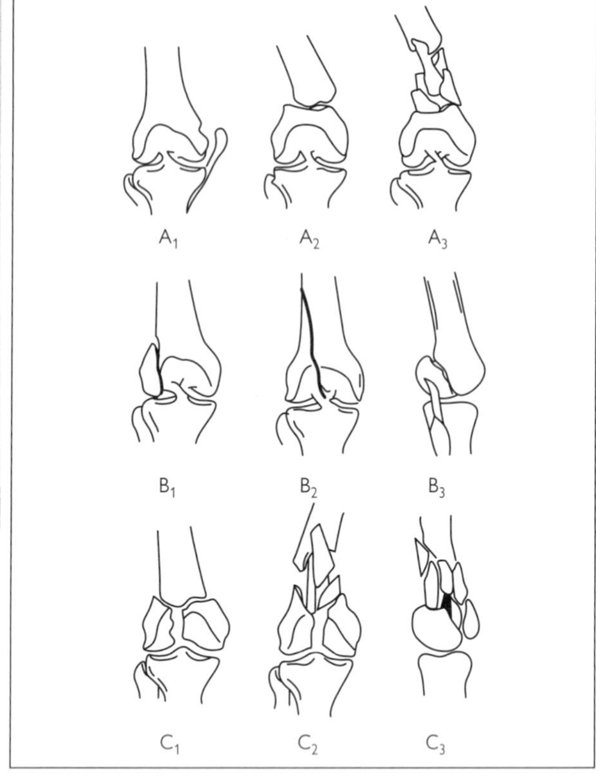

Figure 4.12 Distal femur fracture classification.

be taken to ensure that the splint reaches as high in the hip/femur as possible to immobilize the fractured segment. If the splint is too low, the splint will only act as a lever arm to further displace the fracture.

Admit and Discharge Guidelines

- Request an orthopaedic consult in the emergency department for: Distal femur fractures.
- Adults will all need admission or transfer with operative fixation.
- Children will need consultation or transfer and admission for surgery.

Definitive Treatment

- Surgical fixation is the gold standard for skeletally mature patients. Fixation of these fractures is extremely difficult. There

are many options, including ORIF, submuscular plating, medial/lateral plating, IM nail for Type A fractures, dynamic compression screws, and fixed angle devices.
- Children can be treated with closed reduction and percutaneous pinning (preferred), two-screw lag fixation of the fragment proximal to the growth plate, prolonged traction then casting, exfix, or submuscular plating.

The Knee Extensor Mechanism

The extension of the knee is generated through powerful forces concentrated around the anterior knee. The contraction is from the large quadriceps muscle group, whose force is then applied to the patella through the quad tendon and retinaculum. The patella acts to glide across the femur and increase the distance from the center of rotation, thus increasing the lever arm. Finally, the patellar tendon transmits the force to the anterior tibia through the tibial tubercle. Any disruption in any part of this chain can cause extensor mechanism failure. It should be noted that the typical test of the extensor mechanism, the straight leg raise, could have false negative results if the retinaculum is still intact. In other words, if the leg is passively straightened, it can usually be held straight by the patient by using their retinaculum even with a completely torn tendon. More sensitive testing is done by actively extending the knee against gravity from a 90 degree flexed position. If the patient is unable to cooperate with testing secondary to pain, intra-articular lidocaine may aid in diagnosis.

Quadriceps Tendon Rupture

Symptoms and Findings

- Pain in anterior thigh usually after jumping/landing incident or direct blow
- Can also be seen with overweight patients and low-energy falls or trauma
- Difficulty with ambulation
- +/− Palpable defect in quad tendon
- Difficulty with straight leg raise/knee extension (see above for comment on straight leg raise)
- Typically age >40
- Extensor lag

Imaging

- AP/lateral knee
- Contralateral AP/lateral knee may aid in diagnosis to look for the natural position of the patella and relative patella baja (translates to low patella) in quad tendon injuries. The Insall-Salvati ratio

is a measure of the patellar positioning. It is the maximal length of the patella divided by the distance from the inferior pole of the patella to the top of the tibial tubercule on the lateral radiograph. The normal Insall-Salvati ratio is 1 (Figure 4.13a-A). The ratio is decreased with patella alta (Figure 4.13a-B) and increased with patella baja.

- MRI will definitively diagnose this injury, but is often unnecessary if a good physical exam is done.

Classification

- Complete: Extensor mechanism does not work.
- Partial: Extensor mechanism intact but extensor lag or other concerning symptoms

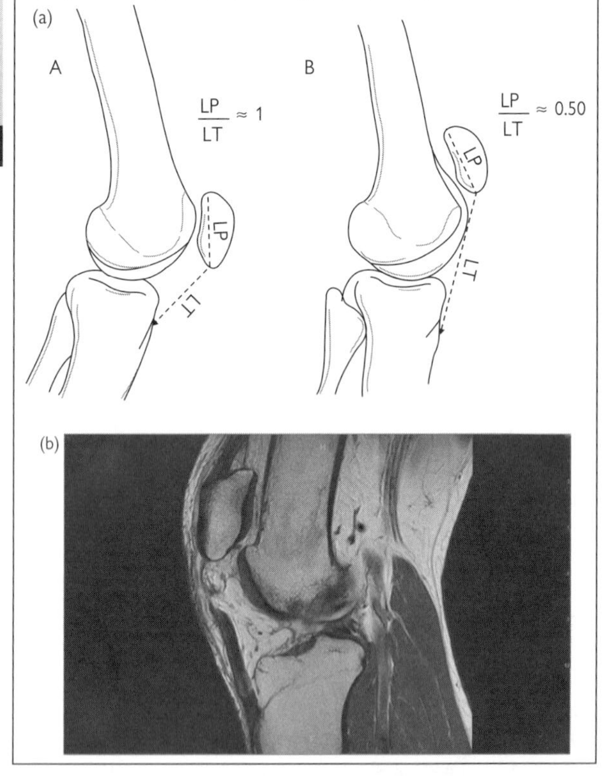

Figure 4.13 Disruption of the extensor mechanism. a. Diagram demonstrating the calculation of the Insall-Salvati ratio. b. MRI demonstrating disruption of the patellar tendon.

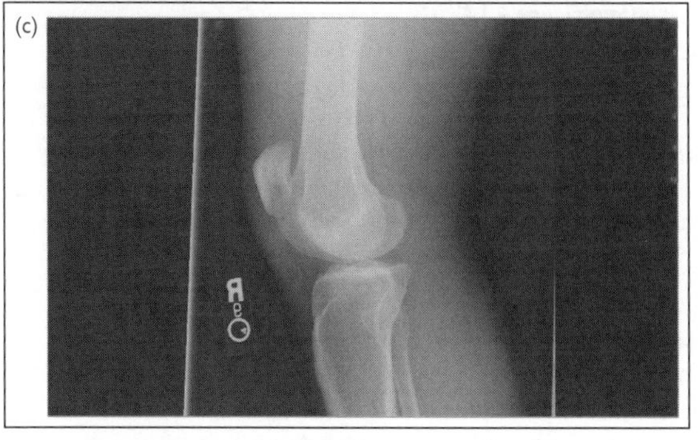

Figure 4.13 (Continued) c. Patellar tendon rupture—patella alta.

Primary Stabilization and Management

- Knee immobilizer

Admit and Discharge Guidelines

- Request an orthopaedic consult in the emergency department for: Complete tears. Complete tears will require surgery and it is best done within 48–72 hours to allow for primary repair.
- If orthopaedic follow-up is immediately available in this time frame, then patients can be discharged, WBAT (weight-bearing as tolerated) in an immobilizer without consultation in the ER.
- Refer to an orthopaedic surgeon for outpatient follow-up for: All tears both complete and partial. WBAT in knee immobilizer with early return to motion as decided upon by the treating orthopaedist.

Definitive Treatment

- Partial tears are treated with immobilization and early return to motion
- Complete tears are treated with primary surgical repair with augmentation as needed.
- Chronic tears will likely need surgical repair with augmentation. There is no urgency to repair chronic tears.

Patella Fracture

Symptoms and Findings

- Pain after direct blow, fall, or dashboard injury (direct mechanism), or jumping/landing injury, or knee flexion against a fully contracted quadriceps (indirect mechanism).
- May have intact or defective extensor mechanism
- Tenderness to direct palpation +/− defect and crepitus
- Significant knee effusion

Imaging (See Figure 4.14)

- AP/lateral knee
- Sunrise view should not be done acutely (unless AP and lateral are negative for fx) as the flexion of the knee can cause displacement of nondisplaced fractures.
- CT can be considered if warranted for operative planning. This is unlikely.

Classification

- Classification is largely related to mechanism of injury.
- Transverse: Medial to lateral split from indirect injury. May be displaced or nondisplaced. May have some comminution but is less often a feature.
- Stellate: Comminuted fracture from direct blow often with a large transverse component. Has more likelihood of articular damage.
- Marginal: Usually vertical fracture of a medial or lateral portion of the bone not directly needed for extensor function.
- Vertical: Rare in isolation
- Sleeve: Unique to children, a sleeve fracture is an avulsion of the inferior pole of the patella. What makes this different is that a large (>50%) segment of the articular cartilage from the underside of the patella is avulsed with the distal pole and is not seen on radiographs. In contrast to many other pediatric fractures, this pattern needs open reduction and fixation.

Primary Stabilization and Management

- Knee immobilizer with the knee in extension. May be WBAT in immobilizer because a well-fitted immobilizer will eliminate the extensor mechanism.

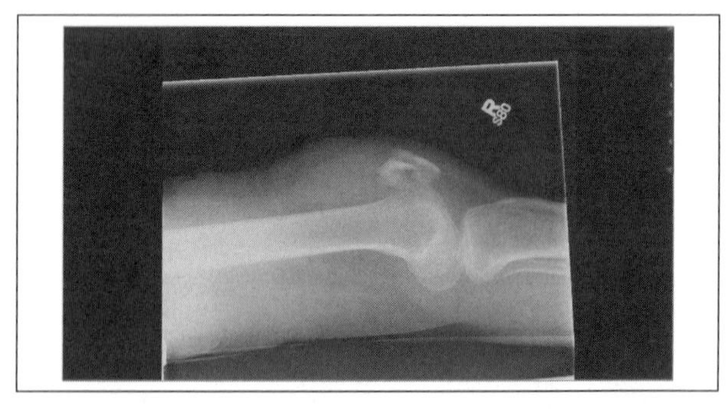

Figure 4.14 Patella fracture seen on lateral X-ray.

- If immobilizer can't be fit properly, a long leg splint or long leg cast can maintain knee extension. Patient should be NWB (non-weight-bearing) if long leg split/cast is applied.
- Children or adults with minimal swelling with non- or minimally displaced fractures can be treated definitively with long leg casting in extension. Some recommend cylinder casting; however, it is the authors' opinion that a cylinder cast is extremely difficult to apply. The difficulty in putting on the cylinder cast is to avoid skin problems at the ankle; in most cases this requires an experienced cast technician.

Admit and Discharge Guidelines

- Request an orthopaedic consult in the emergency department for: Open fracture.
- Refer to an orthopaedic surgeon for outpatient follow-up for:
  - All other patellar fractures
  - Displaced fractures or fractures that disrupt the extensor mechanism; sleeve fractures should be seen within a few days for operative planning.
  - Nondisplaced or minimally displaced fractures, for follow-up in 2 weeks.

Definitive Treatment

- Non- or minimally displaced: Nonoperative treatment in a brace or cast that holds the knee in extension
- Displaced fracture, fracture that disrupts the extensor mechanism, or sleeve fractures: Acute surgical repair

Patella Dislocation

Symptoms and Findings

- Commonly a sporting injury. May be secondary to anatomic variation or acute injury to stabilizing structures in the knee.
- Commonly presents reduced, but may be dislocated
- Pain is a hallmark.
- Bloody effusion and swelling often present
- Positive patella apprehension test
- Positive patella tilt
- Tenderness on medial femoral condyle may represent medial patella-femoral ligament tear.

Imaging (See Figure 4.15)

- Knee XR usually only demonstrates effusion, but it is required to rule out other sources of knee pain. May demonstrate high Q-angle or incompetent trochlea or lateral femoral condyle, all associated with increased occurrence of dislocation.

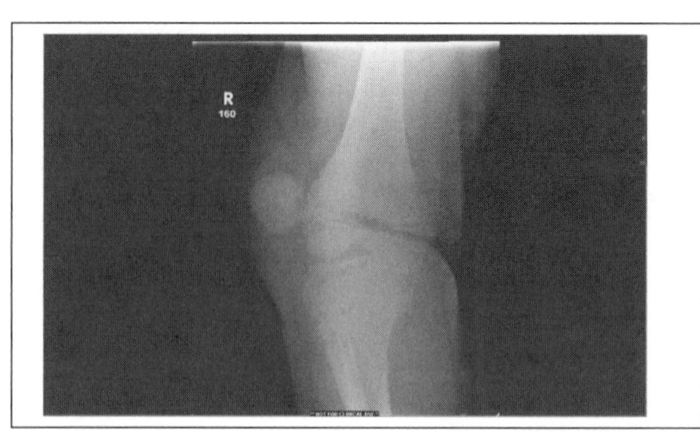

Figure 4.15 Patella dislocation.

- MRI has classic bone bruise pattern on lateral femoral condyle and patella but is generally not required in the acute setting.

Classification

- Reduced versus fixed dislocation
- First time versus chronic dislocator

Primary Stabilization and Management

- Reduced: Knee immobilizer or patellar support brace
- Dislocated: Primary reduction in the ED. Can use intra-articular analgesics; generally responds to direct manipulation or manipulation of knee joint into extension
- Knee immobilizer after reduction.

Admit and Discharge Guidelines

- Request an orthopaedic consult in the emergency department for: Unreducible patella.
- Refer to an orthopaedic surgeon for outpatient follow-up for suspected dislocation. Send patient with knee MRI or prescription for one before specialist appointment if this is the working diagnosis.

Definitive Treatment

- First-time dislocators will often rehab and return to sport with a brace to prevent dislocation or subluxation.
- Recurrent dislocators may require a surgical procedure(s) to correct the Q-angle, femoral condyle, trochlea, or medial patellofemoral ligament (MPFL), depending on the cause of the dislocations.

- Dislocations can sometimes cause loose bodies and cartilage damage to the patella; if symptomatic these may need to be addressed surgically.

Patellar Tendon Rupture

Symptoms and Findings

- Pain in anterior proximal leg, knee, usually after jumping/landing incident when the knee is in flexed position or direct blow. Most often occurs in athletic activity. Requires force of 17.5 times body weight to tear (Zernicke et al. 1977).
- Difficulty with ambulation
- +/− Palpable defect in patellar tendon
- Difficulty with straight leg raise/knee extension (see above for comment on straight leg raise).
- Typically age <40
- Extensor lag

Imaging (See Figure 4.13b–c)

- AP/lateral of the knee.
  - Possible patella alta (compare to other side if question). Insall-Salvati <0.80
  - Possible effusion
  - Possible cortical avulsion from inferior pole of the patella
- MRI only if diagnosis is in question or concern for concomitant injuries (uncommon)
- Ultrasound can make diagnosis if ultrasonographer experienced in musculoskeletal ultrasound

Classification

- No accepted classification system. Described based on location in tendon and whether cortical avulsion.

Primary Stabilization and Management

- Knee immobilizer and WBAT

Admit and Discharge Guidelines

- Request an orthopaedic consult in the emergency department for: Open injury or no orthopaedic surgeon available to patient in next 3–7 days.
- Refer to an orthopaedic surgeon for outpatient follow-up for: Outpatient repair of tendon within 7 days.

Definitive Treatment

- Primary repair/reconstruction for acute complete tears
- Delayed reconstruction for chronic or missed injuries

Knee Injuries

Acute Anterior Cruciate Ligament (ACL), Posterior Cruciate Ligament (PCL), Posterior Lateral Corner (PLC), or Medial or Lateral Collateral Ligament (MCL or LCL) Rupture

Acute injuries to one of the collateral or cruciate ligaments in the knee are common athletic injuries. It is important to note that injury to two or more of the ligaments at the same time is a knee dislocation until proven otherwise and should be evaluated and treated with the same emergency as a knee dislocation.

The knee joint has little bony stability. Without the soft tissue support, it is analogous to a greased ball bearing in a saucer. There are many soft tissue stabilizers to the knee, many of which have multiple roles depending on the angle of flexion of the joint and how it is being used. The intricacies of the knee joint are beyond the scope of this text; the important structures frequently injured will be covered.

Symptoms and Findings

- These injuries occur primarily from athletic pursuits and secondarily to trauma. High-energy trauma should raise one's suspicion for knee dislocation.
- Pain almost always occurs with the injury and in the acute setting.
- Bloody effusion and knee swelling are hallmarks.
- Decreased range of motion due to swelling and pain
- A sense of instability or "giving way" with weight bearing is commonly present.
- ACL: Positive Lachman maneuver, positive pivot shift
- PCL: Positive posterior drawer test, increased dial test at 90 degrees
- PLC: Increased dial test at 30 degrees, reverse pivot shift, ER recurvatum test, and posterolateral drawer
- LCL: Increase in lateral opening with varus stress at 30 degrees
- MCL: Increase in medial opening with valgus stress at 30 degrees

Imaging

- Obtain at least AP/lateral knee
- Common to have significant effusion in all injuries
- Asymmetry of the clear spaces and any subluxation may give clues to a possible injury.
- Segond fracture classic with ACL tear
- Children's ligaments are often stronger than their bone, and therefore their ligament injury presents as a bony avulsion at the insertion or origin of the ligament.
- MRI is gold standard to evaluate ligaments and should be obtained after a complete exam.

Classification

The following scale generically grades ligament injuries:

- Grade I: Tearing of some of the fibers of the ligament. Pain, mild swelling. No joint instability, minimal to no ligament laxity.
- Grade II: Tearing of a portion of the ligament. Significant pain and swelling. Increased laxity with a firm end point.
- Grade III: Complete tearing of the ligament. Significant pain and swelling (although there are reports that suggest that a Grade II injury often hurts more than a Grade III). Increased laxity with a soft or no end point.

Primary Stabilization and Management

- There are custom-made ligament specific braces on the market to help stabilize the ligamentously injured knee. These are best used by experienced sports medicine or orthopaedic surgery practitioners.
- The best acute treatment for a single ligament injury to the knee is a well-fitted knee immobilizer or hinged knee brace if available, WBAT.

Admit and Discharge Guidelines

- Request an orthopaedic consult in the emergency department for:
  - Multiple ligamentous injury (see knee dislocation)
  - Open injuries
  - Irreducible locked knees
- Refer to an orthopaedic surgeon for outpatient follow-up for:
  - Uncomplicated single ligament injuries

Definitive Treatment

- PCL injuries, PLC injuries, and some collateral injuries often need to be fixed to return stability to the knee for normal everyday living.
- ACL and some collateral injuries are commonly fixed to allow return to sport but are not a necessity for daily living in all patients.

Knee Dislocations

An orthopaedic emergency

Symptoms and Findings

- Most common in high-energy trauma; can be low-energy mechanism less commonly
- HIGH occurrence of vascular injury to popliteal vessels and risk of loss of limb. High occurrence of peroneal nerve injuries.
- Dislocation requires >1 ligament tear (at least one cruciate and one collateral)
- May have concomitant fracture, osteochondral defect, meniscal tear, or extensor mechanism injuries
- Pain and swelling always occur.
- MAY be spontaneously reduced or still dislocated

- Patient unable to bear weight
- Evaluate vascular status **immediately** by exam. Any question of difference in pulses requires obtaining ankle brachial index on both legs.
- Any frank difference in vascular status or ABIs that are different get further testing (CTA, magnetic resonance angiography (MRA), angiography—whichever is the most immediately available). Vascular injury is a thrombus or intimal tear. Incidence is high, ranging from 10% to 65%.
- Complete neuro exam is necessary at the time of presentation (10%–45% incidence of peroneal nerve injury).
- Evaluate for possible compartment syndrome.
- Once immediate limb-threatening injuries are ruled out, then a complete ligamentous knee exam should be done.

Imaging (See Figure 4.16a–b)

- AP/lateral knee
- AP/lateral femur
- AP/lateral tibia

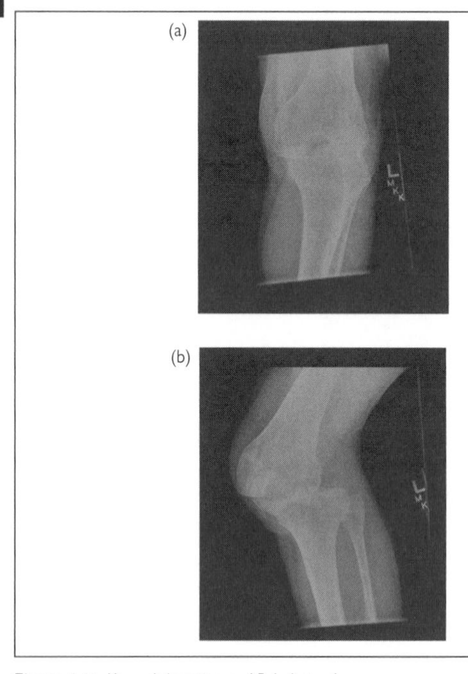

Figure 4.16 Knee dislocation. a. AP. b. Lateral.

Classification

- Dislocated on presentation versus reduced
- Direction of the tibial plateau dislocated in relation to the femoral condyles (anterior vs. posterior)

Knee Dislocation (KD) Classification

Type	Ligament torn	Comments
KD I	One cruciate ligament torn (also one or both collaterals)	Commonly ACL + PLC
KD II	Both ACL and PCL torn; collateral ligaments intact	Rare
KD IIIM	ACL, PCL, and MCL torn	
KD IIIL	ACL, PCL, and LCL torn	Peroneal nerve injury common
KD IV	ACL, PCL, MCL, and LCL torn	Very unstable
KD V	Fracture dislocation	
C (added to above)	Associated arterial injury	
N (added to above)	Associated nerve injury	
Example: KDIVCN = All four ligaments torn + nerve injury + vascular injury		

Primary Stabilization and Management

Consult Orthopaedics on All Knee Dislocations!

- Unreduced knee: Reduce in ED with longitudinal traction and reevaluate neurovascular status after 5 minutes.
- Vascular injury: Immediate orthopaedic and vascular surgery consult with revascularization and stabilization surgeries urgently
- Reduced, stable, vascular intact: Well-fitted knee immobilizer or long leg well-padded splint
- Reduced, unstable, vascular intact: Custom long leg splint; if still unreduced in splint, stabilization surgery (external fixator)

Admit and Discharge Guidelines

- Request an orthopaedic consult in the emergency department for:
  - Every knee dislocation or reduced multiligamentous knee injuries
  - Patients who do not have dedicated imaging of their vessels, who need 24 hour admissions for observation to rule out intimal flap
- Refer to an orthopaedic surgeon for outpatient follow-up for: NONE (all should be admitted).

Definitive Treatment

- There are little to no nonoperative indications for multiligamentous knee injuries. They are unstable, and some reconstruction and repair is necessary to return stability. The order and timing of the procedures is at the surgeon's discretion and will often take more than one procedure over many months.

Acute Meniscal Tear versus Bucket Handle, Locked Knee

Meniscal tears are a very common orthopaedic complaint. Many are chronic attritional tears of the periphery of the meniscus. Although painful, this is not an acute injury. Alternatively, trauma and sporting activities can lead to acute meniscal tears, and in worst case scenario the torn piece of meniscus can flip itself into a position that blocks knee motion and causes significant pain. Any permutations between these extremes are possible.

Symptoms and Findings

- Pain, difficulty with ambulation
- Knee swelling, possible bloody in effusion
- Pain localized to one side of the knee
- Positive McMurray test
- Positive Apley test
- Tender to palpation at medial or lateral joint line
- May have mechanical block to extension (if difficult to determine mechanical block to extension vs. pain preventing extension, intra-articular lidocaine can be used for diagnostic purposes)
- May have snapping, popping, or intermittent "catching"

Imaging

- AP/lateral of knee
  - Effusion likely finding
  - Joint space narrowing on weight-bearing films, especially if chronic
- MRI is gold standard for diagnosis, but not always needed in the acute setting.

Classification

- Meniscal tears are classified anatomically:
  - Medial versus lateral meniscus
  - Anterior horn, posterior horn, middle
  - Radial, longitudinal, oblique, vertical, bucket handle, displaced bucket handle
  - Zone of tear: red-red, red-white, white-white

Primary Stabilization and Management

- If the knee is not locked due to displaced bucket handle: RICE + meds
  - Rest—Can use crutches for a short period if necessary
  - Ice—Pain and swelling control
  - Compression—Ace wrap helps symptomatically. Knee immobilizers are useful with acute injury.
  - Elevation—Helps with pain and swelling
  - Meds: Anti-inflammatory medications are best at targeting the pain of a meniscal tear.

Admit and Discharge Guidelines

- Request an orthopaedic consult in the emergency department for: Locked knee.
- Refer to an orthopaedic surgeon for outpatient follow-up: All others.

Definitive Treatment

- Many meniscal tears can be treated nonoperatively with RICE + Meds, physical therapy, bracing, and activity modification. Younger patients with a tear from an acute injury are more likely to require repair.
- For tears that fail conservative management or are causing locking, surgical intervention is indicated and may consist of debridement versus repair versus transplantation.

Tibial Spine Avulsion Fracture

Symptoms and Findings

- The tibial spine is the location of the attachment of the cruciate ligaments. An avulsion is analogous to a cruciate tear. This injury is more common in skeletally immature individuals.
- They occur and present with the same findings and history as an ACL tear (see Acute Anterior Cruciate Ligament (ACL), Posterior Cruciate Ligament (PCL), Posterior Lateral Corner (PLC), or Medial or Lateral Collateral Ligament (MCL or LCL) Rupture)

Imaging (See Figure 4.17)

- AP/lateral knee demonstrates fracture.
- AP/lateral femur and tibia should be obtained to rule out concomitant injury.
- Postcast AP/lateral knee is crucial.

Classification

- Type 1: Nondisplaced
- Type 2: Displaced with posterior hinge. A—displaced vertically, B—rotated

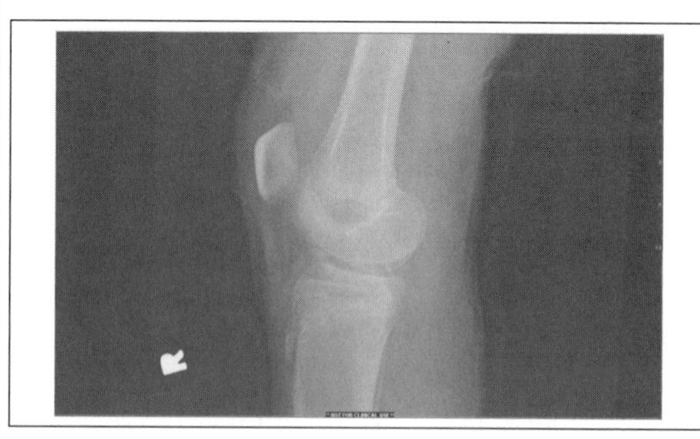

Figure 4.17 Tibial spine avulsion.

- Type 3: Completely displaced

Primary Stabilization and Management

- Immobilization in extension using a straight long leg cast (child) or splint (adult).

Admit and Discharge Guidelines

- Request an orthopaedic consult in the emergency department for: Type 3 or irreducible Type 2.
- Refer to an orthopaedic surgeon for outpatient follow-up for: Type 1 treated in a cast or splint in the ED.
- Type 2 that reduces with extension casting or splinting

Definitive Treatment

- Type 1 fractures can usually be treated in extension immobilization to union.
- Type 3 will likely need surgical reduction and fixation.
- Type 2 depends on the amount of reduction achieved with extension casting and will be surgeon/patient dependent.

Tibial Tuberosity Avulsion Fracture

Symptoms and Findings

- Most common in children and especially boys 12–17 years old, this fracture is analogous to a patellar tendon avulsion in adults. It occurs due to excess force on the patellar tendon, usually in knee flexion, and instead of the tendon tearing, the bone and growth plate fail.
- Anterior knee, proximal tibial pain
- Tenderness localized to the tibial tuberosity and proximal tibia
- Significant swelling and knee effusion can be seen.
- Pain with ROM or activity

- Extensor lag or disruption of extensor mechanism

Imaging (See Figure 4.18)

- AP/lateral knee will provide the diagnosis. A lateral of the contralateral knee can be obtained to determine the baseline size of the tibial tuberosity physis if the diagnosis is equivocal.
- AP/lateral tibia and femur should be obtained to rule out concomitant injury.

Classification (See Figure 4.19)

Type	Description
IA	Fx through primary ossification center, non to min displaced
IB	Fx through primary ossification center, displaced
IIA	Fx through secondary ossification center, non to min displaced
IIB	Fx through secondary ossification center, displaced or comminuted
IIIA	Fx through tibial plateau and knee joint, solitary fracture fragment
IIIB	Fx through tibial plateau and knee joint, comminuted

Primary Stabilization and Management

- IA: Long leg casting with knee in extension, NWB
- IB, IIA, IIB, IIIA, IIIB—Immobilize in extension with long leg splint and refer for surgery

Admit and Discharge Guidelines

- Request an orthopaedic consult in the emergency department for: Open fractures or any question of compartment syndrome.
- Refer to an orthopaedic surgeon for outpatient follow-up: If pediatric orthopaedics is immediately available, long leg splints and referral for surgical intervention is appropriate.

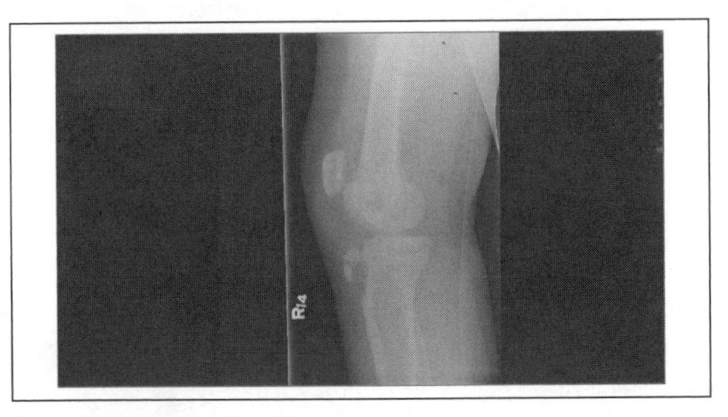

Figure 4.18 Tibial tubercle avulsion.

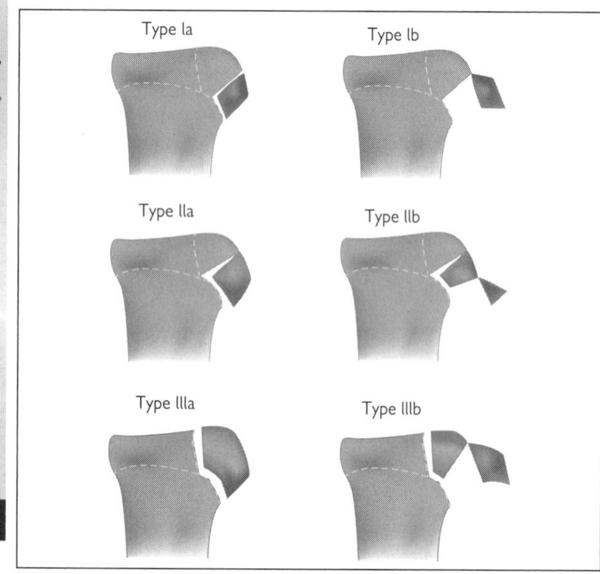

Figure 4.19 Classification of tibial tubercle avulsions.

Definitive Treatment

- IA: Long leg straight knee cast for 4–6 weeks
- All others: ORIF

Corner Fracture

A corner fracture is a specific entity caused by shaking or twisting a child. It is highly suspicious for child abuse. **EVERY** corner fracture warrants a complete child abuse workup regardless of the social situation of the child. Associated injuries include: Bucket handle fractures of the proximal tibial or proximal humeral metaphysis, rib fractures, subdural hematomas, and visceral injury. Old healing fractures in these suspicious areas, especially multiple fractures of different ages or untreated fractures, are also highly suspicious for nonaccidental trauma (i.e. abuse).

Symptoms and Findings

- History usually unclear or inconsistent with injury
- Child has pain and swelling locally

Imaging (See Figure 4.20)

- AP/lateral knee
- Complete skeletal survey to look for associated injuries

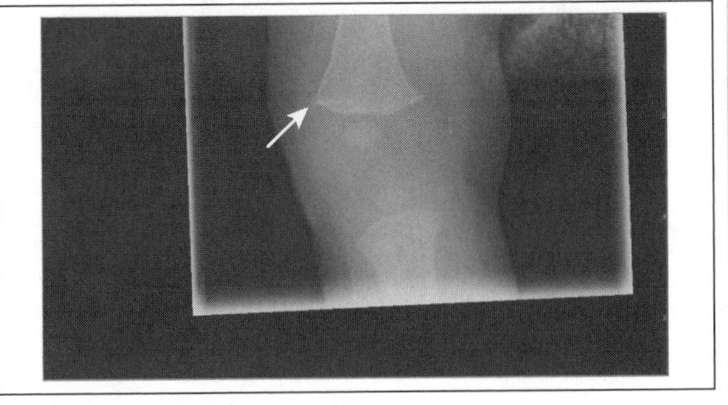

Figure 4.20 Corner fracture of the femur.

Classification

- Describe anatomically

Primary Stabilization and Treatment Management

- Long leg casting, bent knee, ankle at 90 degrees

Admit, Refer, and Discharge Guidelines

- All patients should have complete Child Protective Services/social work evaluation and possible admission for safety until evaluation is complete.

Definitive Treatment

- Casting is definitive treatment.

Salter-Harris Fracture of Proximal Tibia

Symptoms and Findings

- History usually involves a varus or valgus stress applied to knee.
- Proximal tibial pain
- Swelling and tenderness to palpation medially or laterally
- Inability to bear weight
- Can feel like a ligament tear with varus or valgus stress examination

Imaging

- AP/lateral knee may show acute fracture line; however, in a true SHI, may not show any findings
- AP varus or valgus stress examination *may* be required to demonstrate opening of the physis.
- AP/lateral tibia, three-view ankle should be performed for full evaluation if fractured.

Classification

- Classified anatomically:
  - Hyperextension: Apex of angulation is posterior. Highest risk of vascular injury.
  - Varus: Apex lateral (often SHII)
  - Valgus: Apex medial (often SHII)
  - Flexion: Apex anterior (often SHI)

Primary Stabilization and Management

- Hyperextension: Reduction and long leg casting in flexion
- Varus/Valgus/Flexion: Reduction and long leg casting in extension
- NWB

Admit and Discharge Guidelines

- Request an orthopaedic consult in the emergency department for: Reductions that are not within the scope of ED provider's ability.
- Reduction should occur under procedure sedation or general anesthesia, and portable c-arm should be available.
- More than 2–3 attempts are discouraged due to ongoing damage to the growth plate.
- Refer to an orthopaedic surgeon for outpatient follow-up for: Casted, reduced fractures

Definitive Treatment

- Well-reduced and stable fractures can be treated nonoperatively.
- Unstable fractures will need to be pinned across the growth plate in the OR. Hyperextension tends to be unstable.

Tibial Plateau Fractures

Acute tibial plateau fractures in adults are associated with significant swelling and the risk of compartment syndrome.

Symptoms and Findings

- Usually the result of high-energy trauma in healthy individuals
- Significant pain and swelling associated
- Pain with weight bearing
- Decreased ROM
- Joint effusion
- These fractures have a high incidence of popliteal vessel damage and peroneal nerve injury.

- Thorough vascular and nerve examinations must be done and documented at the first presentation.
- Any minor difference in pulse exam between the legs should prompt an ankle brachial index (ABI) examination (ratio of the ankle systolic blood pressure to the brachial systolic blood pressure).
- Any significant difference in pulses or the ABI warrants immediate vascular examination with angio or CTA and a vascular surgery consultation.
- Medial-sided fractures **are analogous to knee dislocations** and should be treated as an **emergency**. Immediate evaluation of the vascular status of the limb is crucial (see "Knee Dislocations").

Imaging (See Figure 4.21)

- AP/lateral knee, AP/lateral tibia, and AP/lateral femur are all required.
- Three-view ankle is recommended.
- Foot series needed if planning skeletal traction through calcaneal pin
- CT scan proximal tibia for operative planning after reduction and splinting

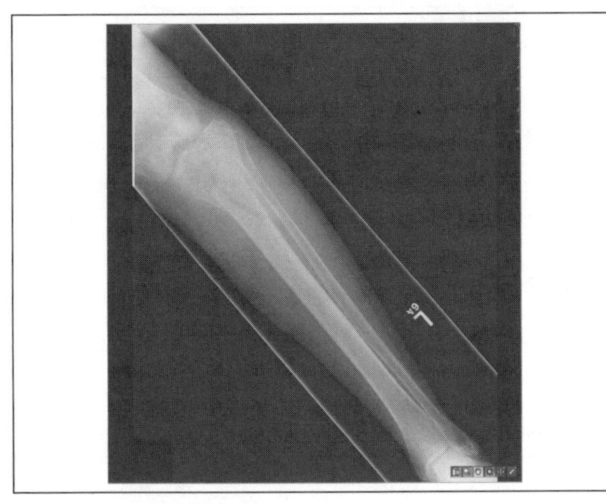

Figure 4.21 Tibial plateau fracture.

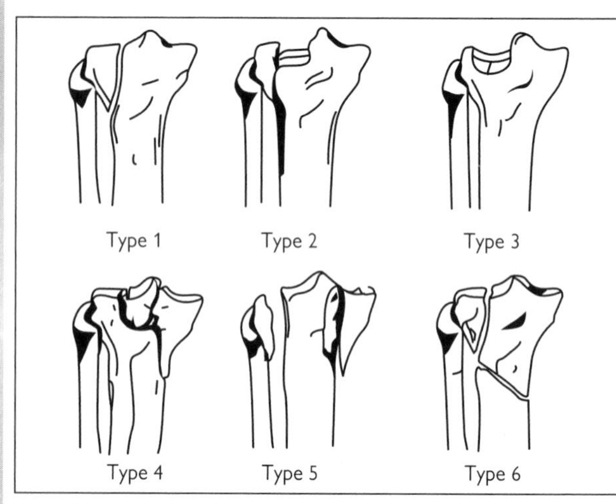

Figure 4.22 Schatzker classification.

Classification (See Figure 4.22)

Classified by the Schatzker system:

Type	Description
I	Lateral split only
II	Lateral split depression combined
III	Lateral depression only
IV	Medial-sided fracture (high incidence of vascular injury)
V	Bicondylar fracture
VI	Any plateau fracture with NO metaphy seal–diaphyseal connection

Primary Stabilization and Management

- Gentle reduction and stabilization in a well-padded long leg splint with the knee at or near full extension and the ankle at 90 degrees
- Many will get same-day or next-day stabilization with external fixation; however, this is not required and is up to the orthopaedic surgeon. Limb shortening is the most common indication for external fixation.
- Skeletal traction with calcaneal pins is an option; however, splinting is the authors' preferred treatment.

Admit and Discharge Guidelines

- Request an orthopaedic consult in the emergency department for: All acute tibial plateau fractures. All acute fractures should be admitted for 24–48 hours for compartment checks, pain control, and elevation.
- Refer to an orthopaedic surgeon for outpatient follow-up: Subacute fracture with no displacement and no swelling. For example: Patient X has had pain in the knee after a minor fall 5 days ago. Patient X is tired of the pain and finally visits the ED. Patient X is diagnosed with a nondisplaced Shatzker I and is treated with hinged knee brace or knee immobilizer and referred to orthopaedics for follow-up.

Definitive Treatment

- Most tibial plateau fractures have some element of displacement at the joint line. These are best treated with ORIF. High-energy injuries with significant soft tissue envelope compromise are best treated with staged ORIF after external fixation to allow for soft tissue healing.
- Some minimally displaced fractures or nondisplaced fractures can be treated in casting after splinting versus hinged knee brace after splinting.
- These fractures are occasionally seen in the nonambulatory population after low-energy injuries (fall during transfer or from wheelchair). These can often be treated nonoperatively.

Injuries to the Lower Leg

Tibial Shaft Fractures

Acute tibial shaft fractures in adults are associated with significant swelling and the risk of compartment syndrome. This occurs in some children as well with less frequency.

Symptoms and Findings

- Most are the result of trauma. Generally high-energy in the healthy population.
- Significant deformity and angulation are often present with displaced fractures.
- Significant swelling
- Tenderness at the fracture site
- A complete neuro and vascular exam at the time of presentation is a must to help with the subsequent evaluations for compartment syndrome.

Imaging (See Figure 4.23)

- AP/lateral knee—Required
- AP/lateral tibial—Required
- Three-view ankle—Required

Classification

Typically described anatomically:

- Proximal, middle, or distal third is a key component.
- Nondisplaced, minimally displaced, or displaced
- Nonangulated versus angulated and direction of apex of angulation
- Fracture description: Transverse, oblique, spiral, butterfly, comminuted, segmental
- Open versus Closed

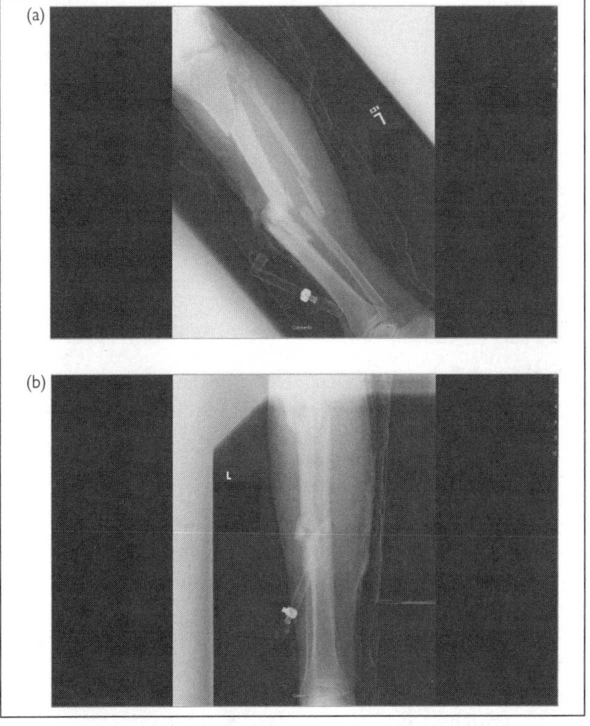

Figure 4.23 Tibial shaft fracture seen on a. AP X-ray. b. Lateral X-ray.

Primary Stabilization and Management

- Adults: Reduction and alignment of fracture with subsequent long leg well-padded splint, knee in extension, and ankle at 90 degrees.
- Children: Conscious sedation with reduction and bivalved long leg bent knee casting.

Admit and Discharge Guidelines

- Request an orthopaedic consult in the emergency department for: All adult tibial shaft fractures and most pediatric tibial shaft fractures, especially those pediatric tibial shaft fractures in need of significant reduction.
- Pediatric patients need education on elevation and the signs and symptoms of compartment syndrome before they are discharged after appropriate treatment and a plan for returning to the ED if these develop.
- Tolerances for discharge after reduction and casting:

Parameter	Acceptable
Cortical contact	>50%
Varus/valgus angulation (tibial plateau to tibial plafond)	<5–10 degrees
Anterior or posterior angulation	<10–15 degrees
Internal or external rotation	<5–7 degrees
Shortening	10–15 mm

- Comminution or <50% cortical contact has been shown to increase nonunion rates in nonoperatively managed tibial shaft fractures.
- Refer to an orthopaedic surgeon for outpatient follow-up for: Nondisplaced pediatric fractures treated appropriately with casting, and reduced and casted fractures for which the social situation allows for discharge home with patient/family monitoring for compartment syndrome.

Definitive Treatment

- Tibial shaft fractures in pediatric patients can often be managed nonoperatively with cast treatment. When they cannot, there are operative techniques that can be used, including IM nails, plating, and external fixation. In the growing child, this often means two surgeries since implants are often explanted in children after union.
- Tibial shaft fractures in adults are often treated with surgery, although arguments can be made for nonoperative treatment. Nonoperative treatment must meet the above stated criteria, and the patient must be able to be nonweight bearing for up to 4 months while the fracture heals. However, operative treatment,

especially with IM nailing in a stable fracture, allows for weight bearing immediately, and often better reduction.

Plafond or Pilon Fractures

The most distal of the tibia fractures is a tibial shaft, typically metaphyseal, fracture with one or more fracture lines extending into the articular surface of the ankle. Often caused by force directed proximally from the talus into the tibia, these injuries can be severe and cause significant destruction. Pilon fractures may be life altering for the patient.

Symptoms and Findings

- Significant ankle pain
- Deformity
- Significant swelling, fracture blisters
- Tenderness at the fracture site
- Neuro changes distally not uncommon
- Unable to bear weight
- Twenty-five percent will be open injuries
- Seventy-five percent will have an associated fibular fracture
- Fifty percent will have some sort of vascular damage
- Spine, pelvis, contralateral injuries all very common

Imaging (See Figure 4.24)

- Three-view foot—look for calcaneus or talus fractures
- Three-view ankle—evaluate articular surface.

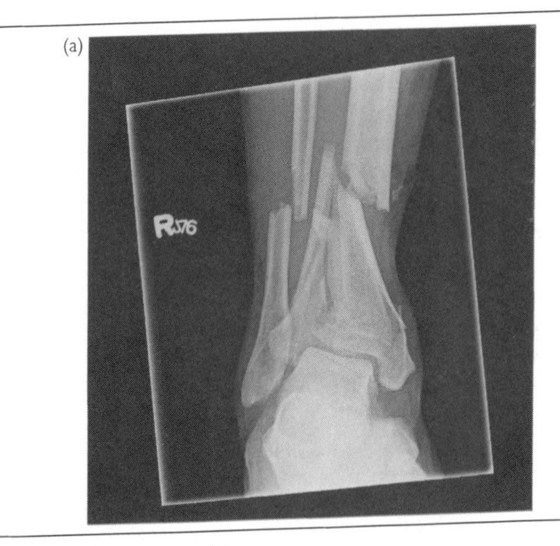

Figure 4.24 Pilon fracture. a. AP.

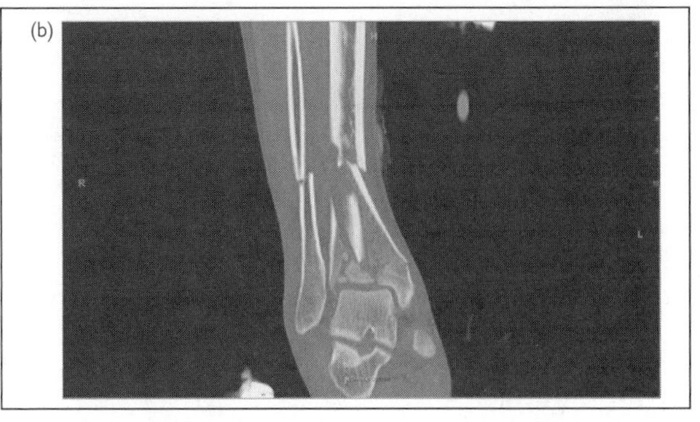

Figure 4.24 (Continued) b. Pilon CT coronal.

- AP/lateral tib fib—evaluate extent of tibia and fibular fractures.
- AP/lateral knee—evaluate for proximal injuries.
- Trauma AP pelvis
- Postreduction/splinting CT, CTA if abnormal vascular exam

Classification (See Figure 4.25)

- Many classifications exist, OA, Topliss, Ruedi-Allgower. None are perfect. Ruedi-Allgower most commonly used historically. Topliss newer and attempts to classify based on pattern typically associated with low versus high energy.
- Ruedi-Allgower:

1. No comminution or joint line displacement
2. Some displacement but no comminution or impaction
3. Comminution and/or impaction

Primary Stabilization and Management

- Focus on soft tissue rest—Reduce as best as possible with axial traction and splint in short leg bulky cotton splint.
- Elevate
- Ice
- NWB

Admit and Discharge Guidelines

- Request an orthopaedic consult in the emergency department for: All—almost all require surgical treatment. Almost all acute fractures should be watched as inpatients for risk of compartment syndrome. Many acute fractures will be treated with external fixation.
- Refer to an orthopaedic surgeon for outpatient follow-up for: Not applicable, all patients should be admitted.

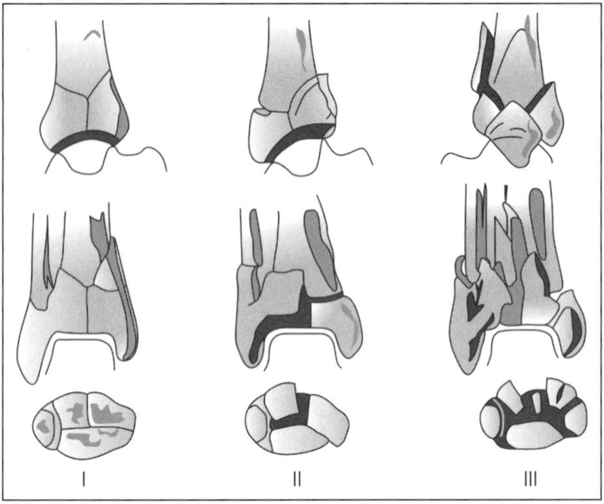

Figure 4.25 Pilon fracture classification.

Definitive Treatment

- True pilon fractures are treated with multiple surgeries. Many will be placed in external fixation to allow for soft tissue healing for up to 2 weeks and then brought back for staged anterior and posterior plating. It is normal for these injuries to take 4 months or more to heal; upward of 10% of people will have an arthrodesis of the ankle joint due to continued pain.

Fibular Shaft Fractures

A true isolated fibular shaft fracture is important to distinguish from a fibular shaft fracture as a manifestation of another injury as the treatment is drastically different. This is usually feasible to do by obtaining a good history and doing a thorough exam. The important injury to rule out is the Maisonneuve fracture (see page 219). A midshaft fibular fracture associated with a tibial shaft fracture can be treated as insignificant.

The fibular shaft above the strong ligaments of the ankle and below the proximal tib-fib joint is a non-weight-bearing bone whose main purpose is to serve as an attachment for muscles and ligaments. Because of this, the fibular shaft is the bone of choice for harvest in vascularized bone grafting since losing a portion of the fibular shaft has low morbidity.

Symptoms and Findings

- Isolated fibular shaft fractures are due to a direct blow or force isolated to the fibula such as a gun shot. This is analogous to the "night stick fracture" in the upper extremity.

- Direct contact is the only mechanism to cause a fracture of the fibula without ankle injury or tibial injury since they are tightly linked. It is analogous to trying to break a hard pretzel: it is possible to do, but it will always break in two places. One needs to definitively rule out the second injury.
- Isolated fibular shaft fractures will have pain and swelling +/− ecchymoses at the fracture site.
- The patient will be tender at the fracture site. Rule out tenderness at the ankle and fibular head.
- The patient may or may not be able to bear weight.
- Proximal fractures may have peroneal nerve involvement, and a thorough neurovascular exam needs to be completed.

Imaging

- AP/lateral tib-fib—Crucial for diagnosis
- AP/lateral knee—Evaluate for proximal tibial injures, fibular head dislocation
- Three-view ankle—Crucial to rule out Maisonneuve fracture

Classification

- Classified descriptively:
  - Transverse, oblique, spiral, greenstick
  - Displaced, nondisplaced
  - Angulation, rotation, shortening
  - Segmental, comminuted
  - Open versus closed
  - Associated injuries

Primary Stabilization and Management

- A true direct-blow fibular shaft fracture with no associated injury does not require reduction or stabilization.
- Make sure to specifically check EHL and eversion function to assess peroneal nerve function as it crosses directly over the proximal fibula.
- These fractures are stable and in a non-weight-bearing bone.
- Treatment includes RICE: Rest, ice, compression (gentle compression with an Ace wrap can help), and elevation.
- Crutches for the immediate postinjury period can help with pain control.
- Cam walker boot (as long as does not directly press on the fracture) can help to rest the ankle and therefore minimize movement of the muscles that attach to the fibula and, therefore, motion of the fragments.

Admit and Discharge Guidelines

- Request an orthopaedic consult in the emergency department for: Open fractures. If Maisonneuve fracture can not be ruled out.

- Refer to an orthopaedic surgeon or nonop orthopaedist for outpatient follow-up for: All others.

Definitive Treatment

- WBAT with symptomatic treatment is definitive treatment. Definitive union of the fracture is not necessary for successful treatment. A painless nonunion is as successful as painless union.

Fibular Head Dislocation

Dislocation of the proximal tibiofibular joint occurs due to direct trauma, extreme twisting in sports such as soccer, or a fracture of the tibial shaft with an intact fibula. It can be missed unless the examiner is specifically looking for it. The joint is inherently stable and dislocation is rare.

Symptoms and Findings

- Pain at or near lateral knee
- May have locking popping
- Pain with palpation of the fibular head
- May have generalized ligamentous laxity
- Pain limits weight bearing
- Ankle motion may cause knee pain
- May have peroneal nerve injury

Imaging

- AP/lateral knee
- AP/lateral tib-fib
- Three-view ankle

Classification

Type	Mechanism	Comment
Subluxation	Any	Treat symptomatically, RICE, cast 2–3 weeks
Anterolateral	Fall, knee flexed, foot inverted, and plantar flexed	Most common dislocation (85%)
Posteromedial	Direct trauma (e.g., car bumper)	Next most common (10%), peroneal nerve injury frequent
Superior	High-energy ankle injury	Evaluate knee stability; often has concomitant injuries

Primary Stabilization and Management

- Dislocations: Proper analgesia, knee flexion, reversal of injury pattern reduces head with palpable pop. Immobilization after reduction.

- Subluxation: Immobilization in long leg splint.
- Irreducible dislocation: Immobilization in long leg splint with open reduction in the OR.

Admit and Discharge Guidelines

- Request an orthopaedic consult in the emergency department for: Irreducible dislocation, open injury.
- Refer to an orthopaedic surgeon for outpatient follow-up for: Reduced injury or subluxation with appropriate immobilization.

Definitive Treatment

- Often closed reduction with immobilization is the definitive treatment. Since the fibular head is the attachment site for many ligaments, it is important to assess knee stability and treat any laxity after the appropriate healing of the tib fib joint has finished.
- Nerve injuries most commonly transient and should be watched for 3 months before studies ordered.
- Open treatment with reduction and fixation is reserved for irreducible dislocations or recurrent dislocations.

Injuries about the Ankle

High Ankle Sprains

The stability of the ankle joint is comprised of both bony and soft tissue stabilizers. Additionally, the tibia and fibula bones need to maintain their relationship to provide ankle stability. This is done by ligaments at the level of the ankle and a large syndesmotic ligament running the length of the bones. Typical ankle sprains do not involve this ligament (see "Ligamentous Injuries of the Ankle"), but "high ankle sprains" are severe enough to tear this ligament and can destabilize the entire ankle joint.

Symptoms and Findings

- Often occur in sporting events
- Commonly a rotational injury
- Pain at ankle and through calf
- Positive "squeeze test" can support diagnosis, but a negative test does not rule it out.
- May have symptoms of ankle "giving out" or instability
- May have swelling and bruising at ankle

Imaging (See Figure 4.26)

- Three-view ankle—Evaluate for any fractures and syndesmotic widening.
- AP/lateral tib-fib—Pay close attention for proximal fibular fractures.

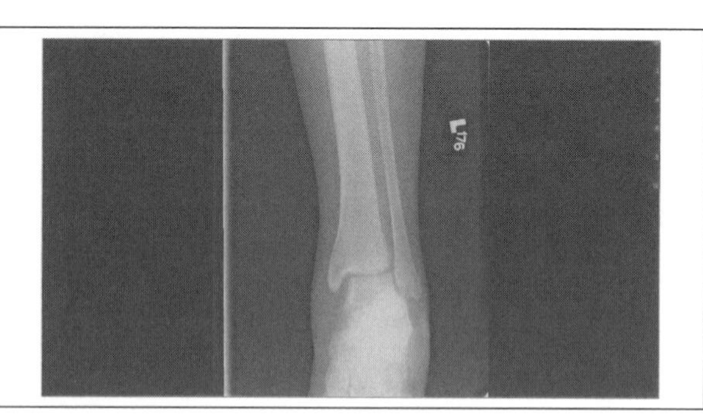

Figure 4.26 High ankle sprain with widened mortise.

- Ankle external rotation stress radiograph: If NO widening is seen on initial mortise and there is suspicion for high ankle sprain, then a mortise view of the ankle with external rotation applied to the foot can be taken. Widening in the nonstress mortise or in this view confirms the diagnosis.
- CT may be required to quantify the amount of the distal fibular displacement in relation to the distal tibia.

Classification

- Stable versus unstable (>5 mm displacement)

Primary Stabilization and Management

- Short leg well-padded splint with the ankle at 90 degrees
- NWB

Admit and Discharge Guidelines

- Request an orthopaedic consult in the emergency department for: Nonreducible or unstable ankle dislocation.
- Refer to an orthopaedic surgeon for outpatient follow-up: Any unstable high ankle sprain cleared for discharge by the orthopaedic consultant.
- Refer to orthopaedic surgeon or nonoperative sports medicine physician for: High ankle sprain that are stable after splinting.

Definitive Treatment

- Stable injuries are treated with cast immobilization for up to 6 weeks. Some may start a progressive PT regimen before the 6 weeks, depending on degree of injury and patient.
- Unstable injuries or high-level athletes may require operative fixation with screws or flexible syndesmotic fixation devices.

Maisonneuve

A Maisonneuve fracture is a high ankle sprain in which the force that causes the injury exits the syndesmosis by traveling through the proximal fibula, leading to a fracture of the fibular shaft. Therefore, the distal segment of the fibula is no longer attached by bony or soft tissue attachments to the tibia, thus causing a destabilized ankle joint.

Symptoms and Findings

- Similar to high ankle sprains, they occur in external rotational injuries. Different fracture patterns based on pronation versus supination
- Pain in fibular shaft, calf, and ankle
- +/− Swelling, ecchymosis at proximal fibular fracture
- Often has positive "squeeze test" (pain with squeezing the fibula and tibia together in a location near the ankle)
- Ankle instability

Imaging (See Figure 4.27)

- Three-view ankle—Evaluate syndesmotic widening, medial mal fx, ankle ligament avulsions.
- AP/lateral tib-fib—Evaluate fibular fracture (proximal third)
- AP/lateral knee—Evaluate fibular head, proximal tibia, knee.

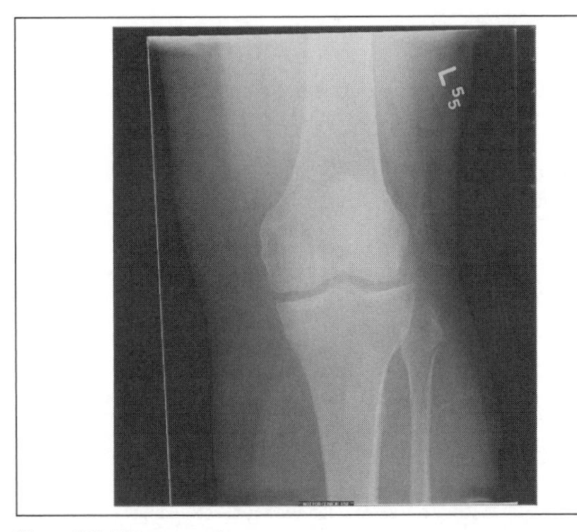

Figure 4.27 Maisonneuve fracture.

Classification

- The Maisonneuve fracture is likely a variant of the typical pattern of ankle fractures. Maisonneuve fractures are a very specific type of fracture, and the medial-sided and concomitant injuries should be described in order to further classify.

Primary Stabilization and Management

- Proper analgesia and reduction/manipulation of fractures
- Short leg well-padded splint
- NWB

Admit and Discharge Guidelines

- Request an orthopaedic consult in the emergency department for: Irreducible dislocation, open fractures, fractures that cannot be stabilized with a splint. The majority of Maisonneuve fractures are unstable and will require operative fixation. The patient should not be referred out of the ED without a good ankle splint as improper splinting may lead to dislocation and significant problems while awaiting outpatient follow-up.
- Refer to an orthopaedic surgeon for outpatient follow-up for: All patients who are not admitted.

Definitive Treatment

Operative fixation is usually the preferred choice for these fractures as they are unstable injuries. This is especially true if the medial and/or posterior malleolus is broken. ORIF of ankle fractures with syndesmotic fixation is preferred. Typically the proximal third fibular fracture is ignored.

Ankle Fractures

Ankle fractures are a grouping of injuries that occur due to rotation of the foot in relation to the leg in such a way that causes the bones to break and ligaments to tear. Although often similar in radiographic appearance and symptoms, these injuries have a wide variety of treatment options, and therefore it is important to understand the differences between them. The classifications used attempt to try and delineate this.

Symptoms and Findings

- Pain in the ankle, medial, lateral, or both
- Swelling, often significant
- Limited ability to bear weight; some patients can with pain, others can not
- Often neurovascular intact unless significant swelling or deformity
- May have dislocation and significant deformity
- Can occur in any age group
- Typically the history involves a fall with a twisting injury, but this is not necessary for diagnosis.

Imaging (See Figure 4.28)

- Three-view foot
- Three-view ankle
- AP/lateral tib-fib
- AP/lateral knee to evaluate proximal fibula and tib-fib joint
- External rotation stress mortise view (when needed)

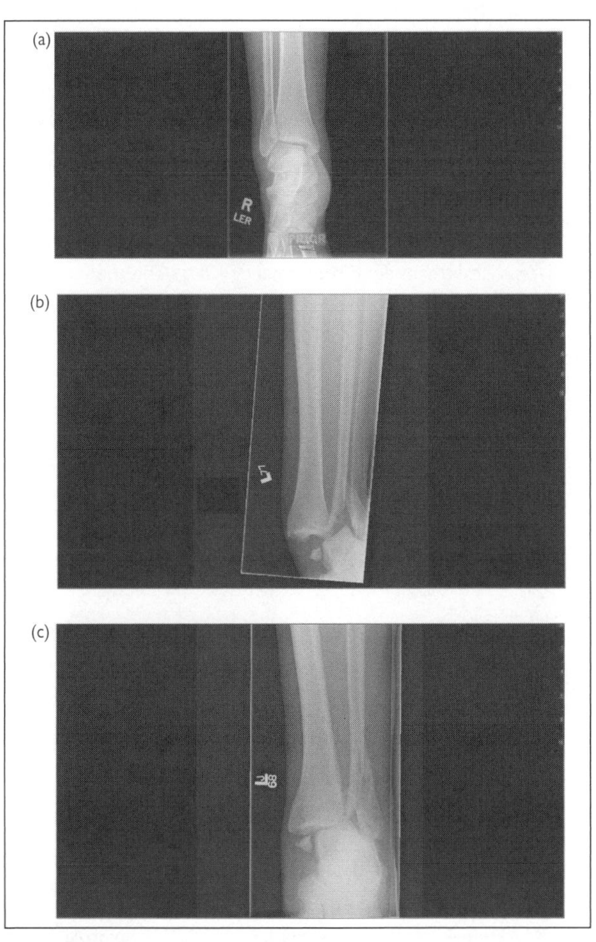

Figure 4.28 Common ankle fractures. a. Lateral malleolus ankle fracture at the joint line on AP X-ray b. Trimalleolar ankle fracture/dislocation seen on AP X-ray. c. Bimalleolar ankle fracture seen on AP X-ray.

Classification (See Figure 4.29)

- Two main classification schemes exist. Both can guide treatment; the Weber system is simpler but less informative, and Lauge-Hansen, more complex and more informative.

Weber—Based on location of fibular fracture with no significance given to other injuries:

- A: Fibular fracture below the ankle joint line
- B: Fibular fracture originating at the joint line
- C: Fibular fracture above the joint line

Weber A fractures are almost always stable, and when isolated, injuries do not require surgical treatment. Weber B fractures can be stable or unstable depending upon other injuries. They are unstable if the medial malleous is also broken. They are unstable if the deltoid ligament is torn. In the event that there is no joint space widening on initial radiographs and the medial mal is intact, an external rotation mortise stress X-ray is taken to determine if the medial soft tissues are intact. If it unstable (i.e., clear space widening medially with stress), surgical treatment is recommended. If it stable, nonoperative treatment may be preferred. Some surgeons prefer to do their own stress views when the patient is seen in the clinic; check with the consultant to see if the specialized imaging is needed in the emergency setting. Weber C fractures represent an unstable pattern with syndesmotic injuries, and surgical treatment is preferred.

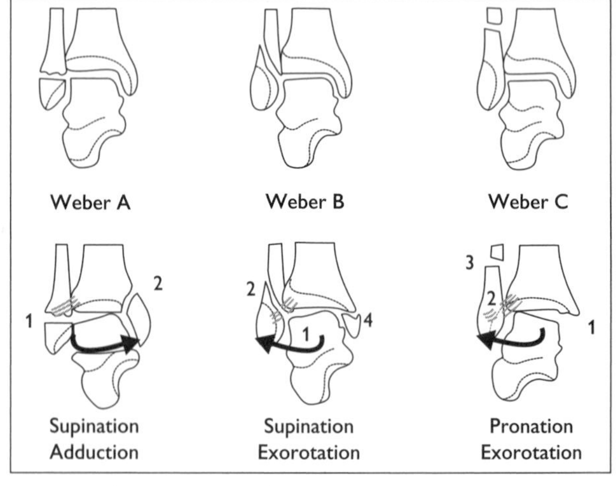

Figure 4.29 Weber ankle fracture classification system and Lauge-Hansen ankle classification systems.

Lauge Hansen—First word describes the position of the foot and the second the direction of the foots movement:

- Supination Adduction
  - Stage 1: Transverse fracture of lateral malleolus, at or below the level of anterior talofibular ligament, or a tear of lateral collateral ligament structures with the anterior talofibular ligament disrupted most often, and frequently the calcaneofibular ligament also being torn. (Nonoperative)
  - Stage 2: Oblique fracture of medial malleolus (surgery recommended)
- Supination-External (Eversion) Rotation: 40%–70% of all ankle fractures
  - Stage 1: Rupture of anterior inferior tibiofibular ligament (nonoperative)
  - Stage 2: Oblique fracture or spiral fracture of the lateral malleolus (nonoperative-stress X-ray negative)
  - Stage 3: Rupture of posttibiofibular ligament or fracture of posterior malleolus of tibia (surgery recommended)
  - Stage 4: Transverse (sometimes oblique) fracture of medial malleolus (surgery recommended)
- Pronation Abduction: Less than 5% of ankle fractures
  - Stage 1: Rupture of the deltoid ligament or transverse fracture of the medial malleolus (surgery recommended for fracture or unstable deltoid injury)
  - Stage 2: Rupture of the anterior and posterior inferior tibiotalofibular ligaments or bony avulsion
  - Stage 3: Oblique fracture of the fibula at the level of the syndesmosis (surgery recommended)
- Pronation Eversion
  - Stage 1: Rupture of the deltoid ligament or transverse fracture of the medial malleolus (surgery recommended for fracture or unstable deltoid injury)
  - Stage 2: Rupture of the anterior inferior tibiotalofibular ligaments or bony avulsion
  - Stage 3: Spiral/oblique fracture of the fibula above the level of the syndesmosis (surgery recommended)
  - Stage 4: Rupture of the posterior inferior tibiofibular ligament or fracture of the posterior malleolus (surgery recommended)
- Pronation Dorsiflexion
  - Stage 1: Fracture of the medial malleolus (surgery recommended for displaced; no operation if reduced)
  - Stage 2: Fracture of the anterior lip of the tibia (surgery recommended for displaced, no operation if reduced)

- Stage 3: Fracture of the supramalleolar aspect of the fibula (surgery recommended)
- Stage 4: Rupture of the posterior inferior tibiofibular ligament or fracture of the posterior malleolus (surgery recommended)

Primary Stabilization and Management

- Despite the large variation in presentations, the initial treatment for ankle fractures is very similar. After initial radiographs, the ankle joint and fracture fragments should be closed reduced and immobilized.
- Adequate analgesia is key; intra-articular block, ankle block, or conscious sedation is commonly used.
- Reduction usually involves a reversal of the injury forces (see above). Commonly this is axial traction, anterior force on the foot, and medial translation of the foot.
- Flexing the knee will reduce the force of the gastroc/soleus and greatly aid in reduction and splinting.
- Children can be casted in a reduced position, short leg cast, ankle at 90 degrees, and bivalved. Families should be educated about the signs and symptoms of compartment syndrome although this is less likely in this injury.
- Adults should be splinted in a well-padded bulky short leg splint, ankle at 90 degrees.
- Fractures that cannot be splinted at 90 degrees because of recurrent dislocation with dorsiflexion should be splinted in plantar flexion and admitted to an orthopaedic service for fixation.
- All splints and especially casts should be checked with the hip flexed and the knee bent to 90 degrees to ensure there is no impingement in the popliteal space as this can cause serious pressure problems.

Admit and Discharge Guidelines

- Request an orthopaedic consult in the emergency department for:
  - Irreducible or unstable fracture. If the reduction of the ankle joint cannot be maintained in the splint, the patient CANNOT be discharged. Be very cautious of any trimalleolus fractures, as these can be very unstable.
- Refer to an orthopaedic surgeon for outpatient follow-up for: Stable, well-reduced fractures. Surgical fractures should be seen within 7–10 days of injury.
- NWB for all ankle fractures

Definitive Treatment

- The classification scheme gives a good perspective on what needs to be fixed surgically versus closed reduced and casted.

- For nonoperative fractures the definitive treatment is usually: Splinting until swelling subsides → Casting or walking boot until radiographic evidence of healing (6–12 weeks) → Progressive rehabilitation
- Surgically treated fractures follow a similar pattern: Initial splinting → Surgical fixation (ORIF, ex-fix) → Immobilization and limited weight bearing until healed (6–12 weeks) → Progressive rehabilitation

Triplane and Tillaux Fractures

Triplane and tillaux (see below) fractures are unique variants of an ankle fracture that occur in children and adolescents due to their unique growth plate characteristics.

Triplane fracture refers to a Salter-Harris fracture through the distal tibial physis that occurs in the frontal, lateral, and transverse planes. There is a vertical fracture through the epiphysis, a horizontal fracture through the physis, and an oblique fracture through the metaphysis (see Figure 4.30). This occurs because the growth plate closes from a medial to lateral direction, leaving the lateral physis but

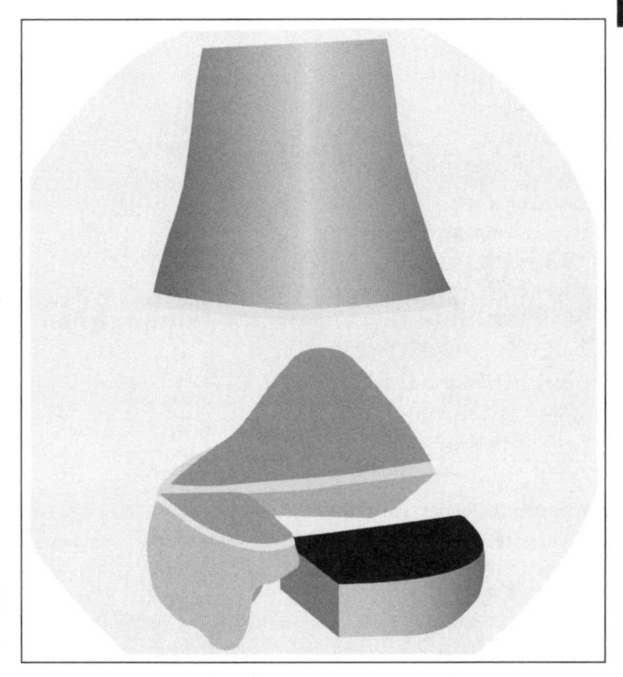

Figure 4.30 Three-part triplane fracture.

not the medial physis prone to injury in an 18-month window often in the early teenage years.

Tillaux fractures are another variant of the Salter-Harris-type ankle fracture. There are only two fracture lines: one vertical through the epiphysis and a horizontal fracture starting at the vertical line and progressing laterally through the physis. The resulting anteriorolateral distal epiphyseal fragment is secured through strong ligaments to the fibula and is often displaced.

Symptoms and Findings

- Pain in the ankle, medial, lateral, or both
- Swelling, often significant
- Limited ability to bear weight; some patients can with pain, others cannot
- Often neurovascular intact unless significant swelling or deformity
- Dislocation and significant deformity less common
- Limited age group
- Typically the history involves a fall with a twisting injury, but this is not necessary for diagnosis.

Imaging (See Figures 4.31 and 4.32)

- Three-view foot
- Three-view ankle
- AP/lateral tibia
- CT ankle often needed to definitively diagnose

Classification

- Based on number of fragments
- Two-part: Fracture through physis stops at vertical fracture line.
- Three-part: Physeal fracture line completely across, giving two fragments, one proximal, one distal

Primary Stabilization and Management

- Conscious sedation with reduction
- Long leg bivalved cast
- NWB

Admit and Discharge Guidelines

- Request an orthopaedic consult in the emergency department for: Reduction beyond scope of provider.
- Refer to an orthopaedic surgeon for outpatient follow-up: Patients should be followed for continued evaluation and healing as outpatients. Residual joint displacement of more than 2 mm will require ORIF.

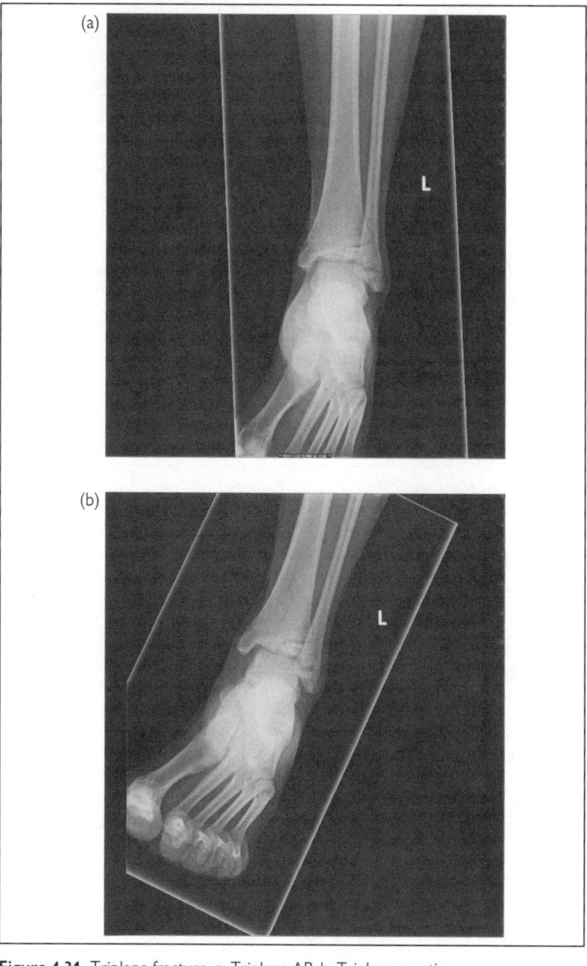

Figure 4.31 Triplane fracture. a. Triplane AP. b. Triplane-mortise.

Ankle Dislocations

The ankle joint is a stable joint with very strong ligamentous support. Therefore, true dislocation without fracture is rare and usually involves high-energy forces.

Symptoms and Findings

- Pain
- Swelling with possible fracture blisters

Figure 4.31 (Continued) c. Triplane-lateral.

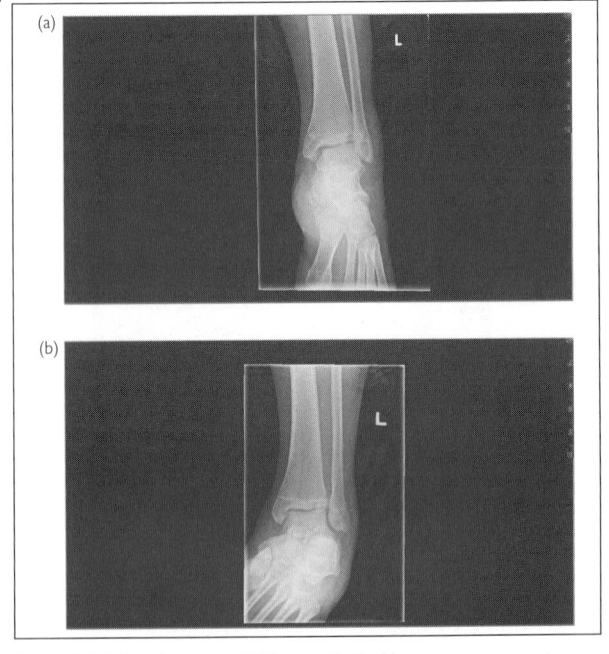

Figure 4.32 Tillaux fracture. a. AP X-ray. b. Mortise X-ray

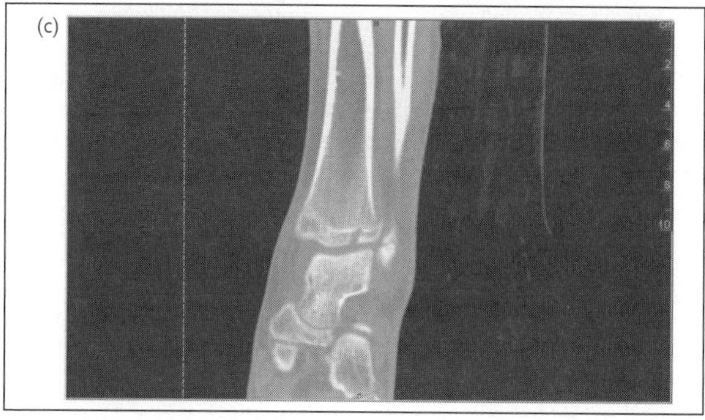

Figure 4.32 (Continued) c. Coronal CT.

- Inability to bear weight
- Significant deformity in unreduced dislocations
- Tenderness medially and laterally over the ligamentous structures of the ankle
- May have a history of prior ankle sprains (incompetent ligaments) or ligamentous laxity

Imaging

- Three-view ankle
- Three-view foot
- AP/lateral knee

Classification

- Classified in the direction of the dislocation:
- Anterior—Foot fixed, tibia driven backwards
- Posterior—Most common, usually associated with syndesmotic disruption
- Superior—Pilon-type force drives talus into syndesmosis
- Lateral—Only occurs with fracture

Primary Stabilization and Management

- Adequate analgesia usually requires a block (intra-articular or ankle block) and/or conscious sedation.
- Reduction of dislocation with axial traction and plantar flexion
- Splint with short leg bulky cotton splint with the ankle at 90 degrees

Admit and Discharge Guidelines

- Request an orthopaedic consult in the emergency department for:
  - Most dislocations, but especially irreducible joint
  - Open injury

- Any dislocation in which the reduction cannot be well maintained with a splint.
- Refer to an orthopaedic surgeon for outpatient follow-up: All ankle dislocations should all be evaluated by an orthopaedist to assess stability. One expects ligamentous healing at 6 weeks after injury given adequate immobilization.

Definitive Treatment

These injuries are rare, and an algorithm for definitive treatment is not agreed upon. Most will get a period of nonoperative management to allow for ligamentous healing. Residual laxity, instability, or repeat dislocation may be indications for operative repair or reconstruction of the torn ligaments. An MRI may be ordered in the postacute phase if surgery is being considered, to fully assess the ligaments involved.

Ligamentous Injuries of the Ankle

Strains and sprains of the ankle ligaments are extremely common. These injuries are a variable constellation of injuries ranging from minor inconveniences to significant tears requiring reconstructive procedures.

The ankle is stabilized by numerous ligaments, including:

- Syndesmosis: Large strong connection between the fibula and tibia
- Anterior inferior tibialfibular ligament (AITFL): Strong connection between tibia and fibula at the level of the mortise
- Posterior inferior tibialfibular ligament (PITFL): Strong posterior connection between the tibia and fibula. This is the ligament that causes the posterior malleolus to be avulsed in trimal fractures.
- Anterior talofibular ligament (ATFL): Lateral stabilizer of ankle
- Posterior talofibular ligament (PTFL): Lateral stabilizer of ankle
- Calcaneofibular ligament (CFL): Lateral stabilizer of the ankle
- Deltoid ligament: Large fan-shaped medial stabilizer of the ankle

Symptoms and Findings

- The symptoms vary based on the degree of injury and ligaments injured.
- Pain and swelling over the injured ligament
- Possible ecchymosis
- Possible feeling of "giving way" or instability
- +/− Able to bear weight
- Position of the foot at the time of injury and mechanism of injury are great guides to which ligament(s) are injured.

Test	Result	Significance
Squeeze test	Usually negative	Positive indicates syndesmosis injury.
Talar tilt	+ Talar tilt in 20 degrees plantar flexion	CFL and ATFL injured
Anterior drawer	Anterior subluxation and rotation around deltoid with anterior applied force with foot in 10–15 degrees plantar flexion	ATFL injured

Imaging

- Three-view ankle to rule out fractures
- Three-view foot to rule out fractures, avulsions
- AP/lateral tib-fib if any concern for high ankle sprain (see above)

Classification

Ligament injuries are generically graded by the following scale:

- Grade I: Tearing of some of the fibers of the ligament. Pain, mild swelling. No joint instability, minimal to no ligament laxity.
- Grade II: Tearing of a portion of the ligament. Significant pain and swelling. Increased laxity with a firm end point.
- Grade III: Complete tearing of the ligament. Significant pain and swelling (although there are reports that suggest a Grade II injury often hurts more than a Grade III). Increased laxity with a soft or no end point.

Primary Stabilization and Management

- Grade I injury: Aircast or gentle compressive wrap, crutches, and partial weight bearing for 3–10 days, elevation, ice, anti-inflammatory medications. Gradual return to sport after 10 days to 2 weeks.
- Grade II: Short leg splint, crutches, and partial weight bearing for up to 2 weeks, elevation, ice, anti-inflammatory medications. Progressive rehab and return to sport after 10 days to 2 weeks.
- Grade III: Short leg splint, crutches, non-weight-bearing, ice, elevation. May need 2–6 weeks to heal ligaments. May require reconstruction if injury leads to recurrent instability or sprains.

Admit and Discharge Guidelines

- Request an orthopaedic consult in the emergency department for: Open injuries, dislocations.
- Refer to an orthopaedic surgeon for outpatient follow-up: Severe Grade II, Grade III, and any unstable ankle should be referred to a sports medicine physician or orthopaedic surgeon for continued follow-up and treatment. Grade I and many Grade II usually do not require close follow-up unless diabetic, then all need close follow-up.

Definitive Treatment

- The ligaments of the ankle usually heal well with proper immobilization and rest. However, recurrent instability or sprains are indications for operative fixation. Ligaments that are stretched but not torn can sometimes be imbricated onto themselves and tightened, while other injuries require using graft material to reconstruct the normal anatomy. Physical therapy referral with range of motion, strengthening, and proprioceptive training is crucial for recovery.

Achilles Tendon Rupture

This is a very common injury. The tendon is predisposed to injury because of its blood supply. The blood supply originates both distally and proximally and is directed toward the center of the tendon. However, the central portion does not have a dedicated blood supply and therefore is a watershed area.

Symptoms and Findings

- Patients feel an acute pop with pain.
- Many will describe it as feeling like someone hit them with a bat in the back of their ankle or fell on them.
- Most common in "weekend warriors" who play sports or exercise only occasionally
- Often unable to walk
- Significant swelling and ecchymosis
- Tender to palpation over the tendon
- Often will have a palpable defect
- Positive Thompson test (squeeze the calf muscles, and the foot does not plantar flex against gravity). Plantaris muscle if intact can give a false negative result.

Imaging

- Three-view ankle to rule out fractures
- Diagnosis made clinically but if questionable, outpatient MRI can aid in the diagnosis.
- MSK ultrasound can be extremely helpful in the diagnosis and is inexpensive, quick, and painless for the patient. Ultrasound is highly dependant on the experience and skill of the technician performing the exam.

Classification

- Location of the tear is described anatomically.
- Complete versus incomplete tear
- Retracted or intact tendon

Primary Stabilization and Management

- Short leg splint in plantar flexion

- Cam walker boot in plantar flexion
- Non-weight-bearing

Admit and Discharge Guidelines

- Request an orthopaedic consult in the emergency department for: Open injuries.
- Refer to an orthopaedic surgeon for outpatient follow-up: All others. Follow-up needs to be arranged quickly as surgical treatment becomes more difficult with time.

Definitive Treatment

- The treatment of Achilles ruptures is either with surgical repair and splinting or splinting alone. Many factors can guide the treatment including patient's baseline function, medical comorbidities, diabetes, smoking status, and other factors influencing wound healing. Surgical repair is best done within 2 weeks of injury.

Injuries about the Foot

Talus Fractures

The talus is a complex bone with many articular interactions, namely with the tibia, fibula, calcaneus, and navicular. Because the surface area of the talus is largely articular, perforating blood vessels are limited to a relatively small area in the neck, medial surface, and posterior process, and, therefore, fractures of the talus may jeopardize the blood flow and cause AVN. Even small malalignment of the talus can alter joint motion significantly and cause a poor functional outcome.

Symptoms and Findings

- Hind foot or ankle pain
- Swelling. This can be significant in talar fracture dislocations and often has tenting or compromise of the skin.
- Pain with weight bearing or unable to bear weight
- Usually involves significant trauma, fall from height, or MVC.
- 50%–80% incidence of remote fracture. Pay special attention to spine, contralateral extremity, and pelvis.
- Displaced fractures puts neurovascular (NV) bundle at risk and complete NV exam needs to be performed.
- Posterior displaced fractures can put the skin at risk.

Imaging (See Figure 4.33)

- Three-view foot
- Three-view ankle
- CT usually required
- Canale view (plantar flexed, pronated, beam 75 degrees from horizontal) shows the neck of the talus.

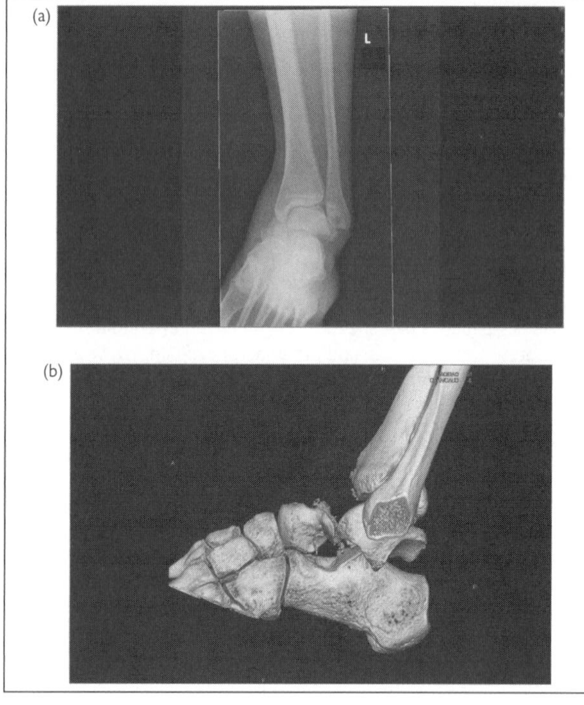

Figure 4.33 Talus fracture. a. AP X-ray b. 3D CT reconstruction.

- Broden: Internal rotation, sequential images from 10–40 degrees to demonstrate the facets

Classification

- Talar neck fractures (50% of talus fractures) use the Hawkins classification:
  - Hawkins 1: Nondisplaced, congruent joints (easily missed; risk of AVN <14%)
  - Hawkins 2: Minimal to complete displacement with resulting subtalar subluxation or dislocation (Risk of AVN 20%–50%)
  - Hawkins 3: Displaced with subtalar and ankle mortise subluxations or dislocations. High rate of open fracture (risk of AVN 70%–100%)
  - Hawkins 4: Displaced with subtalar, ankle mortise, and talonavicualr subluxations or dislocations (risk of AVN approaches 100%)

- Talar body fractures (~30% of talus fractures) are intra-articular and sometimes occur with neck fractures.
  - Described anatomically based on CT
- Talar head fractures (<10% of talus fractures) are rarely isolated. These are always intra-articular. They are either sheer or compressive in nature. They are more common in elderly osteoporotic bone.
  - Described anatomically based on CT
- Posterior process of the talus—Usually avulsion fractures from ligamentous process and should be treated like a ligament injury.
- Lateral process of the talus: Incidence on the rise with the gain in popularity of snowboarding
  - 1: Large fragment, single fracture line
  - 2: Large fragment comminuted
  - 3: Small avulsion-type fragment
- Osteochondral dessicans (OCD) lesions of the talar dome: Described anatomically based on location and size

Primary Stabilization and Management

- Talar neck:
  - Hawkins 1: Short leg bulky cotton splint with the ankle at 90 degrees, NWB
  - Hawkins 2–4: Closed reduction of dislocations with splinting in bulky cotton splint, NWB. If closed reduction of dislocations cannot be completed in the ED, OR closed versus open reduction should be done urgently.
- Talar body, head, lateral process: Closed reduction with splinting in bulky cotton splint, NWB
- Posterior process avulsion: Treat like ligamentous injury of ankle

Admit and Discharge Guidelines

- Request an orthopaedic consult in the emergency department for: Talar neck, talar body, talar head, and lateral process fractures. Nondisplaced Hawkins 1 can be immobilized and discharged whereas almost all others should be admitted for elevation and compartment checks.
- Refer to an orthopaedic surgeon for outpatient follow-up: Posterior process avulsion fractures, talar dome OCD

Definitive Treatment

- The treatment of talus fractures is difficult and often requires operative fixation.
- The talus is extremely vulnerable to AVN, infection, and posttraumatic arthritis. Expert care and reduction of dislocations can help to minimize these complications.
- Talar neck:

- Hawkins 1 with <2 mm displacement may do well with either closed treatment in a cast or percutaneous fixation
- Hawkins 2–4: ORIF
- Talar body: ORIF
- Talar head: ORIF
- Lateral process: Fragment <2 mm: closed treatment, >2 mm ORIF
- OCD: May be asymptomatic after initial painful period and not require treatment. Larger OCDs can be reduced and fixed open versus arthroscopically. OCDs without a salvageable fragment may require osteochondral autograft transfer system (OATS) versus microfracture versus autologous chondrocyte implantation (ACI).

Talus/Subtalus Dislocations

Symptoms and Findings

- Usually the result of high-energy injuries
- Significant deformity present
- Pain
- Swelling
- Inability to bear weight
- May have NV changes; may resolve with reduction

Imaging

- Three-view foot
- Three-view ankle
- Postreduction CT to evaluate for fractures and loose bodies

Classification (See Figure 4.34)

- Describe the direction of the foot relative to the leg: Anterior, posterior, medial, lateral.
- Medial most common
- Lateral most likely to be open
- Anterior/posterior extremely rare

Primary Stabilization and Management

- Adequate analgesia through regional block or conscious sedation
- Flex the hip and knee.
- Axial traction
- Exaggerate deformity before reducing; palpable clunk signifies reduction and should be confirmed with radiographs.
- Splint in short leg bulky splint with the ankle at 90 degrees

Admit and Discharge Guidelines

- Request an orthopaedic consult in the emergency department: Irreducible joint or dislocation emergency

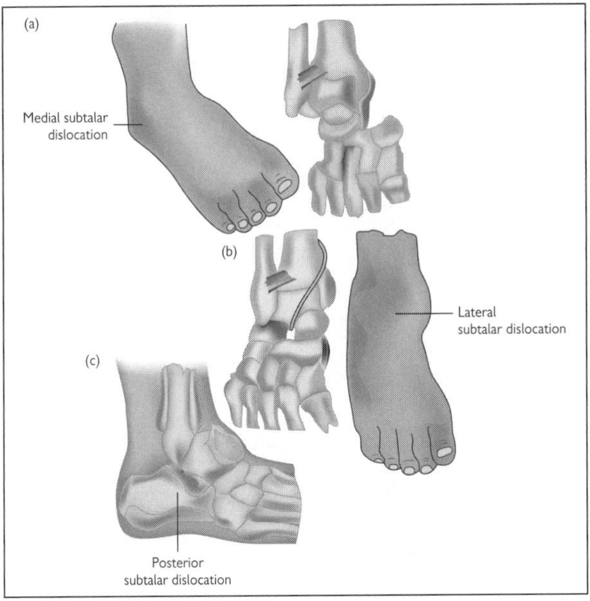

(a) Medial subtalar dislocation
(b) Lateral subtalar dislocation
(c) Posterior subtalar dislocation

Figure 4.34 Subtalar dislocation. a. Medial subtalar dislocation. b. Lateral subtalar dislocation. c. Posterior subtalar dislocation.

providers do not feel comfortable attempting reduction.
Any open injury.

- Refer to an orthopaedic surgeon for outpatient follow-up:
  Reduced stable splinted injury.

Definitive Treatment

Reduced dislocations can be treated with a splint transitioned to cast with initial non-weight-bearing and progressive increase in weight bearing at 2–4 weeks.

Irreducible dislocations or open injuries will often require open reduction and temporary fixation. They will be non-weight-bearing for upward of 6 weeks.

Calcaneus Fractures

Calcaneus fractures are often the result of high-energy injuries. These often a result of falls from heights, MVCs, or a crush or blow out of the calcaneus. There is a high prevalence of concomitant injury of the pelvis, spine, contralateral calcaneus, and ipsilateral or contralateral tibial plateau. Specific attention should be given to these areas. Calcaneus fractures may lead to compartment syndrome of the foot.

Calcaneus fractures often lead to severe disability and significant loss of time from work even if properly treated.

Symptoms and Findings

- High-energy mechanism common, falls or jumping from height a common cause
- Pain
- Significant swelling of the hind/midfoot common
- Can lead to compartment syndrome of the foot
- Unable to bear weight
- Pain with palpation

Imaging (See Figure 4.35)

- Three-view ankle
- Three-view foot—Bohler's angle and angle of Gissane

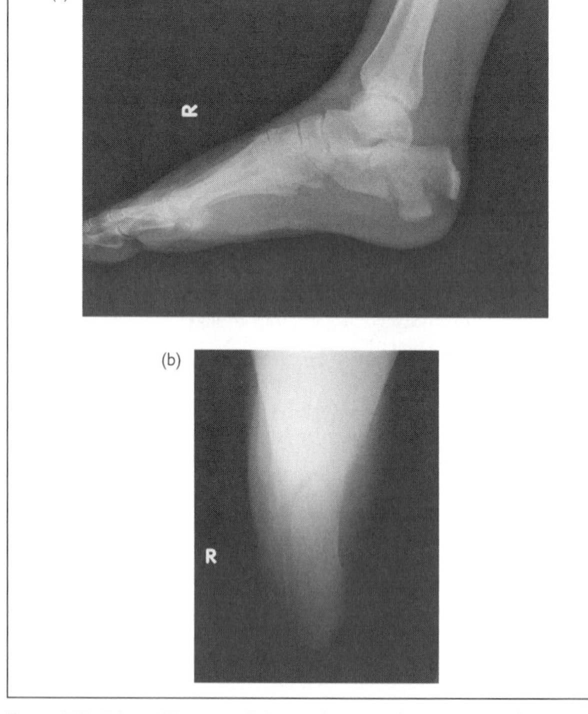

Figure 4.35 Calcaneal fracture a. Calcaneus fracture on lateral X-ray. b. Calcaneus fracture on Harris heel view.

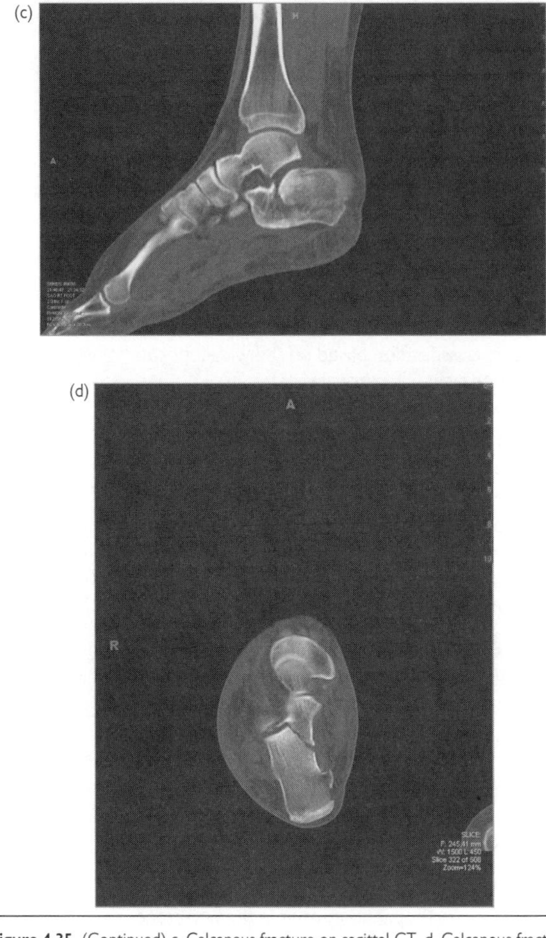

Figure 4.35 (Continued) c. Calcaneus fracture on sagittal CT. d. Calcaneus fracture on axial CT.

- Harris heel view
- CT foot

Classification (See Figure 4.36)

- Intra-articular (70%) versus extra-articular (30%)
- Essex-Lopresti subclassified intra-articular to:
  - Tongue type (50%)

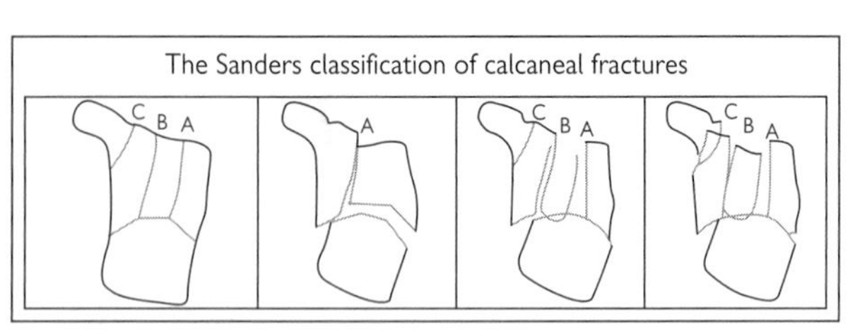

Figure 4.36 Sanders classification.

  - Joint depression (35%)
  - Unclassified (15%)
- Sanders classification based on the widest portion of the posterior facet CT coronal view:
  - Type I: Nondisplaced
  - Type II: Two-part
  - Type III: Three-part
  - Type IV: Comminuted
  - Types II and III further subdivided by adding A, B, or C to describe where the fracture lines are. Area A is the most lateral third, B is the middle third, and C the most medial third.
  - Prognosis worsens as degree increases.

Primary Stabilization and Management

- No true reduction maneuver is generally possible.
- Short leg bulky cotton splint
- NWB
- Ice
- Elevation
- Admit for compartment checks for 24–48 hours
- Pain control

Admit and Discharge Guidelines

- Request an orthopaedic consult in the emergency department for: All acute fractures.
- Refer to an orthopaedic surgeon for outpatient follow-up: Chronic fractures.

Definitive Treatment

The decision to operate or treat nonoperatively is difficult in calcaneal fractures. All types have operative treatment possibilities ranging from percutaneous pinning to large open reductions with internal fixation. The results of treating calcaneal fractures both operatively and non-operatively are fraught with complications including: chronic pain, difficulty with ambulation, inability to wear stock footwear, malunion, nonunion. For operative fixation the complications are the same but infection and wound healing problems are added on and can be significant especially in the elderly, smokers, or diabetics.

Tarsal Bone Fractures

Isolated cuboid or cuneiform fractures are rare. If a cuboid or cuneiform fracture is found, care should be taken to rule out associated injuries such as Chopart (dislocation of the talonavicular and calcaneocuboid) fracture dislocation, navicular fracture, or tarso-metatarsal (TMT) (Lisfranc) injuries.

Symptoms and Findings

- Midfoot pain
- Swelling
- Ecchymosis
- Pain with weight bearing
- Tenderness at the fracture site

Imaging

- Three-view foot
- CT foot to rule out other injuries

Classification

- Described anatomically

Primary Stabilization and Management

- Short leg splint
- NWB

Admit and Discharge Guidelines

- Request an orthopaedic consult in the emergency department for: Multiple fractures or dislocations, skin compromise, open injuries.
- Refer to an orthopaedic surgeon for outpatient follow-up: True isolated cuboid or cuneiform fractures are usually non- or minimally displaced and can be splinted and referred to orthopaedic surgery follow-up.

Definitive Treatment

- The definitive treatment is not 100% agreed upon since these are rare injuries. Patients often do well with splint l cast, NWB treatment. Gross displacement or a lack of healing may prompt ORIF.

Navicular Fractures

The navicular is considered the keystone to the medial arch of the foot. Injury to the navicular is rare, but when present it can lead to significant midfoot and hindfoot dysfunction. The blood supply to the navicular makes it prone to AVN and stress fracture.

Symptoms and Findings

- Fracture from high-energy axial loading (falls, MVC, crush injuries)
- Stress fracture typically occurs in runners, jumpers, hurdlers, or sprinters
- Pain in medial midfoot below ankle
- Pain with weight bearing and push-off
- Tenderness localized over navicular
- Swelling typically medial midfoot

Imaging (See Figure 4.37)

- Three-view foot
- External oblique foot
- Thin-cut CT if question about fracture pattern
- Bone scan or MRI if concern for occult stress fracture

Classification

- Avulsion versus Body fractures
- Avulsion (50%)
  - Dorsal: Deltoid ligament
  - Posterior/medial: Posterior tibial tendon or spring ligament avulsions
- Body (50%)
  - Type I: Coronal fracture line, usually minimal comminution
  - Type II: Dorsolateral-to-plantarmedial fracture line, often has talo-navicular (TN) subluxation or dislocation, + forefoot malalignment (adduction/medial translation)

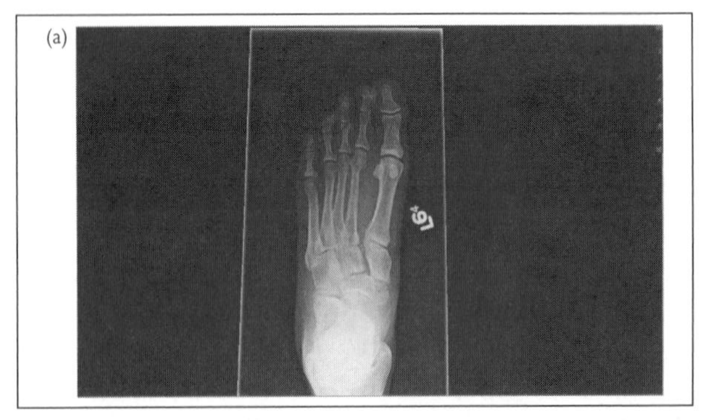

Figure 4.37 Navicular fracture a. Navicular fracture X-ray with Lisfranc injury.

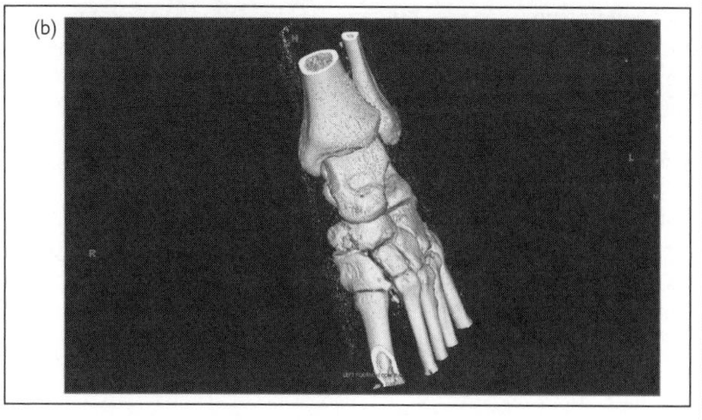

Figure 4.37 (Continued) b. Navicular fracture 3D CT scan.

- Type III: navicular-cuboid (NC) joint subluxation, middle or lateral navicular comminution, high rate of associated fractures

Primary Stabilization and Management

- Adequate analgesia through ankle block, systemic pain meds
- Reduction of deformity
- Splint in short leg, well-padded splint
- NWB

Admit and Discharge Guidelines

- Request an orthopaedic consult in the emergency department for fractures with one of the following surgical indications:
  - Open
  - Impending compartment syndrome
  - Irreducible
  - Displacement
  - Joint incongruity >1 mm
  - Lateral column involvement
  - Skin at risk
- Refer to an orthopaedic surgeon for outpatient follow-up for: Reduced fractures, small avulsion fractures

Definitive Treatment

The majority of the navicular body fractures are treated with ORIF since minor variations in the bone are not well tolerated. The contraindications to surgical treatment of an otherwise indicated fracture are the typical foot contraindications, including vasculopathy, diabetic, heavy smokers, or grossly noncompliant. In these situations, and in many avulsion fractures, the best outcome will be with nonsurgical management initially in a splint and transitioning to a cast. Fractures with severe comminution are difficult problems and often lead to treatment with medial column fusions, external fixation with delayed reconstruction, or other radical treatments. A majority of patients can expect to have long-term stiffness and some degree of midfoot discomfort or pain.

Lisfranc Fractures

The Lisfranc joint is the key to midfoot–forefoot stability. Injuries to this joint or the supporting structures (Lisfranc ligament-lateral surface of the medial cuneiform to the plantar surface of the base of the second metatarsal) can have a significant impact on foot stability and function.

Symptoms and Findings

- Pain and tenderness midfoot
- Swelling midfoot
- +/– Ecchymosis
- Plantar ecchymosis common in ligamentous-only, lower-energy injury
- Chandelier or apprehension signs (stress examination) can be used when diagnosis in question
- High prevalence open injuries
- Malaligned, widened, rotated, abducted forefoot
- Risk of compartment syndrome, skin compromise, intermetatarsal artery injury
- Most are high-energy mechanism: Motor vehicle collision (MVC), motorcycle collision (MCC), fall, crush
- Low-energy occurs during sports (commonly) when axial load applied to plantar flexed, rotated foot. Historically happens when rider falls from horse, and the foot gets stuck in the stirrup. Now more common in windsurfers and occasionally football players.
- Also known as "bunk bed" injury since it can occur when jumping off the top bunk onto a plantar flexed foot.

Imaging (See Figure 4.38)

- Three-view foot
- Bilateral (for comparison), weight-bearing AP with beam tangential to the TMT joints if question of diagnosis
- Stress radiographs can be performed under regional, local, or general sedation if questions remain and the patient unable to bear weight.
- Plain radiography can diagnosis the majority of these injuries.
- CT needed to plan surgery or for diagnosis of avulsions
- MRI can be useful to diagnose ligamentous injury but should not be first-line imaging study.

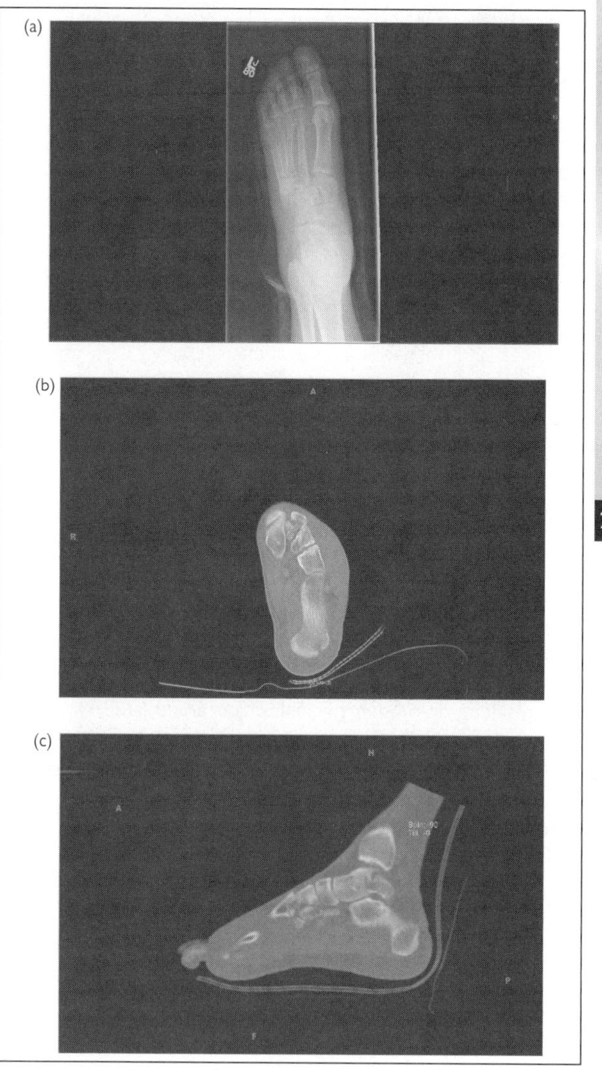

Figure 4.38 Lisfranc injury. a. AP X-ray.b. Axial CT. c. Sagittal CT.

Classification

- Descriptive, anatomic.
- Describe fracture lines.
- Describe avulsions based on anatomic location.
- Describe displacement, subluxation.

Primary Stabilization and Management

- Reduction of any dislocations/subluxations (exaggerate the deformity then return foot to neutral position).
- Short leg bulky cotton splint with ankle at 90 degrees
- NWB

Admit and Discharge Guidelines

- Request an orthopaedic consult in the emergency department for: High-energy injuries, open injuries, irreducible injuries.
- Because of the risk of compartment syndrome, all high-energy injuries should be admitted, elevated, and have serial examinations after closed reduction and splinting.
- Refer to an orthopaedic surgeon for outpatient follow-up: Low-energy injuries that are reduced and splinted well. Follow-up should be within 10 days from injury in case operative planning needs to be done. Low-energy injuries that cannot be confirmed using plain radiography or CT can get an outpatient MRI in a splint while awaiting orthopaedic referral.

Definitive Treatment

Fractures in this location are treated with ORIF. Ligament tears with resulting dislocation or subluxation are treated with ORIF. Well-aligned joints with ligament sprains/strains are treated with 6 weeks of short leg NWB casting and reassessment. If still painful in the midfoot at that time, ORIF can be considered.

Metatarsal Fractures (1–4) (Excluding Lisfranc Injuries)

Metatarsal fractures are relatively common injuries but are well tolerated and generally do not require perfect alignment or reduction for successful outcomes, with a few notable exceptions. The fifth metatarsal is a special circumstance and will be discussed elsewhere. There is a high incidence of concomitant foot injuries, and the whole foot needs to be examined when metatarsal fractures are found.

Symptoms and Findings

- Pain at the fracture site
- Swelling
- Ecchymosis often present in the forefoot
- Occur due to high-energy injury (MVC, crush, fall)

- Can less often occur in lower-energy twisting injuries
- Pain with weight bearing

Imaging (See Figure 4.39)

- Three-view foot
- Weight-bearing radiographs with comparison films of the contralateral side if the diagnosis is in question

Classification

- Classified anatomically
- Location of fracture: First versus second versus third, and so on. Head, neck, shaft, base
- Description of fracture: Transverse, spiral, oblique, comminuted, displaced, shortened, and so forth
- Intra-articular versus extra-articular

Primary Stabilization and Management

- Displaced: Local or regional analgesia and reduction
- Short leg, well-padded splint with the ankle at 90 degrees
- NWB

Admit and Discharge Guidelines

- Request an orthopaedic consult in the emergency department for: Open injuries, question of compartment syndrome, and any first or second metatarsal base fractures or dislocations (Lisfranc injuries).
- Refer to an orthopaedic surgeon for outpatient follow-up: All others. Referrals should be in the 7–10 day range in case operative reduction is indicated.

Definitive Treatment

- Metatarsal fractures in the 1–4 metatarsals are usually stable fractures with minimal displacement. The typically heal well with

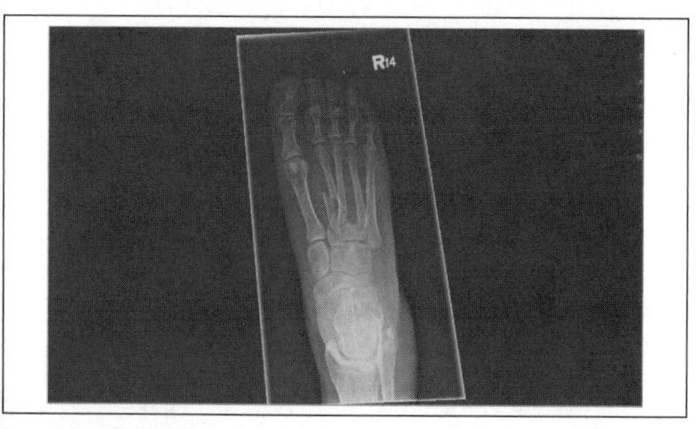

Figure 4.39 Multiple metatarsal fractures: AP.

nonoperative treatment. Nonoperative treatment usually is 1–2 weeks immobilization and NWB. Immobilization continues, and WB can be progressed as tolerated until 4–6 weeks, where immobilization ends.
- Surgical indications are not clearly elucidated. The most important relationship to correct is the position of the metatarsal heads. Angulation of more than 10 degrees or shortening of 3 mm may affect weight bearing across the heads and therefore may be an indication for operative reduction and fixation.

Fifth Metatarsal Fractures

Jones, Pseudo Jones, and diaphyseal fractures are unique fractures of the base of the fifth metatarsal. They differ slightly in the location of the fracture but significantly in the treatment of the fractures, and therefore it is important to be able to distinguish between them. The fifth metatarsal is unique due to its above average range of motion at the TMT joint and the fact that it has the insertions of both the peroneal brevis tendon and the plantar aponeurosis at the base of the fifth. There is a watershed area of blood supply at the base of the fifth metadiaphyseal region that is prone to nonunion. The fifth metatarsal diaphyseal fracture usually represents a stress fracture.

Symptoms and Findings

- Pain, usually lateral midfoot
- Swelling
- Occasional ecchymosis
- Tenderness at the fracture site
- Usually able to bear weight, but with pain laterally
- Commonly fractured through indirect trauma

Imaging (See Figure 4.40)

- Three-view foot

Classification (See Figure 4.41)

- Zone I: Any fracture proximal to the level of the fourth/fifth metatarsal joint (avulsion fracture or Pseudo-Jones)
- Zone II: The metadiaphyseal region starting at the level of the fourth/fifth metatarsal joint (Jones fracture)
- Zone III: Diaphyseal region (stress fractures)
- Distal shaft fractures (dancer's fracture): Analogous to a boxer's fracture in the hand

Primary Stabilization and Management

- Zone I: Can be treated in a soft wrap, walking boot, or temporary splint. Limited weight bearing, ice, elevation also helps. Patients usually tolerate a progressive return to WBAT quickly, and no further treatment is needed afterward.

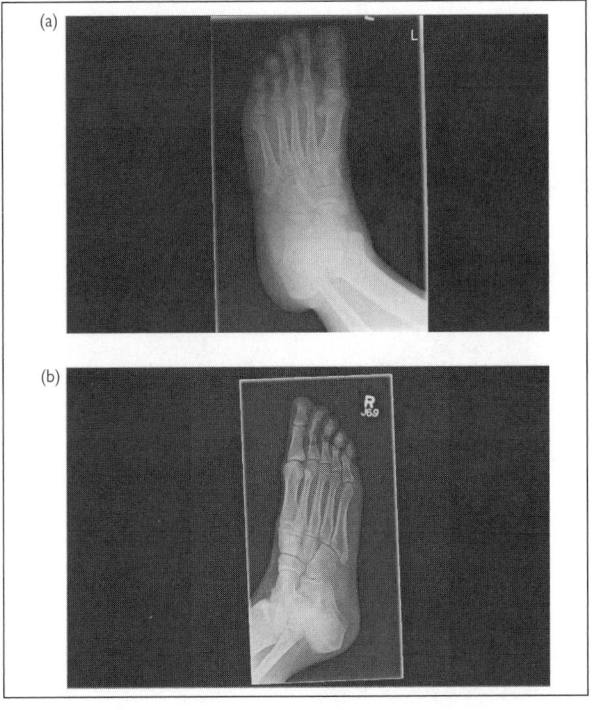

Figure 4.40 Fifth metatarsal fractures. a. 5th Metatarsal base fracture on AP X-ray. b. 5th Metatarsal avulsion fracture on AP X-ray.

- Zone II: Short leg splint, NWB
- Zone III: Short leg splint, NWB
- Distal fifth metatarsal shaft: Short leg splint NWB
- These fractures usually do not require reduction maneuvers, Zone III being the occasional exception. Local analgesia versus ankle block will help accomplish reduction.

Admit and Discharge Guidelines

- Request an orthopaedic consult in the emergency department for: Open injuries.
- Refer to an orthopaedic surgeon for outpatient follow-up: All others, especially Zone II or III fractures, distal shaft fractures.

Definitive Treatment

- Zone I: Progressive return to WBAT. Large displaced avulsions (>1 cm) can sometimes be amenable to ORIF.

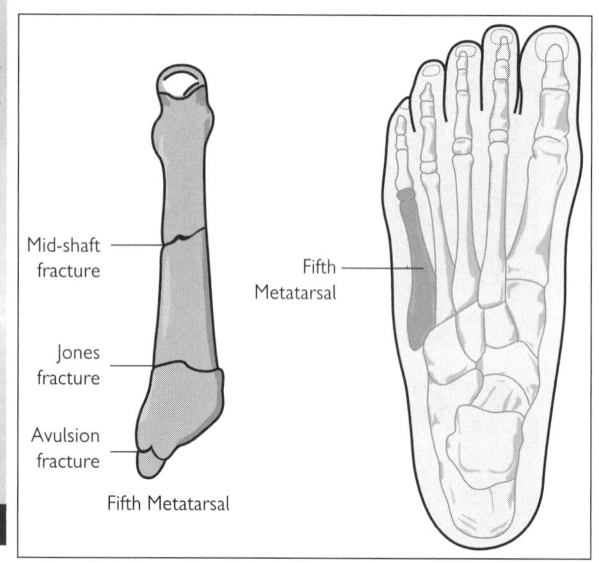

Figure 4.41 Fifth metatarsal fracture classification.

- Zone II: Six weeks of NWB in short leg cast and then progressive rehab. Surgical fixation for nonunions, grossly displaced, or high-level athletes
- Zone III: Six weeks of NWB in short leg cast, operative fixation for evidence of nonunion or grossly displaced
- Distal shaft fractures: NWB in short leg cast. Nonunion and malunion are well tolerated in this region and operative fixation is rarely indicated.

Great Toe Injuries

Because the great toe is physiologically more important and has less stabilizing structures, it is prone to injury. Great toe injuries can range from sprains to fractures and treatment varies widely.

Symptoms and Findings

- Localized pain and swelling
- May have deformity (significant deformity is common in dislocations)
- May have instability (+ anterior drawer is common in plantar plate injuries)
- Pain with weight bearing

Imaging (See Figure 4.42)

- Dedicated AP/oblique/lateral of toe
- Three-view foot

Classification

- Described anatomically: Location, type of fracture, displacement, alignment, open or closed, subluxation or dislocations, intra versus extra-articular.
- Turf toe is an injury to the plantar plate ligament system with axial loading and dorsiflexion of the toe
- Examine all dorsiflexion injuries for plantar plate injuries:
  - Imaging may show retraction or lateral subluxation of the sesamoids.
  - Disruption of the plantar plate or sesamoids disrupts the windlass mechanism required for arch support.

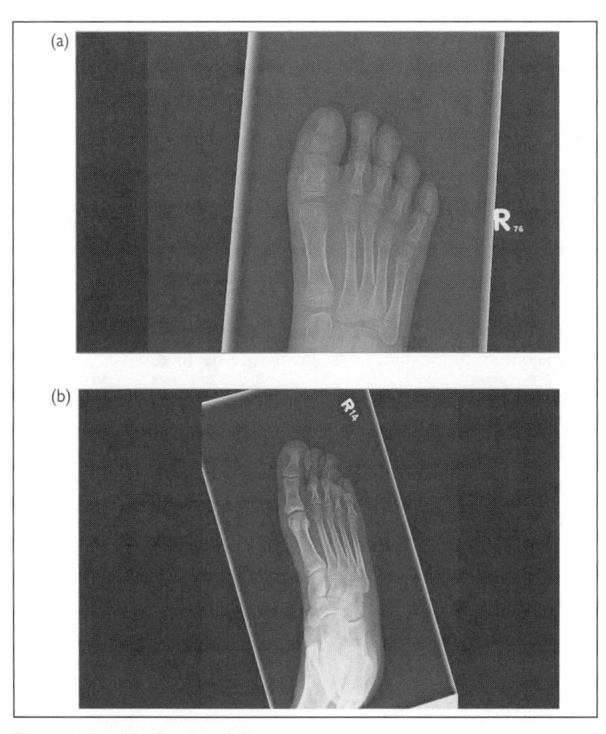

Figure 4.42 a. & b. Great toe fractures.

Primary Stabilization and Management

- Local or regional analgesia
- Reduction of dislocations or deformities
- Buddy taping
- Hard sole shoe
- WBAT is usually possible in a rigid-sole shoe.

Admit and Discharge Guidelines

- Request an orthopaedic consult in the emergency department for: Irreducible dislocation, open injury.
- Refer to an orthopaedic surgeon for outpatient follow-up: Intra-articular fractures, turf toe, unstable dislocations, and displaced or unstable fractures.

Definitive Treatment

- Turf toe: Often can be treated nonoperatively with a period of healing followed by footwear modification. Severe tears (Grade III) or tears with joint interposition may need operative reconstruction.
- Dislocations: These are often stable after reduction and can be treated nonoperatively. Delayed reconstruction can be an option for recurrent dislocations or instability.
- Fractures: Fractures of the proximal phalanx of the great toe are subjected to significant strain from ligamentous and tendinous attachments. Stable nondisplaced fractures do well with nonoperative treatment while displaced fractures are often best closed reduced and pinned. Distal phalangeal fractures or tuft fractures are usually best treated nonoperatively.
- Intra-articular fractures with >1 mm step-off may require surgical realignment of the joint surface with internal fixation.

Lesser Toe Injuries

The lesser toes are physiologically less significant than the great toe, and injuries to these toes are generally well tolerated.

Symptoms and Findings

- Pain at the injury site
- Swelling
- Tenderness
- +/− Ecchymosis
- Deformity is common in dislocations.

Imaging

- Three-view of the toe in question
- Three-view foot

Classification

- Described anatomically: Location, type of fracture, displacement, alignment, open or closed, subluxation or dislocations, intra- versus extra-articular.

Primary Stabilization and Management

- Local versus regional analgesia
- Reduction of deformity or dislocation
- Buddy taping
- Hard sole shoe
- WBAT
- Significant nail bed injuries require removal of the nail and direct repair of the nail bed primarily.
- Subungual hematoma of >25% surface area requires drainage.

Admit and Discharge Guidelines

- Request an orthopaedic consult in the emergency department for: Irreducible dislocation. Open injuries.
- Refer to an orthopaedic surgeon for outpatient follow-up: Intra-articular fractures, significant residual deformity after reduction.

Sesamoid Fractures

The sesamoid bones are imbedded in the short flexor mechanism and are crucial for load displacement around the metatarsal head with weight bearing and ligament functionality.

Symptoms and Findings

- Pain
- Plantar based swelling and/or ecchymosis
- Stiffness of the MTP (metatarsophalangeal) joint
- Occurs either through: Direct trauma, overuse, and stress fracture, or fracture dislocation of the MTP.

Imaging (See Figure 4.43)

- AP/lateral/sesamoid view
- Contralateral views or CT can help distinguish acute fracture from bipartite sesamoids.

Classification

- Described anatomically including location and displacement

Primary Stabilization and Management

- Strapping of the toe in neutral to slight flexion
- Rigid short leg splinting out to beyond the end of the toe
- NWB

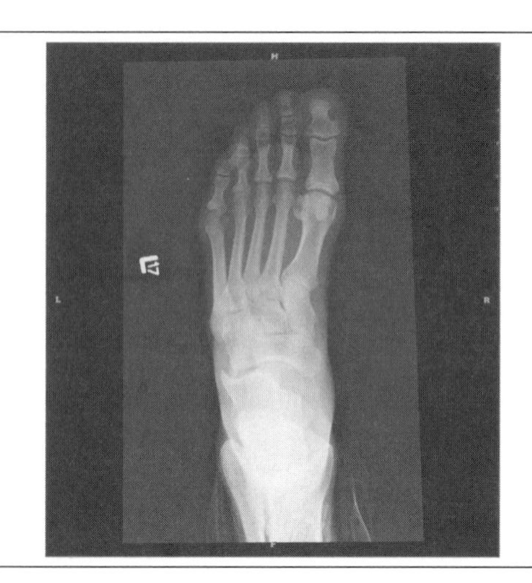

Figure 4.43 Sesamoid fractures.

Admit and Discharge Guidelines

- Request an orthopaedic consult in the emergency department for: Open injury.
- Refer to an orthopaedic surgeon for outpatient follow-up for: All others.

Definitive Treatment

Nonoperative treatment is the mainstay and first-line choice. Treatment consists of initial splinting with transition to rigid foot orthothoses and limited motion of the MTP joint. Pain often takes 4–6 months to resolve completely. Nonunions are common. Painful nonunions are treated with ORIF with bone grafting versus sesamoidectomy.

Subacute Lower Extremity Injuries

- *Hip/Knee/Ankle/Foot Osteoarthritis*: Often debilitating from pain and a lack of range sof motion, this is a very gradual condition and is best referred to a total joint specialist (hips and knees) or a foot and ankle surgeon (ankle and feet). Radiographs demonstrate joint space narrowing, subchondral sclerosis, osteophyte formation, and subchondral cysts. Fractured osteophytes are not considered broken bones. Anti-inflammatory medications, ice, and rest are the best first-line therapies. See Chapter 7: Septic Arthritis section, page 298 regarding ruling out septic arthritis.

- *Greater Tronchanteric Bursitis*: This condition is often confused with hip or lumbar pathology. Diagnosis is easily made with reproduction of the patient's pain from direct palpation over the greater trochanter. Treatment is a steroid injection in the greater trochanteric bursae done as an outpatient.
- *Groin Strains*: Common in athletes, these are often strains of the adductor muscles or tendons. They are best treated as an outpatient with activity modification, ice, anti-inflammatories, and a gentle compression dressing.
- *Prepatellar Bursitis*: Common after prolonged kneeling (roofers, gardeners, etc.) or contusion to the knee. The prepatellar bursae swell significantly and can cause significant deformity and discoloration. Initial therapy is anti-inflammatory medication, ice, elevation, and gentle compression. Failure of these modalities can prompt needle aspiration as an outpatient. If there are concerns of septic bursitis, an oral antibiotic can be given.
- *IT Band Syndrome*: Caused by friction of the IT band over the underlying structures of the distal femur, this syndrome is marked by pain, tenderness, and often a snapping sensation. Thomas and Ober tests can aid in the diagnosis. Activity modification, physical therapy, stretching, heat, and anti-inflammatories are the first-line treatments with a referral to the appropriate provider for follow-up.
- *Plantar Fascitis*: Caused by traction injury to the origin of the plantar fascia, this injury can cause significant discomfort. Patients have pain in the proximal plantar aspect of the foot and may or may not have osteophytes visible on foot radiographs. Initial treatment consists of oral anti-inflatories, ice, and most importantly, night splinting to stretch the plantar fascia. Injections should be avoided.
- *Toe Deformities Including Bunion, Hammertoe, and Curly Toe*: Although these can be significant deformities, they rarely have an indication for acute or urgent treatment. Many of the problems associated with these deformities involve using standard footwear or cosmetic concerns. They should be referred to a foot and ankle specialist.

Chapter 5

Pelvic Injuries

Derek Papp*

*Derek Papp, MD, Chief Resident, Department of Orthopaedic Surgery, Johns Hopkins, Baltimore

Introduction

Evaluation of the patient with a pelvic ring injury can be intimidating as the consequence of misdiagnosis or inadequate treatment may lead to death. Most physicians recognize this and respond to pelvic ring injuries aggressively with fluids, blood products, early orthopaedic consultation, and/or pelvic binder placement. The first step is to understand the differences between commonly encountered pelvic fractures and to recognize those that require urgent intervention.

The incidence of pelvic fractures that present with hemodynamic instability is actually low. After blunt trauma, pelvic fractures account for only 3% of skeletal injury, and only 10% of pelvic fractures seen at Level 1 trauma centers present with both mechanical and hemodynamic instability (White et al., 2008). Others have estimated that the overall mortality from high-energy pelvic ring injuries is between 6% and 35% (Hak et al., 2009). Distinguishing low-energy, mechanically stable injuries from high-energy, unstable fractures allows proper treatment, both at initial evaluation and in later management of the fracture.

It is important to understand the relevant anatomy and what injury patterns should cause concern. The human pelvis consists of three bones, the two innominate bones (ilium, ischium, and pubis) and the sacrum. It is commonly thought of as a ring, with the two innominate bones articulating with the sacrum posteriorly and with each other anteriorly at the pubic symphysis. Stability of the pelvis depends on the ligamentous attachments between the sacrum and the innominates. The strongest of these, the posterior sacroiliac ligaments, require a substantial amount of force before being disrupted. Other important ligamentous attachments include the anterior sacroiliac ligament, sacrotuberous ligament, and the sacrospinous ligament. A series of iliolumbar ligaments run from the transverse processes of the fourth and fifth lumbar vertebrae to the iliac crest. These can be avulsed after significant trauma and typically present as fracture of the L4 or L5 spinous process on plain radiograph when disrupted. The acetabulum, or socket of the hip joint, lies at the confluence of the pubis, ilium, and ishium. Fractures of the acetabulum, while in the pelvis, are not considered "pelvis fractures" per se and should be differentiated as such when describing the fracture pattern.

A number of major vascular structures pass through and lie upon the walls of the pelvis, and injuries to these vessels can cause significant blood loss. The common iliac artery splits within the pelvis. While the external iliac courses out of the pelvis over the pelvic brim, the internal iliac artery dips and runs close to the sacroiliac joint. Its branches course both anteriorly and posteriorly within the pelvis. The branches most commonly identified as causing hemorrhage include the pudendal anteriorly and the superior gluteal posteriorly (*White*). Arterial bleeding causes only 10%–15% of pelvic hemorrhages however; bleeding from the fracture site and disruption of the posterior pelvic venous plexus more commonly result in intrapelvic bleeding (*White*).

While pelvic fractures can cause concern, the treating physician is encouraged to approach the injury like any other trauma. Remember the ABCs of trauma care, perform a thorough history and physical examination, and recognize associated injuries.

Physical Examination of the Pelvis

Surface Anatomy

- See Figure 5.1

Radiographs

- All: AP pelvis (see Figure 5.2)
- Hip injuries: AP and lateral hip
- Pelvic ring injuries: Inlet/outlet
- Acetabular injuries: Judet views

Neurovascular Exam

- Diagram showing cutaneous innervation of the pelvis, lower extremities and perianal region(see Figure 5.3) (see also Chapter 6)

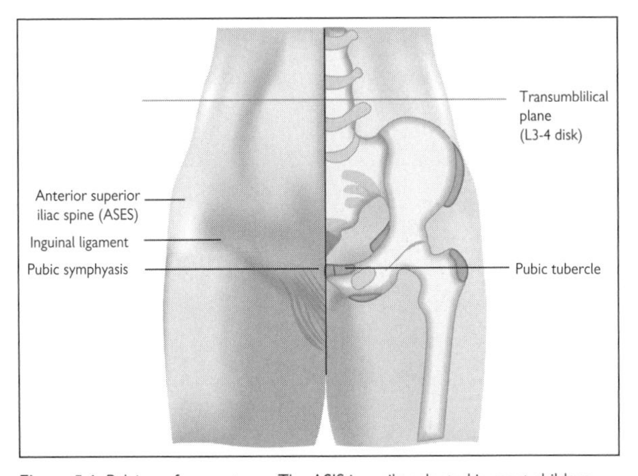

Figure 5.1 Pelvic surface anatomy. The ASIS is easily palpated in most children and nonobese adults. The PSIS can also routinely be felt. Laterally, the greater trochanter is the most prominent and lateral bony landmark and is also the correct location to center a pelvic binder.

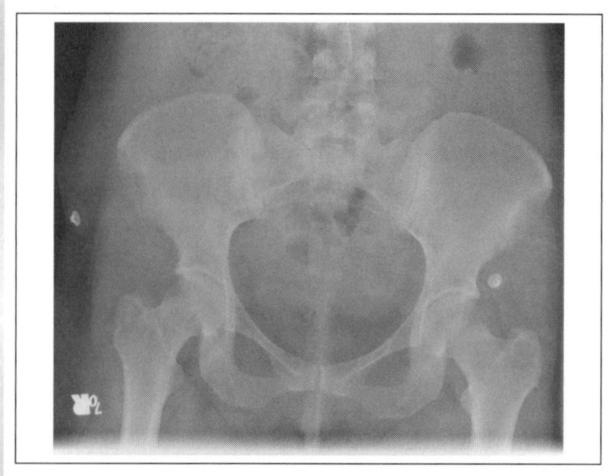

Figure 5.2 A normal AP pelvis radiograph.

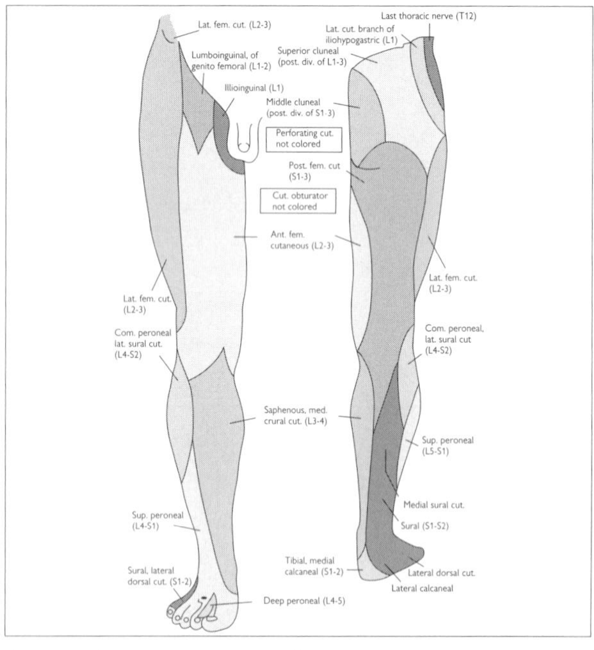

Figure 5.3 Cutaneous innervation of the lower extremity.

- Vascular exam (use of Doppler for difficult to palpate pulses); capillary refill. The following pulses should be checked routinely: Femoral, dorsalis pedis, posterior tibial

Motor Exam

- Detailed motor exam of lower extremities important, **especially L5 nerve root**, which travels just anterior to the SI joint
- Motor strength is graded from 0 to 5:
  - Grade 0: No movement
  - Grade 1: Flicker of movement only
  - Grade 2: Movement with gravity eliminated
  - Grade 3: Movement against gravity
  - Grade 4: Movement against resistance
  - Grade 5: Normal power

Motion	Muscle	Nerve	Root
Hip flexion	Iliopsoas	Femoral	L2–3
Hip extension	Hamstrings	Sciatic	L5-S2
Knee flexion	Hamstrings	Sciatic	L5-S2
Knee extension	Quadriceps	Femoral	L2–4
Ankle dorsiflexion	Tibialis anterior	Deep peroneal	L4
Ankle plantarflexion	Gastrocsoleus	Tibial	S1
Ankle eversion	Peroneus longus	Superficial peroneal	L5-S1
Great toe extension	Extensor hallucis longus	Deep peroneal	L5

Documenting the Physical Exam

- ABCs
  - A formal trauma evaluation should be performed, especially in patients that sustain high-energy injury.
- Skin
  - Example: Skin is intact without breaks, edema, or ecchymosis
  - Watch for the Morel-Lavallee lesion, a closed internal degloving injury about the pelvis and/or greater trochanter of the femur that results in large subcutaneous, serosanguinous fluid collections. It is associated with high-energy injury.
- Genitalia
  - Injuries to vagina and/or rectum are common with pelvic injuries.
  - Digital rectal examination for gross blood or high-riding prostate. Should examine for resting/active tone and perianal

sensation as high-energy pelvic injuries can be associated with spine injury.
    - Examine vaginal walls for gross blood with associated laceration.
    - Blood at penile meatus associated with urethral injury. Call urology for retrograde urethrogram.
- Palpation/Deformity
    - **Defer** mechanical test of pelvic stability to treating orthopaedic surgeon. Performed by compressing anterior superior iliac spines (ASIS) together. Some argue that repeated tests for stability could break up intrapelvic clot and cause hemodynamic instability.
    - Patients with muscle tears will have pain at muscular origins.
- Pulses
    - Example: 2+ dorsalis pedis and posterior tib pulses
- Sensation
    - Example: Sensation intact to light touch in the superficial peroneal, deep peroneal, and tibial nerve distributions.
- Motor
    - Example: Strength is 5/5 in the iliopsoas, hamstrings, quadriceps, tibialis anterior, gastrocsoleus, and extensor hallicus longus.

Imaging of Pelvic Injuries

Initial Radiograph

- An AP pelvis is part of the ATLS screening radiographic series (the other imaging studies are X-rays of the c-spine, chest, and a FAST abdominal ultrasound). This initial radiograph should be performed in any patient with an injury mechanism suspicious for pelvic fracture.

Specific Fracture Patterns

Hip Fractures

- If a hip fracture is seen or suspected based on the initial radiograph, AP and lateral hip views of the involved side should be performed. See Chapter 4.

Acetabulum Fractures

- If an acetabulum fracture is seen or suspected based on the initial radiograph, Judet views (iliac oblique and obturator oblique views) of the involved side should be performed. AP and lateral hip views of the involved side should also be performed to rule out associated femoral neck or peritrochanteric hip fracture

Pelvic Ring Fractures

- If a pelvic ring fracture is seen or suspected based on the initial radiograph, inlet and outlet views should be performed.

CT Imaging

- When fractures of the acetabulum or pelvis are identified, CT imaging should generally be performed if available. Isolated fractures of the proximal femur do not require CT scans. Scanners with 3D reconstruction are particularly useful.
- CTA (CT angiography) can be useful in identifying continued pelvic bleeding, which may require embolization.

Pelvic Ring Fractures

Symptoms and Findings

- Dual distribution of incidence and mechanism
  - Younger patients who sustain high-energy trauma from motorcycle or motor vehicle collision or a high-level fall. Other mechanisms include pedestrians struck by a car or other high-speed rapid deceleration injuries.
  - Elderly patients can sustain pelvic fractures from low-level fall from standing
- Patients with high-energy injuries can have:
  - Hypotension, tachycardia, labored breathing
  - Multisystem injuries—evaluate for bleeding and other injuries in the thorax, abdomen, and from other extremity injuries in addition to the pelvis and retroperitoneum. Primary and secondary surveys are crucial! Examine genitalia as described above.
  - Associated soft tissue injuries—Morel-Lavallee lesion as mentioned above (a large subcutaneous, serosanguinous fluid collection)
- Elderly patients present with:
  - Falls from standing or from bed
  - Inability to ambulate or pain with ambulation
  - Associated conditions such as syncope, heart failure, or delirium.
- Tenderness over the pubic symphysis and/or fracture site

Imaging (See Figure 5.4)

- X-rays: AP pelvis and inlet/outlet views
- CT scan
  - Should be performed to further delineate fracture pattern
  - Should be performed to assess for continued bleeding
  - Does not obviate the need for plain radiographs as the fracture will be followed on an outpatient basis with plain radiographs regardless of whether or not the fracture is treated operatively

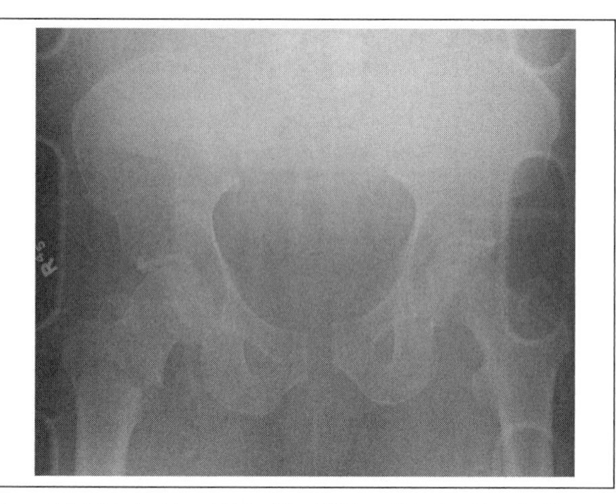

Figure 5.4 Pelvic ring injury. This is of the APC type. Note the widened pubic symphysis and widening of the SI joint posteriorly. This patient also has an intertrochanteric fracture secondary to his high-energy injury.

Classification

- Young-Burgess Classification (see Figure 5.5)
  - Four major categories based on mechanism: Anterior-posterior compression (APC), lateral compression (LC), vertical shear (VS), and combined mechanism (CM). The APC and LC divisions can be graded from I to III.
  - APC injuries caused by force transmission through the pelvis from front to back—such as motorcyclists hitting the gas tank with the front of their pelvis as they get ejected from their seat in a collision.
  - High-level APC injuries—APC II and APC III (the "open book pelvis") are associated with increased transfusion requirement and mortality.
  - LC injuries have lower rates of mortality; the most commonly identified cause of death in patients who sustain LC fractures is closed head injury.
  - Elderly patients who sustain falls from standing often present with pubic rami fractures and a small fracture of the sacrum consistent with an LCI-type injury.

Primary Stabilization and Management

- Formal trauma evaluation based on advanced trauma life support principles
- Application of a pelvic sheet or binder is a simple intervention for a mechanically unstable pelvis in a patient with hemodynamic instability.

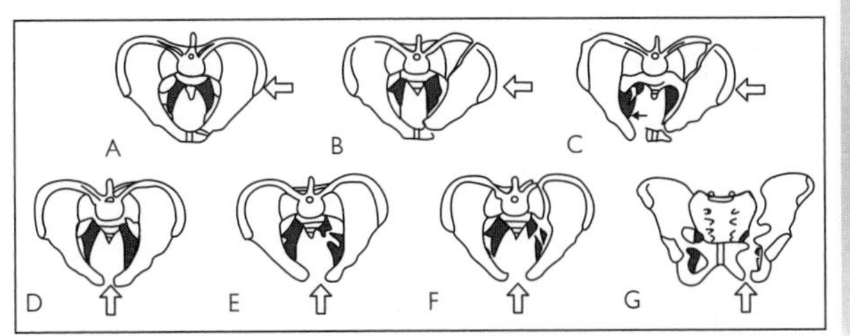

Figure 5.5 The Young-Burgess classification of pelvic ring injuries. A: Lateral compression (LC). I, Rami fracture +/– sacral fracture. B: LCII, Rami fracture and sacral or iliac fractures. C: LCIII, Windswept pelvis with compression fracturing both sides of the pelvis. D: Anterior-posterior compression I: Pubic symphasis diastsis less than 2 cm, posterior ligaments intact. E: APCII, Pubic symphasis diastsis > 2 cm, anterior SI ligaments torn. F: APCIII, Pubic symphasis widened, both anterior and posterior SI ligaments torn. G: Vertical shear.

- A binder's only role is to close down the volume of an APC2 or 3 injury. They are ineffective in VS- or LC-type fractures.
- When possible, the binder or sheet should be removed within 24 hours of placement to avoid skin breakdown; an external fixator can be placed for definitive management.
- If the patient remains unstable, angiography and embolization can be performed. Binders should have a window cut to allow vascular access instead of being removed for angiography.

Admit and Discharge Guidelines

- High-energy pelvic ring fractures require a multidisciplinary approach with attention toward associated injuries, fluid resuscitation, and early operative intervention if the patient does not respond adequately.
- Lower-energy pelvic ring fractures can often be treated nonoperatively; though the patient may need to be admitted to maximize his or her medical health and provide physical therapy.
- Request an orthopaedic consult in the emergency department for all high-energy injuries.
- Consult an orthopaedic surgeon for outpatient follow-up or inpatient cofollow for elderly patients admitted who sustain pubic rami fractures after a fall.

Definitive Treatment

- High-energy injuries with significant displacement may require stabilization with external fixation until the patient is stable and then ORIF (open reduction internal fixation).
- > Two centimeters of diastasis of the anterior pelvis, > 1 cm of vertical displacement, and SI joint dislocation are all indications for ORIF.

- LC1- and Rami-only fractures can be treated by WBAT and follow-up with an orthopaedist.

Acetabulum Fractures

Symptoms and Findings

- Most often occur with high-energy injury where force vector goes through the femur and into the acetabulum; an example would be a patient's knee hitting the dashboard in an MVC.
- Severe pain in groin/hip along with inability to walk
- Sciatic nerve injury in up to one in five fractures involving the posterior wall or column—**examine foot dorsiflexion and eversion.**
- Associated multisystem injuries—evaluate for hemodynamic instability and other injuries in the thorax, abdomen, and other extremity injuries. Primary and secondary surveys are crucial! Examine genitalia as described above.
- Associated soft tissue injuries—the Morel-Lavallee lesion (a large subcutaneous, serosanguinous fluid collection)

Imaging (See Figure 5.6)

- X-rays: AP pelvis, Judet views (obturator and iliac oblique)
  - CT scan should be performed to: Further delineate fracture pattern.

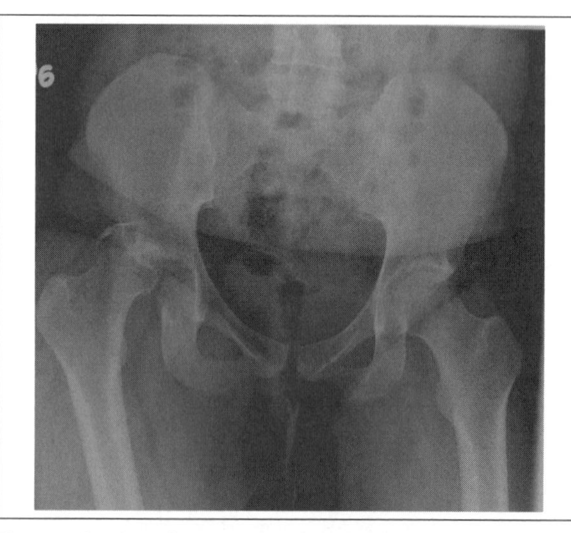

Figure 5.6 Acetabulum fractures with femoral head dislocation.

- Assess for marginal impaction of fracture fragments or osteochondral fragments trapped in the joint.
- CT does not obviate the need for plain radiographs as there will be outpatient follow-up of the fracture with plain radiographs regardless of whether or not the fracture is treated operatively.

Classification (See Figure 5.7)

- Most orthopaedists refer to the Letournel classification scheme. It recognizes 10 common fracture patterns—five simple and five complex. The scheme is useful in determining surgical approaches.

Primary Stabilization and Management

- Formal trauma evaluation based on ATLS principles
- If the femoral head is dislocated, it should be reduced as soon as possible.
- Placement of a femoral skeletal traction pin is often advised, especially if the fracture has an unstable posterior wall

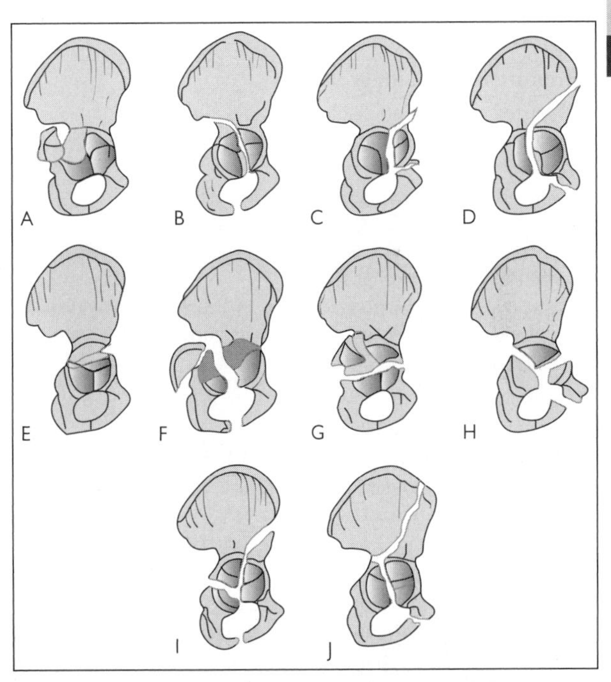

Figure 5.7 Letournel acetabulum classification. A: Posterior wall. B: Posterior column. C: Anterior wall. D: Anterior column. E: Transverse. F: Posterior column, posterior wall. G: Transverse posterior wall. H: T-Type. I: Anterior column posterior hemitransverse. J: Both column.

component and the femoral head tends to fall out of joint, if any bony fragments remain incarcerated within the joint, or if definitive management of the fracture will not be performed in the immediate future.

Admit and Discharge Guidelines

- These are high-energy injuries that typically require intervention from a multidisciplinary team.
- An orthopaedic surgeon should be consulted in the emergency department for the vast majority of acetabulum fractures—nondisplaced fractures should be evaluated by an orthopaedist before allowing weight bearing.

Definitive Treatment

- Many of these injuries will require ORIF. Because of their complexity, most general orthopaedic surgeons refer them to orthopaedic traumatologists.

Sacral Fractures

Symptoms and Findings

- Estimated to occur in 45% of all pelvic fractures
- Usually a high-energy injury such as a motor vehicle or motorcycle collision
- Twenty-five percent associated with neurologic compromise—perform a digital rectal examination.
- Like pelvic and acetabulum, may be associated with soft tissue compromise and/or other injuries.
- Sacral insufficiency fractures occur in 1% of women older than 55. These patients have groin, low-back, and buttock pain along with difficulty walking.

Imaging

- X-rays: Pelvis, inlet/outlet views
- CT scan in stable patients to identify pattern
- MRI or bone scan to demonstrate sacral insufficiency fractures

Classification

- The Denis classification scheme provides a simple means of classifying vertical sacral fractures.
  - Zone 1 injuries occur laterally to the foramen. Zone 2 injuries violate the foramen, and Zone 3 fractures occur medial to the foramen. Incidence of neurologic compromise is lowest in Zone 1 and highest in Zone 3.
  - Zone 3 injuries can be subclassified if any sagittal translation or angulation exists.

- A descriptive classification scheme exists for fractures with transverse components. These are H, U, lambda, and T types.

Primary Stabilization and Management

- Treatment of any associated pelvic ring injuries will usually address the sacral fracture.
- No clear guidelines on neural decompression as results are not reliable
- Treatment of insufficiency fractures should focus on pain control, activity modification.

Admit and Discharge Guidelines

- High-energy pelvic ring fractures require a multidisciplinary approach with attention toward associated injuries, fluid resuscitation, and early operative intervention if the patient does not respond adequately.
- Pelvic insufficiency fractures can be managed with outpatient follow-up if the pain is well controlled and if the patient can ambulate.
- An orthopaedic consult should be called to follow the patient if he or she is admitted to the hospital for pain control or physical therapy.

Pediatric Apophyseal Avulsion Fractures

Symptoms and Findings

- Pain at the muscular origin or insertion after sudden muscle contraction (usually eccentric)
- Associated swelling, pain, weakness. There may be noticeable ecchymosis.
- Occur in children because the muscular attachment to bone is stronger than that of the physis
- Most commonly involve the anterior superior iliac spine—sartorius and tensor fascia lata origin site, and ischial tuberosity—hamstring insertion site
- Other sites include the anterior inferior iliac spine (AIIS), lesser trochanter and iliac crest.

Imaging (See Figure 5.8)

- X-rays: AP pelvis may demonstrate the affected fragment. Oblique images of the pelvis may also show the injury.
- Best imaged by CT scan or MRI. MRI especially useful if the ossification center has not ossified.

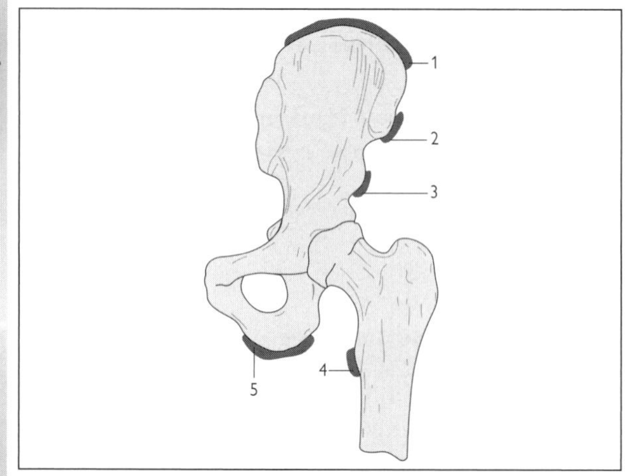

Figure 5.8 The locations of apophyseal avulsion fractures in skeletally immature individuals. 1-Abdominal muscles. 2-Sartorius. 3-Rectus Femoris. 4-Iliopsoas. 5-Hamstrings.

Classification

- McKinney et al. (2009) proposed a modified classification system for all pediatric apophyseal avulsion fractures:
  - Type I: Nondisplaced
  - Type II: Displacement ≤ 2 cm
  - Type III: Displacement > 2 cm
  - Type IV: Symptomatic nonunions or painful exostosis

Primary Stabilization and Management

- Most do not require immobilization but may help assuage patient's pain. Can consider knee immobilizer acutely, especially for hamstring injuries.
- Generally treated nonoperatively. Start with gentle passive range of motion, then 1 week after injury follow with active range of motion. Gentle strengthening may begin 2–3 weeks after injury.

Admit and Discharge Guidelines

- Refer to patient to orthopaedist as an outpatient, especially if he or she presents with a painful nonunion.

Chapter 6

Spine Injuries

Introduction

The potential morbidity of a spinal column injury is catastrophic, as is the consequence of mismanagement. For that reason, definitive management of any suspected spinal cord injury necessitates immediate consultation of a spine surgeon. The primary challenge in the initial evaluation of spinal column injuries is determining whether a bony or ligamentous spinal column injury has harmed—or has the potential to harm—the spinal cord.

Advanced trauma and life support (ATLS) protocols should be followed with any patient presenting after a traumatic injury. Full spine precautions should always be followed until spinal injury can be ruled out. Information about the mechanism of injury should be obtained whenever possible from first responders. Details about how the patient was injured can provide important information about likely spine injuries; for example, sudden deceleration accidents are associated with distracting ligamentous injuries.

While spinal precautions should be followed throughout resuscitation, evaluation for deficits (the "D" of the "ABCDEs" of ATLS) should only commence after the ABCs (airway-breathing-circulation) have been fully addressed. The physical examination of the spine has multiple components and therefore should occur in a step-by-step manner. A useful guide to the spine exam is the American Spinal Injury Association ASIA scale. (see Box 6.1)

Box 6.1 Vocabulary

Incomplete spinal cord injury: Presence of some motor or sensory function more than three segments below the level of injury (ASIA definition).

Complete spinal cord injury: Absence of motor or sensory function more than three segments below the level of injury.

Spinal shock: state of flaccid paralysis, hypotonia, and areflexia that may occur immediately after a spinal cord injury. The end of spinal shock is determined by the return of the bulbocavernosus reflex. The level of injury cannot be determined until after the time period of spinal shock has passed.

Neurogenic shock: Loss of autonomic reflexes; sympathetic outflow is disrupted, resulting in unopposed vagal tone. Patients are hypotensive and bradycardic. Treat with invasive monitoring, fluids, and vasopressors.

Physical Examination of the Spine

Surface Anatomy

- See Figure 6.1

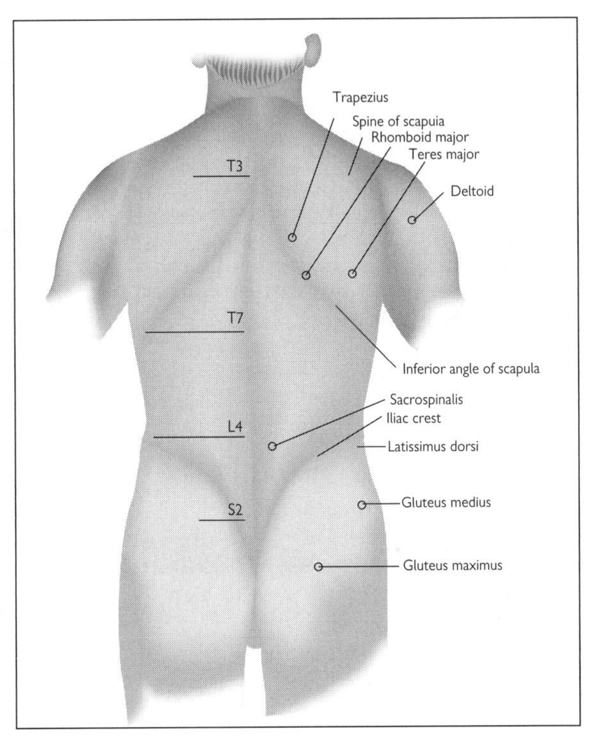

Figure 6.1 Surface anatomy of the spine.

Vascular Exam

- Vascular exam (use of Doppler for difficult to palpate pulses); documented carotid, radial, ulnar, posterior tibialis and dorsalis pedis; capillary refill

Examination of the Spine by Neurologic Level

- The ASIA scale provides a simple reference for the complete spinal cord examination with respect to sensory, motor, and reflex function.
- The patient should be systematically examined from head to toe.

Sensory Exam (See Figure 6.2)

- A true sensory examination should include both pinprick (using the sharp end of a sterilized safety pin) and light touch (using the cotton tip of a swab). In an emergency situation, this detailed examination is often reduced to the gross presence or absence of sensation. While a less comprehensive examination is sometimes necessary in order to determine immediate management, a full sensory examination should be performed as soon as possible in order to accurately assess and document the patient's initial deficit
- Sensation is graded from 0 to 2:
  - Grade 0: No sensation
  - Grade 1: Impaired sensation
  - Grade 2: Normal sensation
- See dermatome diagram on the ASIA scale (Figure 6.2)

Motor Exam (See Table 6.1)

- Motor strength is graded from 0 to 5:
  - Grade 0: No movement
  - Grade 1: Flicker of movement only
  - Grade 2: Movement with gravity eliminated
  - Grade 3: Movement against gravity
  - Grade 4: Movement against resistance
  - Grade 5: Normal power

Reflexes

- Deep tendon reflexes are used to demonstrate the integrity of reflex arcs through a spinal nerve root. They are graded on a five-point scale:
  - Grade 0: Absent
  - Grade 1+: Hypoactive
  - Grade 2+: Normal
  - Grade 3+: Hyperactive without clonus
  - Grade 4+: Hyperactive with clonus

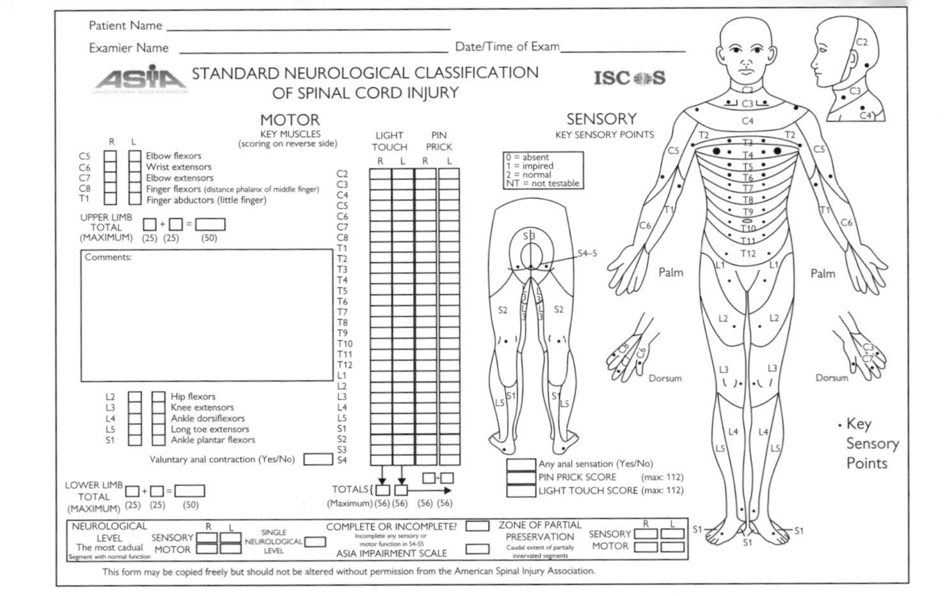

Patient Name

Examier Name   Date/Time of Exam

ASIA

STANDARD NEUROLOGICAL CLASSIFICATION OF SPINAL CORD INJURY

ISCoS

MOTOR

KEY MUSCLES

(scoring on reverse side)

	R	L	
C5			Elbow flexors
C6			Wrist extensors
C7			Elbow extensors
C8			Finger flexors (distance phalanx of middle finger)
T1			Finger abductors (little finger)

UPPER LIMB TOTAL (MAXIMUM) ☐ + ☐ = ☐ (25) (25) (50)

Comments:

	R	L	
L2			Hip flexors
L3			Knee extensors
L4			Ankle dorsiflexors
L5			Long toe extensors
S1			Ankle plantar flexors

Valuntary anal contraction (Yes/No) ☐

LOWER LIMB TOTAL (MAXIMUM) ☐ + ☐ = ☐ (25) (25) (50)

	LIGHT TOUCH R	LIGHT TOUCH L	PIN PRICK R	PIN PRICK L
C2				
C3				
C4				
C5				
C6				
C7				
C8				
T1				
T2				
T3				
T4				
T5				
T6				
T7				
T8				
T9				
T10				
T11				
T12				
L1				
L2				
L3				
L4				
L5				
S1				
S2				
S3				
S4				
TOTALS (Maximum)	(56)	(56)	(56)	(56)

SENSORY

KEY SENSORY POINTS

0 = absent
1 = impired
2 = normal
NT = not testable

☐ Any anal sensation (Yes/No)
☐ PIN PRICK SCORE (max: 112)
☐ LIGHT TOUCH SCORE (max: 112)

NEUROLOGICAL LEVEL The most cadual Segment with normal function — SENSORY R ☐ L ☐; MOTOR R ☐ L ☐

SINGLE NEUROLOGICAL LEVEL ☐

COMPLETE OR INCOMPLETE? ☐ Incomplete any sensory or motor function in S4-S5

ASIA IMPAIRMENT SCALE ☐

ZONE OF PARTIAL PRESERVATION Caudal extent of partially innervated segments — SENSORY R ☐ L ☐; MOTOR R ☐ L ☐

This form may be copied freely but should not be altered without permission from the American Spinal Injury Association.

Figure 6.2 ASIA scale. American Spinal Injury Association: International Standards for Neurological Classification of Spinal Cord Injury, revised 2011, Atlanta, GA. Reprinted with permission.

Table 6.1 Examination of the Spine by Neurologic Level

Root	Reflex	Muscle	Sensation
C5	Biceps	Deltoid Biceps	Lateral arm
C6	Brachioradialis	Wrist extension Biceps	Lateral forearm
C7	Triceps	Wrist flexion Finger extension Triceps	Middle finger
C8	None	Finger flexion Hand intrinsics	Medial forearm
T1	None	Hand intrinsics	Medial arm
L4	Patellar	Anterior tibialis	Medial leg and foot
L5	None	Extensor hallucis longus	Lateral leg and dorsum of foot
S1	Achilles	Peroneus longus and brevis	Lateral foot

Source: Hoppenfeld, S., Thomas, H., & Hutton R., *Physical Examination of the Spine and Extremities*. Prentice Hall, 1976.

- The bulbocavernosus reflex is the most distal spinally mediated reflex. It is a reflexive contraction of the anal sphincter elicited by a gentle tug on a Foley catheter. If no catheter has been placed, the reflex may be elicited by gently squeezing the glans penis in men or the clitoris in women. Absence of the bulbocavernosus reflex immediately after a spinal cord injury indicates the presence of spinal shock (see Box 6.2).

Box 6.2 Case Scenario: Importance of the Bulbocavernosus Reflex

An 18-year-old patient presents to the emergency department 30 minutes after sustaining a cervical spine fracture-dislocation. On examination, he has a complete cord injury at the C5 level.

If the bulbocavernosus reflex is absent:
The patient's injury level cannot be determined as he is still in spinal shock. The patient has hope of recovering function after spinal shock is no longer present.

If the bulbocavernosus reflex is present:
The patient has a C5 spinal cord injury.

- The presence of certain reflexes indicates an upper motor neuron injury. These reflexes are present at birth but go away usually with the first few months and are considered abnormal over 2 years old. In 90% of adults, the presence of the reflexes listed below should be considered pathologic (1 of 10 adults may have a false positive).
  - Hoffman's sign: Flexion of the thumb or index finger at the DIP joint when the nail of the long finger is flicked.
  - Babinski's reflex: Up-going toes (dorsiflexion of the great toe with splay of the other toes) when a sharp instrument is run along the sole of the foot from calcaneus to the forefoot

Documenting the Physical Exam

- ABCs
  - With very few exceptions, patients presenting with a spine injury should have a formal trauma evaluation.
- Skin
  - Example: Skin is intact without breaks, edema, or ecchymosis.
- Palpation/Deformity
  - Example: Nontender to palpation of the spine from C1 to sacrum; no step-offs or shifts of spinous processes appreciated.
- Pulses
  - Example: 2+ radial, dorsalis pedis, and posterior tibialis pulses
- Sensation
  - Example: Sensation intact to light touch and pinprick from C1 to S1
- Motor
  - Example: Sketch out below tables for upper and lower extremities (Tables 6.2 and 6.3)
- Reflexes
  - Example: 2+ biceps (C5), brachioradials (C6), triceps (C7), patellar (L4), Achilles (S1)
  - Negative Hoffman and Babinski (toes downgoing)

Table 6.2 Normal Upper Extremity Motor Exam

	Biceps (C5)	Extend wrist (C6)	Triceps (C7)	Grip (C8)	Abduct fingers (T1)
Right	5	5	5	5	5
Left	5	5	5	5	5

Table 6.3 Normal Lower Extremity Motor Exam

	Flex hip (L2)	Extend knee (L3)	Dorsiflex foot (L4)	Extend toes (L5)	Plantarflex foot (S1)
Right	5	5	5	5	5
Left	5	5	5	5	5

- Rectal
  - Example: Intact rectal tone, anal wink, and perianal wink (i.e., bulbocavernosus reflex intact)

Imaging of Spine Injuries

Initial Radiographs

- Cervical spine radiographs are part of the ATLS screening, radiographic series (the other imaging studies are X-rays of the chest and pelvis, and a FAST abdominal ultrasound).
- A lateral, AP, and open-mouth odontoid view should be obtained in patients. According to ATLS criteria, patients who are alert, neurologically normal, without neck pain, and who can voluntarily range their neck without pain, do not require radiographs
- The skull to the first thoracic vertebra must be visualized on the lateral X-ray. If all seven cervical vertebrae cannot be visualized, a swimmer's view must be obtained.

CT Imaging

- According to the ATLS manual, CT imaging should be performed when initial radiographs are suspicious for injury or if a region cannot be adequately visualized. However, the ease and availability of CT imaging has increased their use as the initial assessment tool for possible spine injuries.
- CT imaging is useful in assessing osseous injury and displacement. It is less useful in assessing ligamentous or spinal cord injury. Ligamentous injury can be detected using flexion extension radiographs or MRI imaging. MRI is the preferred modality for evaluation of spinal cord injury.
- While MRI is the preferred modality for assessing spinal cord injury, a CT scan is faster to perform and typically easier to obtain than an MRI. For this reason, patients who present with neurologic deficits will usually undergo a CT scan prior to an MRI. The CT scan allows detection of osseous causes of a neurologic deficit.

- A CT myelogram allows visualization of the space available for the spinal cord through the injection of contrast dye into the subarachnoid space. Because a needle is introduced into the subarachnoid space, it is significantly more invasive than MRI. However, depending on operator experience, it can be performed much more quickly than MRI, and it can be performed on patients in which MRI is contraindicated (e.g., patients with pacemakers).
- CTA (CT angiography) can be useful in identifying injury to the carotid or vertebral arteries.

MRI

- MRI should be performed in patients with neurologic deficits. MRI provides the best imaging of damage to the spinal cord, as well as soft tissues that may be obstructing the spinal cord, such as an epidural hematoma or herniated disk.
- The utility of MRI is somewhat limited by the time it takes to acquire imaging and the need for patients to remain still throughout the procedure.
- The individual or team who will treat the spinal cord injury should be consulted prior to ordering MRI imaging, as specific techniques (e.g., contrast, angiography) may be needed for optimal visualization of the injury pattern.

Initial Treatment

Resuscitation

Any patient with major trauma should be assumed to have a cervical spine injury until proven otherwise. Evaluation and management of the ABCs should be initiated, but stabilization of the spinal cord should be maintained throughout the resuscitation. If a patient requires intubation, stabilization of the cervical spine should be maintained. A jaw thrust maneuver may be required to prevent hyperextension of the cervical spine during intubation. Alternatively, a nasotracheal intubation can be performed, though this requires someone experienced in performing the procedure. Finally, a surgical airway may be established through a cricothyroidotomy.

Assessment and management of circulation is also a top priority. Patients in hypovolemic shock will present with decreased blood pressure and increased heart rate while patients in neurogenic shock demonstrate decreased blood pressure and decreased heart rate. Whether hypovolemic or neurogenic, the initial treatment of shock is volume repletion. However, if a patient is suffering from neurogenic shock, fluid resuscitation should be guided by central venous pressure. Adequate resuscitation is a key component of spinal injury care as decreased blood pressure results in decreased perfusion of the spinal cord, which may worsen an initial spinal injury.

Spine Stabilization

The spine should be maintained in a neutral, supine position with immobilization of the head, neck, and body. Generally, a semirigid plastic collar and a long spine board are utilized. Side head supports and sandbags may also be used.

There is evidence that in rare situations cervical collars may cause additional injury, particularly certain hyperextension and dissociative injury patterns. In these injuries, a cervical collar may distract the vertebral segments beyond their usual anatomic relationship. A cervical collar should not be applied if it appears to distract the patient's cervical spine. One of the main reasons for discomfort with a cervical collar has to do with the size of the collar. An ill-fitting collar is not always an advantage and may put the patient at increased risk. Additionally, if a patient is conscious and holding his or her head in a flexed position or is apprehensive about applying a cervical collar, it may be better to defer cervical collar application until initial radiographs are performed. Certain fracture patterns are more stable in flexion than extension.

It is important to be suspicious for spinal column injury for any patient with antecedent spinal pathology as certain preexisting conditions increase the risk of spine injury after trauma:

- Ankylosing spinal disease (e.g., ankylosing spondylitis or diffuse idiopathic skeletal hyperostosis): Increased risk of fracture with minor trauma
- Down syndrome: Increased incidence of occipito-cervical and atlantoaxial instability due to ligamentous laxity
- Preexisting spinal stenosis: Higher risk of spinal cord injury with hyperflexion or hyperextension
- Os odontoideum: Increased incidence of atlantoaxial joint instability
- Rheumatoid arthritis: Known to cause weakness and instability at C1–C2 from attenuation of the transverse ligament

Clearing the C-spine

The purpose of cervical spine clearance is determining that no cervical spine injuries are present. There is Level-1 evidence supporting clearance of the cervical spine without radiographs, including the NEXUS studies and the Canadian C-spine rule. The protocol advised by the American Trauma and Life Support group is based on this evidence (see Box 6.3).

If a patient cannot be assessed due to intoxication, he or she must be reevaluated at a time when sober. If the patient has a distracting injury, it must be assumed that the patient has a cervical spine injury

Box 6.3 ATLS Protocol

Immediate removal of cervical collar in the awake, alert, sober, and neurologically normal patient who has no tenderness to palpation in the cervical spine and who exhibits full, pain-free range of motion

until another evaluation can be done after the distracting injury has been attended to appropriately.

Initial radiographic assessment of the cervical spine should consist of three radiographic views. In many trauma centers, a multidetector CT scanner has replaced screening radiographs as the initial imaging modality. If imaging demonstrates a fracture or dislocation, the cervical collar must remain in place, and consultation with a spine expert should be sought.

If initial imaging is negative for injury, a determination should be made as to the necessity for MRI imaging. MRI imaging is indicated in patients with unexplained neurologic deficits or suspected ligamentous injury. Ligamentous injury should be suspected if patients have midline, cervical spine tenderness.

Patients who complain of pain without neurologic deficits or suspected ligamentous injury do not require immediate MRI. These patients should be maintained in a high-quality cervical collar and scheduled for follow-up. The cervical collar should be well fitted and well padded, such as a Miami-J or a Philadelphia collar. Patients should not be discharged in the temporary collar placed by EMS or in the emergency department. At the 2-week follow-up examination, radiographs of the cervical spine with or without flexion-extension radiographs should be performed.

Fractures and Dislocations of the Spine

In the emergency setting, the goal of treatment is to appropriately resuscitate the patient and stabilize the spine to allow for optimal clinical outcome. After emergency care, most patients will be transitioned to the care of a surgical trauma service where definitive treatment will be provided. As part of the transition of care, a spine surgery consult is frequently required. The complex and evolving nature of spine injury classifications may hinder rather than facilitate communication. We recommend avoiding eponyms and classification systems when calling a surgical consult in favor of using clear descriptive assessments.

The first step in describing an injury is its location. Anatomically, there are three parts of the spine—cervical, thoracic, and lumbar. In trauma, there are also three broad categories of injury—preaxial (above and including the axis or C2), postaxial (below C2 or C2–C7), and thoracolumbar. We will consider postaxial and thoracolumbar injuries together as they can be described in the same manner.

Preaxial Cervical Injuries

Occipital Condyle Fractures

- Mechanism: High-energy trauma
- Associated injuries: Head trauma, cranial nerve palsies, cervical spine trauma
- Imaging: Often missed on radiographs, usually diagnosed on CT scan
- Initial stabilization: Maintain in collar or stabilize with sandbags. Do not place in traction, but provide inline stabilization.
- Watch out for…atlanto-occipital (a.k.a., occipito-cervical) dislocation.

Atlanto-Occipital Dislocation (See Figure 6.3)

- Mechanism: High-energy trauma
- Associated injuries: Brainstem injuries, subarachnoid hemorrhage, intracranial and vertebral artery disruption
- Imaging: Diagnosed based on the lateral cervical radiograph (see Figure 6.3a). Diagnosis depends upon understanding the normal

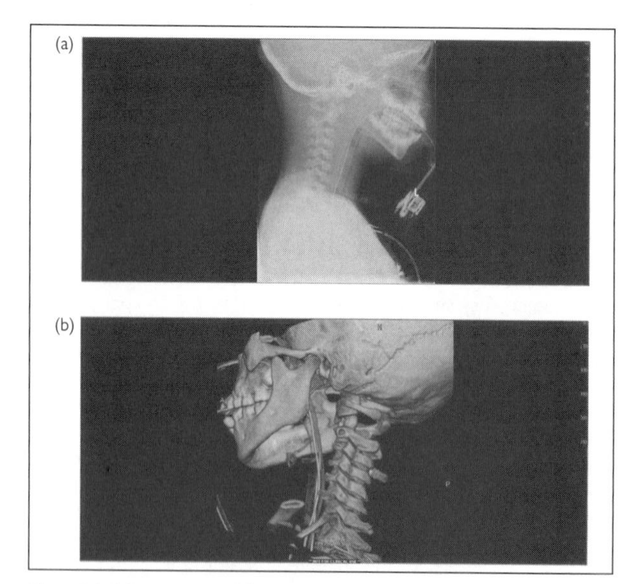

Figure 6.3 Atlanto-occipital dislocation. a. X-ray. b. 3D CT scan.

anatomic relationships between the occiput and atlas. This relationship is calculated by Power's ratio (see Box 6.4), which is usually 1. CT imaging (particularly the sagittal cuts) is also useful for identifying and defining this injury, particularly concomitant occipital condyle fractures (see Figure 6.3b).

- Initial stabilization: Maintain in collar or stabilize with sandbags. Do not distract the cervical spine with an overlarge collar. Any distraction of the head from the body can be fatal and must be avoided.
- Pitfall: Severe dislocations are typically fatal, so most patients who survive have less displacement and thus more easily missed injuries.

Box 6.4 Power's Ratio

Power's ratio is calculated as the distance between the basion to the posterior arch of C1 compared to the distance between the opisthion to the anterior arch of C1. This ratio is usually 1. If the occiput is displaced anteriorly, the distance from the basion to the posterior arch of C1 will lengthen while the distance from the opisthion to the anterior arch of C1 will shorten, leading to a ratio greater than 1.

C1 (Atlas) Fractures (See Figure 6.4)

- Mechanism: Compressive trauma
- Associated injuries: Occipital fractures, other cervical injuries
- Imaging: Disruption of the C1 ring may be seen on the open-mouth odontoid view. These fractures are best seen on CT imaging, particularly axial slices. (See Figure 6.4 for representative axial images.) Because of the ringed structure of the atlas, it frequently breaks in more than one place along the ring and may be in as many as four large parts.
- Initial stabilization: Cervical collar

C2 (Axis) Fractures (See Figure 6.5)

- Mechanism: May be high- or low-energy mechanism. Many fractures involve hyperextension +/− flexion or distraction
- Associated injuries: Atlanto-occipital instability, other cervical injuries, facet dislocations, spinal cord injury, cranial nerve deficits, head trauma
- Types of injuries:
  - C2 traumatic spondylolisthesis: Fracture of the pars interarticularis

SUPERIOR VIEW OF C1 ON C2

Spinous process of the axis

Superior articular facet of the atlas

Posterior arch of the atlas

Dens

Transverse ligament of the atlas

Anterior arch of the atlas

Jefferson Fracture

Lateral Mass Fracture

Bony Avulsion of the Transverse Ligament

Midsubstance Tear of the Transverse Ligament

Figure 6.4 C1 fracture.

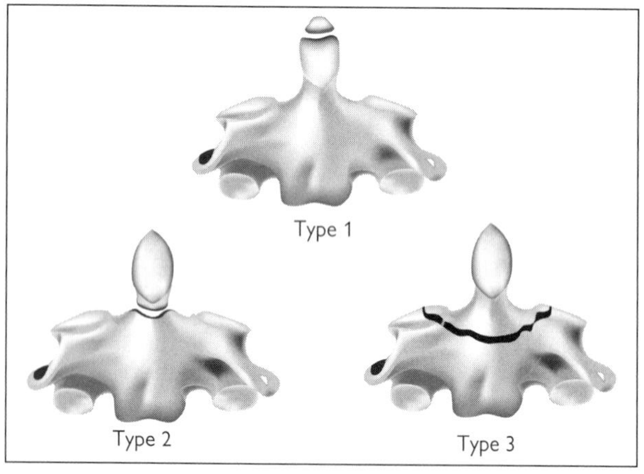

Figure 6.5 C2 fracture.

- Odontoid fracture (see Figure 6.5)
- C2 body fracture

- Imaging: Odontoid fractures may be seen on the open-mouth odontoid view while C2 traumatic spondylolisthesis may be seen on lateral radiographs; CT imaging is usually necessary in C2 fractures as precise fracture location and displacement will dictate definitive management.

- Initial stabilization: Cervical collar
- Watch out for…other cervical spine injuries, which may occur in up to 30% of axis fractures.

Transverse Ligament Disruption

- Mechanism: Trauma, infection, congenital anomaly, inflammatory process
- Imaging: These injuries may be diagnosed on the open-mouth odontoid views if there is significant subluxation of C1 on C2; usually, more advanced imaging is required. If there is a bony avulsion of the ligament, these injuries may be seen with CT imaging. MRI is the study of choice for soft-tissue injuries, such as ligamentous disruption.
- Initial stabilization: Cervical collar
- Watch out for…underlying infectious or inflammatory process.

Postaxial Cervical and Thoracolumbar Injuries

Three-Column Concept (See Figure 6.6)

A preliminary assessment of fracture stability can be made using the three-column concept. While this system was initially developed based on thoracolumbar injuries, its use has been expanded to include the cervical spine. The three-column approach divides

Anterior Column	Middle Column	Posterior Column
Anterior longitudinal ligament Anterior half of vertebral body Anterior annulus fibrosis	Posterior half of vertebral body Posterior annulus/ posterior disk	Spinous process Laminae Facets
Anterior disc	Posterior longitudinal ligament	Pedicles Posterior ligamentous structures: Ligamentum flavum Intraspinous ligaments Supraspinous ligaments
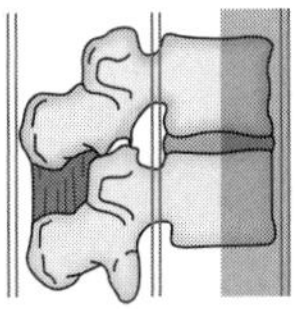 A	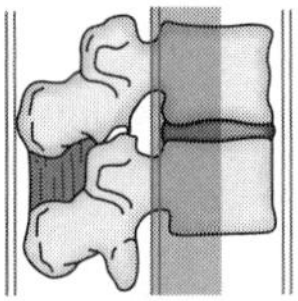 B	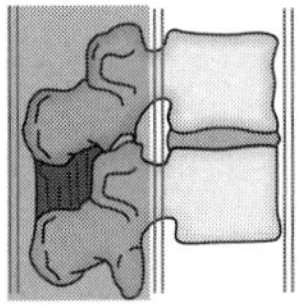 C

Figure 6.6 Three-column drawing demonstrating the anterior, middle, and posterior column.

the spine into three columns: anterior, middle, and posterior (see Box 6.5).

Box 6.5 Three-Column Concept

Anterior column: Anterior vertebral body, anterior annulus fibrosus, and anterior longitudinal ligament
Middle column: Posterior vertebral body, posterior annulus fibrosus, and posterior longitudinal ligament
Posterior column: Posterior ligamentous complex (facet joints, ligamentum flavum, and other posterior elements)

The number of involved columns determines the stability of an injury. Generally, one-column injuries are stable while three-column injuries are by definition unstable, requiring halo-immobilization and/or surgical stabilization. The stability of two-column spinal injuries is more difficult to assess. All two-column injuries should be considered potentially unstable; that is, even injuries that are not unstable at presentation have the potential to become unstable without stabilization. As an example, compression fractures are fractures of the anterior column and are stable whereas burst fractures involve both the anterior and middle column and may not be stable.

Imaging (See Figure 6.7)

Postaxial cervical and thoracolumbar injuries are diagnosed using plain radiographs as well as CT imaging. MRI is generally obtained in spinal cord injury; however, the timing of such imaging should be directed by the spine surgeon involved. The goal of imaging is to identify the region of injury, the number of involved columns, and the degree of displacement of fractures or dislocations. For any postaxial injury, the severity of the deformity (fragmentation, displacement, and loss of vertebral height) and the degree of canal compromise should be assessed. Figure 6.7 shows an anterior column fracture with maintenance of the middle and posterior columns.

Communicating with Consultants

With few exceptions, a spine surgery consult should be requested as soon as a spine injury is diagnosed. Adequate communication about the spine injury allows for optimal patient outcomes. Depending on the injury, patients may require urgent operative reduction or decompression, and surgical intervention may be delayed by poor communication.

Step-by-Step Guide to Calling a Consult

1. Age of patient and mechanism of injury
    - Example: I have a 24-year-old man who was an unrestrained passenger in a high-speed motor vehicle collision.

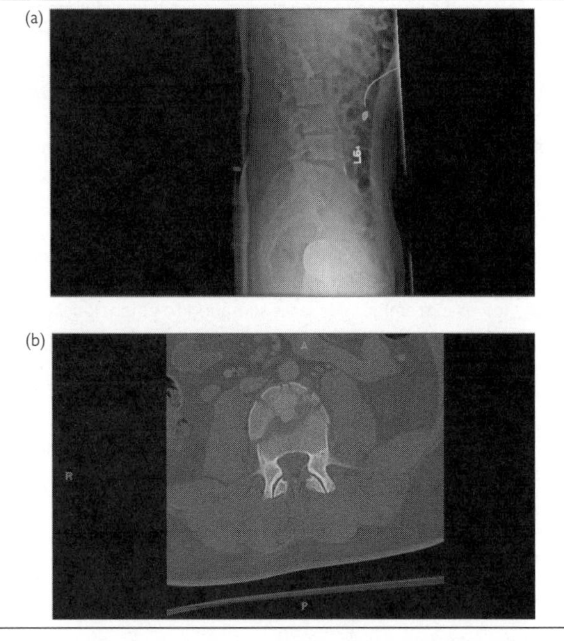

Figure 6.7 L3 fracture. a. Lateral X-ray demonstrating disruption of the anterior column while the middle and posterior columns remain intact. b. Axial CT scan demonstrating fracture of the anterior column with intact middle and posterior column.

2. Description of bony and ligamentous injury
   - Location and type of injury
     - Example: I have a 24-year-old man who was an unrestrained passenger in a high-speed motor vehicle collision *who sustained a fracture of the anterior and middle columns of C6.*
   - Severity of the deformity and/or canal compromise
     - Example: I have a 24-year-old man who was an unrestrained passenger in a high-speed motor vehicle collision who sustained a fracture of the anterior and middle columns of C6 *with posterior displacement of the posterior vertebral body into the spinal canal with approximately 50% of the canal compromised.*
   - Imaging obtained
   - Example: I have a 24-year-old man who was an unrestrained passenger in a high-speed motor vehicle collision who

sustained a fracture of the anterior and middle columns of C6 with posterior displacement of the posterior vertebral body into the spinal canal with approximately 50% of the canal compromised *seen on CT scan of the cervical spine.*

3. Neurologic status
   - Mental status
     - Example: He is alert and oriented with a GCS of 15.
   - Deficits
     - Example: *He is alert and oriented with a GCS of 15.* He is able to fire his deltoids and wrist extensors bilaterally but cannot move his triceps or fingers. He has no motor function in his lower extremities. He lacks sensation to pinprick distal to C6 dermatome.
   - Bulbocavernosus
     - Example: He is able to fire his deltoids and wrist extensors bilaterally but cannot move his triceps or fingers. He has no motor function in his lower extremities. He lacks sensation to pinprick distal to C6 dermatome. *He does not have a bulbocavernosus reflex.*
4. Current management of the spine
   - Collar, sandbags, backboard
     - Example: He is currently wearing the collar he arrived in on EMS transport.
5. Other important details
   - Other injuries and what teams are involved
     - Example: He has a splenic laceration that will require monitoring but no intervention at this time per trauma surgery. We will be consulting plastic surgery for his multiple facial fractures.
   - Interventions performed or about to be performed
     - Example: He had a Foley placed on arrival. He has two large-bore IVs in each arm.
   - Overall stability
     - Example: He is currently stable without a pressor or fluid requirement, and he is mentating appropriately and maintaining his airway.

Classification of Spinal Cord Injuries (Table 6.4)

Spinal cord injuries can be broadly classified into complete and incomplete spinal cord injuries (see Table 6.4). A complete spinal cord injury is defined by the absence of motor or sensory function

Table 6.4 Patterns of Incomplete Spinal Cord Injury

Syndrome	Motor	Sensory	Prognosis
Central cord	Upper extremities more affected than lower extremities	Variable sensory loss	Fair
Anterior cord	Variable motor loss	Variable loss of pain and temperature; preservation of proprioception and deep pressure	Poor
Brown-Sequard	Ipsilateral motor loss	Ipsilateral loss of proprioception and deep pressure; contralateral loss of pain and temperature	Good
Cauda Equina* **(Surgical Emergency)**	Variable loss of lower extremity motor function, reflexes and bowel/bladder function	Saddle anesthesia	Good

* The cauda equina is technically not a part of the spinal cord as it has terminated by this point; however, its importance can not be underemphasized.

more than three segments below the level of injury. An incomplete spinal cord injury is defined by the presence of some motor or sensory function more than three segments below the level of injury (ASIA definition). There are several well-defined patterns of incomplete spinal cord injury, and prognosis varies by pattern.

Summary

Spinal injuries can be devastating. When in doubt, assume a cervical spine injury until a thorough evaluation can be performed. All spinal injuries require consultation in the emergency department.

Chapter 7

Orthopaedic Emergencies and Urgencies

Compartment Syndrome

Introduction

Compartment syndrome is a limb-threatening emergency in which the end capillary perfusion pressure within a limited space (a compartment) is less than the surrounding intracompartmental pressure. Perfusion and tissue function are compromised by increased intracompartmental pressure. If untreated, compartment syndrome causes permanent devastating results, including neurologic deficits, muscle necrosis, ischemic contractures, infection, and delayed fracture healing. Amputation is a common sequla. Compartment syndrome may not only be limb threatening but also life threatening. If muscle ischemia causes significant myoglobinuria, patients may develop acute renal failure.

Compartment syndrome has been found wherever a compartment is present, including any limb segment, abdomen, and buttock. Of all extremity fractures, tibia fractures are most frequently associated with compartment syndrome. However, any fracture may cause a compartment syndrome (as can burns and crush injuries). The mechanism of injury is also important as high-energy injuries with significant damage to soft tissues, particularly of a crushing mechanism, are much more likely to cause compartment syndrome.

Diagnosing Compartment Syndrome

The classic signs of compartment syndrome are the "Five Ps": pain, paresthesias, pallor, paralysis, and pulselessness. While the Five Ps are a useful mnenomic, their applicability to clinical practice is limited as any patient suspected of having compartment syndrome should have undergone fasciotomy before paralysis or pulselessness develops. The most sensitive symptom of compartment syndrome is pain out of proportion to the injury while the most sensitive sign of compartment syndrome is pain with passive stretch.

> Pain out of proportion and/or pain with passive stretch
> =
> Compartment syndrome until proven otherwise

The key to diagnosing compartment syndrome is having a high index of suspicion and performing serial compartment examinations. A compartment check exam should be performed every 4 to 6 hours in any patient with a fracture or injury mechanism worrisome for compartment syndrome. There is a high frequency of morbidity as well as litigation associated with compartment syndrome and, to those ends, it is important to document compartment checks (see Box 7.1). The date and time of the check is particularly important.

Diagnosing compartment syndrome is particularly difficult in children, obtunded or intubated patients, patients with underlying neurologic deficits, or patients with an iatrogenic decrease in neurologic function, that is, regional or spinal analgesia. A clue to an evolving compartment syndrome is increased analgesic (usually narcotic) requirement. Prophylactic fasciotomy may be performed in some cases if an injury is felt to have a high likelihood of developing compartment syndrome, particularly if the patient is intubated or nonresponsive.

Box 7.1 Documenting a Compartment Check in SOAP Note

SUBJECTIVE
Pain on a 10-point scale
Method of analgesia: PO, IV, IM, patch analgesics, patient controlled anesthesia (PCA), epidural

OBJECTIVE
Vitals: Beware of increasing HR and BP
General: Lying comfortably in bed vs. writhing uncontrollably
Mental status: Awake and alert vs. confused and disoriented
Palpation: Compartments feel soft vs. tense
Sensation: Paresthesias or deficits; document any change from initial or previous exams
Circulation: Pulses and capillary refill
Motor: Change in strength or new deficits
Passive stretch: Pain with passive stretch = suspected compartment syndrome

ASSESSMENT#1
No evidence of compartment syndrome at this time

PLAN #1
Continue Q6H compartment checks

ASSESSMENT #2
Examination is concerning for compartment syndrome

PLAN #2
Emergency fasciotomy

ASSESSMENT #3
Examination is equivocal

PLAN #3
Immediate measurement of compartment pressures

Measuring Compartment Pressures

Diagnosing compartment syndrome based on physical examination is an art rather than a science. If the clinical picture is borderline or the examination is equivocal, the compartment pressures can be directly measured using commercially available devices. These devices require training and experience, and the individual checking a compartment pressure should be capable of performing emergent fasciotomy if a compartment pressure reading is consistent with compartment syndrome.

Thresholds for intracompartmental pressures have not been established, with studies supporting a range of absolute pressures (30–50 mmHg) as being consistent with compartment syndrome. The preferred alternative to the absolute measurement is a comparative measurement, whereby the compartment pressure is compared to a patient's diastolic pressure. The critical range of the delta pressure (diastolic pressure − intracompartmental pressure = delta pressure) is greater than 30 mmHg. A delta of 30 or less should prompt immediate fasciotomy.

Treatment of Compartment Syndrome

Compartment syndrome is treated by initially removing all circumferential dressings down to the skin. If, after reevaluation, compartment syndrome exists, this should be followed by open and extensive fasciotomies with decompression of all involved compartments within the limb.

Preventing Compartment Syndrome

Iatrogenic Compartment Syndrome

Compartment syndromes may be created in the treatment of fractures and soft-tissue injuries. Inadequate padding or the application of circumferential splints greatly increases the risk of compartment syndrome. A patient will generally have improvement in his or her pain after immobilization. If pain has increased, the splint should immediately be removed.

IV catheter infiltrates are another iatrogenic cause of compartment syndromes, particularly in the forearm and hand. Initial management should consist of cessation of IV use and elevation. If symptoms continue and compartment syndrome is suspected, a surgical consult should be requested.

Prevention

Elevating an injured extremity may help decrease the incidence of compartment syndrome by decreasing the swelling and thus the intracompartmental pressure of an injured limb. Please note that once compartment syndrome is suspected, the limb should be kept at the level of the heart. Elevating a limb with compartment syndrome could impede arterial flow, which could further compromise an already ischemic limb. Similarly, icing the injured limb can help decrease swelling (though ice should never be applied directly to the skin as it may cause a hypothermia injury). Avoid the placement of arterial or intravenous lines in injured extremities as they increase swelling within the extremity. Isolated soft-tissue injuries may benefit from placement of a well-padded splint in a protected position; such splints reduce swelling through soft-tissue rest.

Summary Points: Compartment syndrome can happen with a variety of injuries, including burns, crush, or any fracture or dislocation. If compartment syndrome is suspected, the patient in question should have an orthopaedic consultation with appropriate monitoring, compartment pressure evaluation, analgesia, and potentially preparation for a fasciotomy. If not treated aggressively, compartment syndrome can result in loss of limb and lead to other complications.

Open Fractures

Diagnosing an Open Fracture

Definition

An open fracture is defined as a fracture in which there is disruption of the overlying soft tissue and skin such that the fractured region communicates with the outside environment. The belief that bone must be visible for a fracture to be "open" is a common misconception. The primary complication of open fractures is infection because the bone has communicated with the outside environment. A fracture is defined as open independent of the length of time the fractured bone has spent in communication with the outside environment.

Physical Examination

A careful physical examination is the key to diagnosing an open fracture. The patient's skin should be carefully examined for abrasions or lacerations. At times, the only evidence of an open fracture is a tiny pinprick where a spike of fractured bone penetrated the soft tissue and skin. Open fractures are generally due to higher-energy trauma, and it is important to recognize that the soft-tissue injury is much more extensive than the visible skin disruption. It is commonly said that open fractures can be thought of as soft tissue injuries that include a fracture.

Imaging

While open fractures should be diagnosed by physical exam rather than imaging studies, it is helpful to know some of the radiographic findings associated with open fractures. The most obvious radiographic finding is bony penetration beyond the radio-opaque lines indicating the soft tissue of the skin. More subtle findings can include radio-opaque foreign body material within the soft tissue or adjacent to the bone. Additionally, air within the soft tissue may indicate an open fracture, though it can also be seen in other conditions such as gas gangrene.

Classification of Open Fractures

The Gustilo and Anderson classification system is the most widely utilized system for classifying open fractures.

Gustilo and Anderson Classification

- Type I: Puncture wound of less than or equal to 1 cm in length with minimal contamination or muscle crushing
- Type II: Laceration of greater than 1 cm in length with moderate soft tissue damage and crushing and adequate bone coverage with minimal comminution
- Type IIIA: Extensive soft-tissue damage; however, soft-tissue coverage of the bone is adequate. Injury is often due to a high-energy injury with a crushing component; includes massively contaminated wounds and severely comminuted or segmental fractures.
- Type IIIB: Flap coverage required to provide soft-tissue coverage; due to extensive soft-tissue damage, periosteal stripping and bone exposure.
- Type IIIC: Open fracture associated with an arterial injury requiring repair.

It is important to note that not all criteria are needed to become Type II or III. For example, a clean, noncontaminated laceration <1 cm with a "severely comminuted or segmental fracture" is not a Type I because of the skin laceration; it is at least a IIIA because of the segmental or comminuted aspect of the fracture.

An accurate description of the soft-tissue injury is preferred for cross-service communication due to the high interobserver variability of classification systems. The length of the wound should be documented, as should the neurovascular status. Type and degree of contamination should be described. It is important to note any mitigating factors, which might change antibiotic treatment, such as barnyard contamination. Finally, it is most important to document the remainder of the musculoskeletal exam, particularly the neurovascular status. An open fracture with a large wound may be a distracting injury for the physician who fails to note the loss of a pulse amid a particularly impressive fracture.

It is worthwhile to note that many fractures may be associated with a violation of skin integrity. A careful examination should be undertaken to ascertain if the soft-tissue disruption is deep enough that bone is exposed to the open environment.

Initial Management of Open Fractures

Antibiotic Treatment and Tetanus Prophylaxis

All open fractures should be treated with intravenous antibiotics as soon as possible, as prompt delivery of antibiotics decreases the risk of infection. Tetanus prophylaxis should be given in the emergency department according to standard treatment algorithms. If immunization history is unknown or incomplete, tetanus immunoglobulin should also be given. No consensus has yet emerged about the type and duration of antibiotic treatment. There is disagreement in the literature about whether coverage of both gram-positive and gram-negative organisms is required for prophylaxis in all open fractures and whether fractures contaminated with organic material mandate clostridial coverage. The antibiotic recommendations given below are based on commonly accepted treatment algorithms (see Box 7.2). When deciding on antibiotic therapy, it is important to review your local institutional recommendations as sensitivities and resistance varies geographically.

Box 7.2 Antibiotic Recommendations

Type I and II: Cefazolin (1 gram IV) every 8 hours until 24 hours after wound closure

Type III: Cefazolin plus gentamicin (3–5 mg/kg/day) or levofloxacin (500 mg/day)

Farm injuries: Cefazolin plus gentamicin or levofloxacin plus penicillin (2,000,000 units) every 4 hours or metronidazole (500 mg) every 6 hours

Preliminary Irrigation and Debridement

All open fractures require definitive surgical debridement in the operating room. If definitive treatment will be delayed, the wound should be gently irrigated with an isotonic solution, such as sterile saline, and gross contamination should be removed, excluding bony fragments. The wound should be dressed with saline-soaked gauze. A splint should be applied both for comfort and to decrease additional soft-tissue injury from the fracture. Preliminary treatment should never delay the definitive operative management of open fractures, and the treating orthopaedic surgeon should be consulted as quickly as possible. Vascular surgery should be contacted in cases of vascular injury. Plastic surgery should be consulted in cases of soft-tissue loss.

Definitive Management of Open Fractures

Surgical debridement is the standard of care for almost all open fractures. Sharp, centripetal debridement around the open wound should be performed under tourniquet. Management of open fractures often requires a multidisciplinary approach with coordination between orthopaedics and plastic surgery. In Type III injuries, orthopaedics will often assist vascular surgery in stabilization procedures, such as external fixator application, to help maximize the likelihood of a successful vascular outcome. Fractures require adequate blood supply and soft-tissue coverage in order to heal; early involvement of appropriate specialists often determines the success or failure of fracture healing.

Gunshot Injuries

Gunshot wounds are a special subclass of open fractures, as low-velocity injuries (such as those caused by handguns) may not require operative irrigation and debridement. The rationale for nonoperative management is that such injuries have less microbial contamination than other open fractures and less soft-tissue damage than high-velocity gunshot wounds. Low-velocity gunshot wounds may be treated with antibiotics (cefazolin), tetanus treatment, and irrigation of the entry and exit wounds in the emergency department. After these initial steps, the fracture can be treated as a closed fracture.

Gunshot wounds with gross contamination, delayed presentation, and signs of infection, large wounds, bullet fragmentation, bony comminution, and/or multiple projectiles may require formal irrigation and debridement even if the injury was a result of a low-velocity weapon.

All fractures associated with skin integrity violations should be examined closely to determine if an open fracture exists. Open fractures require orthopaedic consultation, and it is up to the consulting physician to determine if operative irrigation and debridement is necessary.

Septic Arthritis

Septic arthritis is not a true orthopaedic emergency in most cases. A septic joint becomes a true emergency when the joint infection spreads, resulting in overwhelming sepsis with a clear threat to the patient's life. Otherwise, the major risk is to the articular cartilage, which is progressively injured due to the bacterial load. Since patients with total joint replacements have no articular cartilage, the need for immediate intervention is lower in this population, but the concern that the infection could spread remains.

For native joints, the diagnosis should be made by needle aspiration by an appropriately trained provider. Surgical irrigation and debridement should then be performed within 12 hours of diagnosis if the aspirate is positive for bacteria or has an abundance of white blood cells.

If a causative microbe has been positively identified from an aspiration, a tailored antibiotic regimen can be started as soon as the diagnosis is made. Total joint replacements that are suspicious for infection should only be aspirated and treated **after** discussion with or by an orthopaedic surgeon, preferably the surgeon who performed the total joint replacement.

Presentation

The signs and symptoms of septic arthritis are nonspecific but when added together can paint a clear picture. Table 7.1 shows the typical symptoms associated with septic arthritis in the order of the most important diagnostically.

Table 7.1 Signs and Symptoms of Septic Arthritis

Sign/symptom	In septic arthritis	Comment
Range of motion	Extremely painful with ***any*** motion. Patients with near-normal motion compared to their contralateral side can be ruled out.	Should be distinguished from pain at the ends of ROM typically associated with degenerative pain
History consistent with septic arthritis	Puncture wounds or local trauma to joint, blood-borne bacteria, recent illness, recent sick contact, recent international travel, or recent dental work are all possible predisposing factors.	Septic arthritis in adults is rare without a reason. Therefore, the history is crucial.
Fever	>101.5°	Normally present unless immunocompromised or early in the disease process
Lab abnormalities	Elevated erythrocyte sedimentation rate (ESR) (will depend on institution's lab normal values) Elevated c-reactive protein (CRP) (will depend on institution's lab normal values) WBC >12 Aspirate with WBC >80 K and left shift (native joint only).	Normally present unless immunocompromised or early in the disease process. CRP rises and falls faster than ESR. Caution: Gout can cause an aspirate to have WBC >100 k.
Imaging	Radiographs often show effusion. CT or MRI rarely indicated.	Radiographs are mostly needed to help rule out other causes of joint pain such as nondisplaced fracture, etc.
Joint effusion	Often present	Nonspecific as it is present in almost all joint injury patterns
Joint erythema	Occasionally present	Nonspecific
Joint "warmth"	Occasionally present	Nonspecific

Children present differently than adults do and can occasionally get septic arthritis without a good reason in their history. Additionally, septic arthritis in children should be distinguished from transient synovitis, which is an inflammatory condition that occurs after a recent illness and presents like septic arthritis of the hip. Transient synovitis is treated symptomatically while septic arthritis may require surgery.

Table 7.2 outlines the likelihood of septic arthritis in a child given certain findings and lab values.

History and Physical Examination

A complete history is crucial to the diagnosis. Many findings are usually present but are often nonspecific. It is good to look for erythema, swelling, warmth, and evidence of a break in the skin or source of infection. Range of motion is one of the best predictors of true florid septic arthritis. Patients with septic arthritis will have intense pain with minimal joint motion and will often stop the examiner after only 5–10 degrees of motion. Patients that can achieve full or near full range of motion (compared with the other side) are very unlikely to have septic arthritis. Radiographs of the affected joint should always be evaluated to ensure that there is no other pathology that could be causing the symptoms.

Joint Aspiration

Joint aspiration is the gold standard of diagnosis, and in some cases, serial aspirations are the preferred treatment. Joint aspiration is joint dependent (please refer to Chapter 9, "Miscellaneous Procedures"). One should always adhere to strict sterile technique in order to prevent seeding a joint and to prevent contaminating the sample. It is a good idea to escort the sample to the lab onself to ensure it does not get lost or ruined, as it is very difficult to reaspirate a joint. Cell count with differential and microscopic evaluation for crystals should always accompany the gram stain and culture if enough fluid is obtained.

Definitive Management

Irrigation and debridement of the joint with microbe-specific systemic antibiotics is the gold standard. More than one I+D may be required.

For patients too ill to tolerate surgery, serial aspirations (q24hr or q12hr) with appropriate systemic antibiotics will decrease bacterial load.

Summary Points

Septic joints should be evaluated as quickly as possible to avoid a spreading systemic illness. Diagnosis is confirmed by needle aspiration under sterile conditions by an experienced and knowledgeable provider. Irrigation and debridement should follow after a septic joint is confirmed.

Table 7.2 Multivariate Predictor

History of fever	Nonweight- Bearing	Erythrocyte sedimentation rate ≥ 40 mm per hr.	Serum white blood cell count > 12,000 cells per mm^3 (12.0×10^9 cells per L)	Predicted Probability of septic arthritis *(percent)*
Yes	Yes	Yes	Yes	99.8
Yes	Yes	Yes	No	97.3
Yes	Yes	No	Yes	95.2
Yes	Yes	No	No	57.8
Yes	No	Yes	Yes	95.5
Yes	No	Yes	No	62.2
Yes	No	No	Yes	44.8
Yes	No	No	No	5.3
No	Yes	Yes	Yes	93.0
No	Yes	Yes	No	48.0
No	Yes	No	Yes	33.8
No	Yes	No	No	3.4
No	No	Yes	Yes	35.3
No	No	Yes	No	3.7
No	No	No	Yes	2.1
No	No	No	No	0.1

Kocher MS, Zurakowski D, Kasser JR. Differentiating between Septic Arthritis and Transient Synovitis of the Hip in Children: An Evidence-Based Clinical Prediction Algorithm. The Journal of Bone and Joint Surgery Am 1999;81: 1662–70.

Dislocated Joints

Dislocated joints constitute orthopaedic emergencies or urgencies for two reasons. First, the joint dislocation can compromise the blood and nutrient flow to the articular cartilage, causing permanent damage incrementally with time (the most sensitive to this is the hip joint). Second, the dislocation can cause significant soft-tissue damage and put the limb at risk (such as a traumatic knee dislocation).

Rarely, there are patients who can voluntarily dislocate and relocate their joints (usually the shoulder). These patients have learned that presenting with a dislocation and being relocated can provide them with some sort of secondary gain such as narcotic prescriptions. Many will visit multiple EDs and providers to ensure they do not get identified as voluntary dislocators. If there is suspicion, the other emergency facilities in the area should be contacted to establish a pattern.

Presentation

Dislocated joints occur most often in the setting of trauma, often high energy. Occasionally, they can occur from low-energy mechanisms such as twisting or falling. Patients will almost always have pain associated with the dislocation and often deformity and a significant decrease in the range of motion of the joint.

History and Physical Examination

A complete history should be obtained. A careful exam should always be done with specific attention to the neurovascular status of the limb and concomitant injuries. It is crucial to inspect the skin as there can sometimes be open injuries, which substantially complicate the treatment.

Definitive Management

Closed reduction of the dislocation within 6 hours of the injury is the preferred technique for almost all dislocations. *Adequate analgesia, sedation, and relaxation are crucial to all relocations.* Appropriate procedural sedation is often required as a brief period of sedation and relaxation of the surrounding muscles is necessary for the relocation. The general technique for reduction usually involves exaggerating the deformity, and applying axial traction with countertraction. Slow steady pressure is far superior to large jerky movements. Often rotation is necessary. Immediately after relocation one should perform a range of motion exam to ensure stability in the reduced position or record the position in which the joint redislocates. If the joint is unstable in any plane, it needs to be immobilized to prevent this motion. Please see the individual injury sections for joint-specific information.

Summary Point

Dislocations of the hip or the knee should be reduced as quickly as possible by the orthopaedic surgeon. Any shoulder dislocation with neurological deficits should be reduced by an orthopaedic surgeon.

Pelvic Fracture with Hemodynamic Instability: A Potentially Life Threatening Emergency

Presentation

These injuries most commonly result from high-energy mechanisms. The patient is often multiply injured, and ATLS trauma resuscitation protocols should be initiated. As a part of the ATLS trauma evaluation, an AP Pelvis radiograph should be performed. If this image shows a widened pubic symphysis, an orthopaedist or someone trained in placing a pelvic binder should be called immediately.

Physical Examination

Evaluate the patient for pelvis stability. This examination should not be repeated multiple times as blood clots may have formed in the pelvic vessels, and repeat manipulation may cause them to break free and increase bleeding.

Initial Treatment

- ATLS ABCs take priority. Managing the pelvic fracture is a part of "C" due to the hemodynamic instability that results from these fractures.
- Anterior-posterior compression pelvic injuries with hemodynamic instability should get a pelvic binder placed immediately. The binder needs to be centered over the greater trochanters NOT the anterior superior iliac spine.
- Repeat radiographs should be done to assess the quality of the reduction of the pelvis once the binder is placed.
- The binder should not be removed before the stabilization procedure. A binder can remain on for 12–24 hours. If femoral artery or vein access is needed, a hole can be cut in most binders and sheets to form a window and allow adequate access without compromising binder function.

Definitive Management

Unstable pelvis injuries often will need ORIF. Many patients will be too ill from their other injuries to get this at the time of presentation. External fixation is used to bridge the gap between presentation and definitive fixation.

Summary Points

Pelvic fractures are associated with high-energy mechanisms and often with other injuries. These patients need a formal trauma evaluation as well as an orthopaedic evaluation. Testing for pelvic stability should be performed by the most experienced practitioner and should be repeated sparingly if it all. A pelvic binder or a sheet should be placed until definitive treatment in the operating room or an external fixation device can be placed.

Acute Cauda Equina Syndrome

Most often caused by postsurgical changes, cauda equina syndrome can have permanent and serious effects ranging from mild numbness to permanent paralysis. Acute cauda equina can also be secondary to trauma, tumors, or other space-occupying lesions.

Presentation

- Saddle anesthesia
- Bowel or bladder dysfunction
- Numbness or weakness in a nerve root distribution
- +/− Pain
- Can occur quickly

Diagnosis

- Often a clinical diagnosis for postoperative patients based on changing neurologic exam
- Trauma patients, tumor patients, or space occupying lesion patients may require imaging of the spine to fully characterize.

Treatment

Urgent/emergent surgical decompression

Summary Point

Cauda equina syndrome can occur quickly and is usually associated with pain along with certain neurological deficits including numbness and weakness in a nerve root distribution, bowel or bladder dysfunction (including urinary retention), or saddle anesthesia. These patients need immediate attention from a spine surgeon for acute surgical decompression.

Pyogenic Flexor Tenosynovitis

Pyogenic flexor tenosynovitis is infection of the flexor tendon sheath in a digit. It is a diagnosis that must be recognized and treated urgently

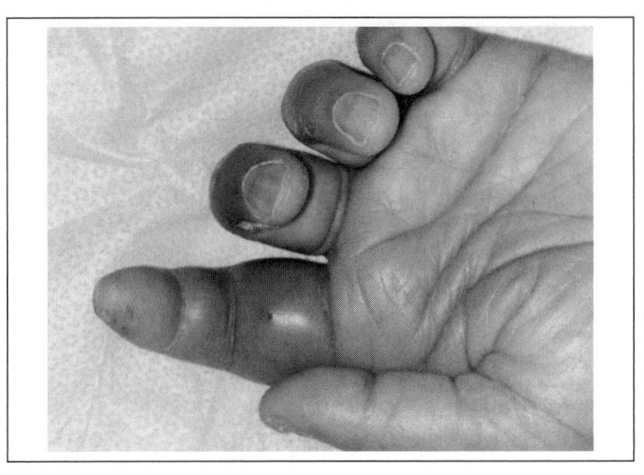

Figure 7.1 Pyogenic flexor tenosynovitis.

to prevent loss of finger function and spread of infection more proximally as well as to avoid systemic infection. Left untreated, this condition will result in tendon necrosis, disruption of the tendon sheath, and digital contracture.

It occurs most often secondary to penetrating injury and introduction of an infectious agent in the closed space of the tendon sheath. The most common organism is *Staphylococcus aureus*. The organism flourishes in the medium of the tenosynovial fluid.

Presentation

The patient presents with complaints of a red, painful, swollen digit and often with a fever (see Figure 7.1).

Physical Examination

The diagnosis can be made clinically if **Kanavel's signs** of flexor tendon sheath infection are present. These include:

1. Finger held in flexed position
2. Fusiform swelling (sausage-like swelling)
3. Tenderness along the flexor tendon sheath
4. Pain with passive finger extension

These signs may be minimally present if the patient has recently been treated with antibiotics or is immunocompromised or has chronic infection.

Aspiration

While the diagnosis can typically be made by clinical exam, if infection is questionable, the flexor sheath can be aspirated and the fluid sent for gram stain and culture.

Definitive Management

Tendon sheath infection requires irrigation and debridement in the operating room within 12–24 hours. Once cultures are obtained, IV antibiotics should be initiated (either cefazolin 1–2 g IV Q8H or clindamycin 600 mg IV Q8H until culture and sensitivity results reported). Patients that present very early—in the first 24 hours of flexor sheath infection—may be able to be managed nonoperatively, with IV antibiotics (either cefazolin 1–2 g IV Q8H or clindamycin 600 mg IV Q8H) and immobilization. If the patient's exam does not show marked improvement within 12 hours, prompt surgical management is indicated. Medical management is not advisable in immunocompromised patients or patients with diabetes, and early surgical intervention is recommended.

Summary Point

Patients with flexor tenosynovitis need urgent evaluation and treatment by a hand specialist because of the rapid progression of untreated disease. If Kanavel's signs are present or the suspicion of disease is high enough, plan for definitive operative treatment and start IV antibiotics after cultures are obtained.

Femoral Neck Fractures in Patients under 60

Femoral neck fractures, especially displaced fractures with an intact capsule, greatly compromise the tenuous blood supply to the femoral head. For patients that are older, this is insignificant because the preferred surgical treatment is a hemiarthroplasty or total hip replacement, both of which replace the femoral head. However, for young patients, it can have disastrous complications if they get AVN of the femoral head. Therefore, prompt treatment and restoration of adequate blood flow to the femoral head is paramount. Studies suggest that permanent damage starts occurring at 6 hours postinjury and builds from there. Rapid transfer of these patients if orthopaedic surgery is not available is crucial. Surgery within 6 hours of injury is the goal. Intertrochanteric fractures are extracapsular, and therefore do not carry the risk of AVN. Please refer to the femoral neck fracture section of Chapter 4 for presentation and initial treatment.

Chapter 8
Analgesia

Introduction

The care of patients with injuries to the musculoskeletal system often requires pain management. Most people will experience some pain with an orthopaedic injury, but the degree of pain felt by patients often varies. Pain associated with these injuries can be intense and, in many cases, will worsen especially when attempting reduction or manipulation. While the degree of discomfort generally lessens after a successful reduction, it is important to note there may be significant residual pain. Pain management must be an important consideration when deciding on how to best treat a patient with an orthopaedic injury.

Systemic Analgesia

There are many different strategies for pain management. One approach is treating with oral systemic analgesia. It is a simple, yet effective treatment method that provides several therapy options. Multimodal analgesia (especially with oral only) is preferred.

Nonopiate Analgesia (see Table 8.1)

- Acetaminophen is a readily available oral analgesic that can be used in most situations as long as there are no contraindications. Hepatic failure or the risk of hepatic toxicity should be considered prior to administration, especially for those patients with a G6PD deficiency or ETOH liver injury. Acetaminophen is a relatively weak analgesic and is often combined with other oral agents for a greater pain-relieving effect.
- Another useful oral medication is tramadol which is a centrally acting opioid with both serotonin and norepinephrine activity. Tramadol has some of the analgesic properties of an opioid medication with fewer side effects. It is a good alternative to acetaminophen, but the two could be used in combination for patients requiring more significant pain relief. It is important to note that both acetaminophen and tramadol are hepatically metabolized, so caution should be exercised in certain populations.
- Nonsteroidal anti-inflammatory agents (NSAIDs) are a growing class of medications commonly used in orthopaedics for postmanipulation pain and outpatient analgesia. NSAIDs should not be used in some patients, including those with renal insufficiency or gastric bleeding disorders. Remember, when prescribing NSAIDs for pain management, the agents in this class vary widely in their analgesic effect, as well as dose and duration of activity. Not all NSAIDs are created equal. Table 8.1

Table 8.1 Nonopioid Pain Management				
Medication	**Recommended dose (mg)**	**Frequency (Q-hr)**	**T1/2 (hr)**	**Max (mg/day)**
Acetaminophen	650–1000	4–6	2–4	4000
Tramadol	50–100	4–6	6	400
Naprosyn	250–500	12	12–17	1375
Ibuprofen	400–800	6–8	2	3200
Ketorolac	10 PO	4–6	5.3	40 (No > 5 days)
	60 × 1 IM or 30 IM 30 × 1 IV or 15 IV	6	5.6	120

lists different NSAIDs and their recommended dosing. Some orthopaedic surgeons avoid NSAIDs in patients with fractures (and fusions) due to concern over compromised bone healing.

- One of the more effective classes of medications used for the treatment of musculoskeletal pain are the opioids. Fortunately, there are several different formulations with varying strengths to allow for better titration of effect. Refer to Chapter 10 for a quick reference table that describes the relative potency between the different opioid agents. Note that individual variation in opioid metabolism can result in unpredictable and abnormal patient responses. This holds true for all opioid analgesics. This chapter briefly touches on some of the more commonly prescribed opioid medications.

Opiates

- Codeine, a prodrug of morphine, is a weak opioid analgesic with less than half the potency of morphine. Codeine's half-life ranges from 2 to 4 hours. Higher doses are often necessary for adequate pain control and are generally not well tolerated.
- Hydrocodone is a very common mild-to-moderate analgesic that is often combined with other substances including nonsteroidal medications or acetaminophen. It is considered less potent than morphine, and its half-life is about 2.5 to 4 hours.
- Morphine is a frequently used agent with a relatively long duration of activity (4–5 hours) and an elimination half-life of about 2 hours. Initial doses typically range from 5 to 10 mg orally, or 2 to 4 mg intravenously, (with careful patient monitoring of vital signs).
- Oxycodone is a moderate analgesic. It has a potent metabolite, oxymorphone, which is about 10 times more potent than

morphine. It has good bioavailability, mild side effects, and half-life of about 2.5 to 3 hours.

- Hydromorphone has an analgesic effect that is 7–11 times that of morphine, depending on IV or oral therapy. It comes in a variety of forms and can be given IM, IV, subcutaneously, or PO. The oral formulation has an onset of about 30 minutes and duration of about 4 hours.
- Methadone, a synthetic opioid agent, has a very long half-life (120 hours in chronic users and 15 hours for one or two doses), has excellent bioavailability orally, and has minimal euphoric side effects. It is most often used in chronic pain and is generally not recommended in the acute pain setting.
- Fentanyl is an extremely effective analgesic medication, with about 80 times more potency than morphine. It is also available in several formulations that allow a variety of administration routes including oral transmucosal (lozenge), IV, and subcutaneous or IM routes (slightly longer onset much like intranasal but very effective in the acute setting with no IV access). It has a rapid onset and a short duration of action (1–2 hours).

- Naloxone administered intravenously or intramuscularly will reverse respiratory depression caused by the administration of excess opioids. It is a pure opioid antagonist.

Fentanyl for Pediatric Patients

One very useful administration route for fentanyl is the intranasal application. While there is limited literature available on this administration route, the data does suggest it can be of benefit in children over the age of 6 months, reducing the need for IV insertion. The dose is typically 1–2 mcg/kg, based on available literature (same dose as the IV form) to a maximum single dose of 50 mcg, but this can be repeated. Analgesic effect usually occurs within 5 minutes and lasts about 1 hour.

Administration

The medication should be drawn up in 1 ml luer-lock syringe. The syringe is held horizontal in the patient's nostril while the head of the patient is reclined at 45 degrees, and the medication is then administered in divided doses between the two nostrils. Repeated doses can be given, but as always, close patient monitoring with appropriate equipment(including pulse oximetry) and personnel should be in place before dosing.

Important Point about Procedural Sedation

If an analgesic is to be given intravenously, especially if repeat doses or multiple agents are to be given, then all necessary precautions should be in place. This should include, but is not limited to, cardiac monitoring, oxygen monitoring, supplemental oxygen, and close observation. Procedural sedation is used in many institutions for complex orthopaedic procedures; the description of how to do it is beyond the scope of this text. Most institutions will have their own protocols, many of which will include the recommendations suggested in this text. In most cases, procedural sedation requires additional well-trained personnel to assist with the analgesia delivery and close patient monitoring.

Local Analgesia

Nerve blocks are another excellent means to provide regional analgesia. There are several types of nerve blocks that will not be discussed in this chapter, but commonly used nerve blocks are reviewed below. Lidocaine or bupivacaine is most often used for local analgesia, lidocaine has the theoretical advantage of being faster onset, while bupivacaine is longer acting. In many cases, a 50–50 mixture of the two agents is used to achieve adequate analgesia.

Digital Blocks

Digital blocks are an excellent way to provide local anesthesia in order to complete a procedure on a finger, thumb, or toe. Introducing an anesthetic agent, such as lidocaine, (or the longer acting bupivacaine) locally, at the base of the digit allows for local anesthesia and the ability to perform manipulation of or procedure on the affected area without pain. Local anesthetics **without** epinephrine are recommended in the hand to avoid vasoconstriction of the arteries and ischemia of soft tissue. Lidocaine is one of several amide anesthetics that can be considered for this method of nerve block. The objective is to provide enough anesthetic agent around the neurovascular bundle without causing injury to the nerve or vessels. It is important to remember to allow enough time for the anesthetic agent to take effect, to get the full benefit of the anesthetic intervention (see Figure 8.1).

Most local anesthetic agents have an onset of activity of 10–30 minutes.

- Two percent lidocaine has an onset of 10–20 minutes and duration of 2–5 hours.
- 0.5% bupivacaine has an onset of 15–30 minutes and duration of 5–15 hours.

When significant crush injury or infection is present, this can impede anesthetic flow to the area, causing a longer time to onset or decreasing overall efficacy. Digital blocks are contraindicated in the setting of an infected injection site, compromised digit circulation, and known allergy to anesthetic agent.

Clean the digit in the usual sterile fashion with appropriate preparation and draping.

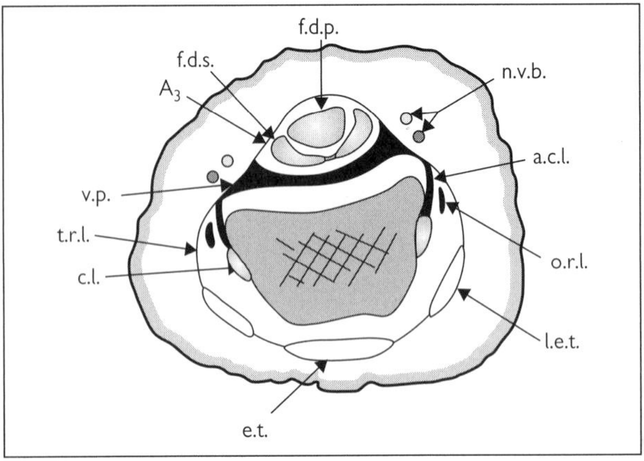

Figure 8.1 Finger anatomy—relationship of digital neurovascular bundle.
f.d.p. flexor digitorum profundus; f.d.s. = flexor digitorum superficialis; A3 = annular pulley; v.p. = volar plate; t.r.l. = transverse retinacular ligament; c.l.= collateral ligament; e.t. = extensor tendon; l.e.t.= lateral band; o.r.l. = oblique retinacular ligament; a.c.l. = accessory collateral ligament; n.v.b. = neurovascular bundle.

Gear List

- One to two percent lidocaine without epinephrine (if longer anesthesia is needed, consider using bupivacaine (0.25%–0.5%).
- Twenty-five- to twenty-seven-gauge needle
- Ten cc syringe

Method 1 (See Figure 8.2)

- Place the patient's hand palm side up. Identify the distal palmar crease in line with the involved digit.
- Palpate the flexor tendons to that digit at this level.
- Advance the needle through the skin at the midaspect of the flexor tendons down to the periosteum and then withdraw the needle 1–2 mm.
- Draw back on the syringe to ensure that the needle is not in an artery and then inject 5–10 cc of lidocaine (or about 5 ml of bupivacaine).
- The anesthetic will pass in the flexor sheath and should affect the radial and ulnar digital nerves.
- Wait at least 10 minutes before attempting a procedure or manipulation.

Method 2 (See Figure 8.3)

- Place the patient's hand palm side down.
- Holding the syringe perpendicular to the digit, advance the needle palmarly through the skin at the web space, just distal to the metacarpophalangeal joint.

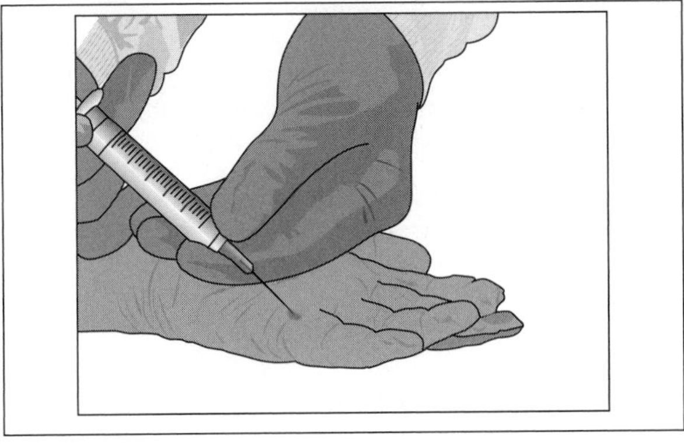

Figure 8.2 Digital block—volar.

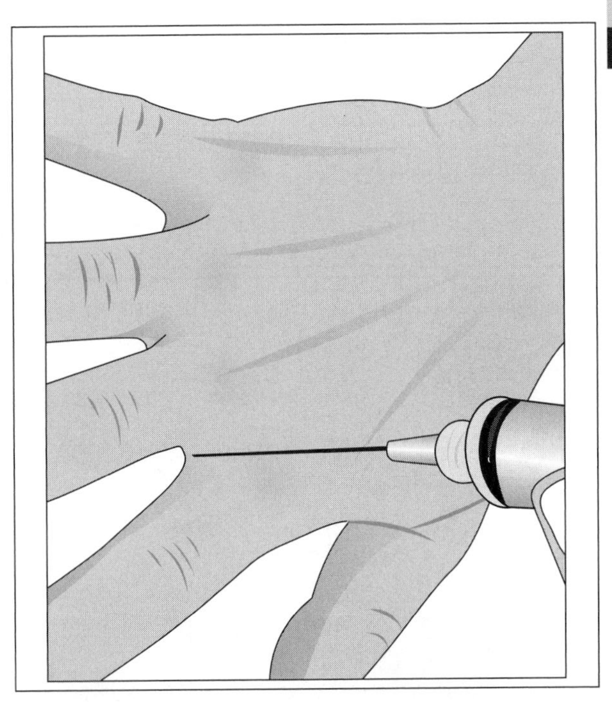

Figure 8.3 Digital block—dorsal.

- Draw back on the syringe to ensure the needle is not in an artery and then slowly inject up to 5 cc of lidocaine or bupivacaine as the needle is withdrawn.
- Repeat this procedure at the other web space of the involved digit; however, as the needle is slowly withdrawn, but before it is removed, advance the needle over the dorsum of the digit in a subcutaneous plane and inject lidocaine as the needle is slowly withdrawn to provide a dorsal "ring block" and anesthesize any dorsal crossing sensory branches.
- Wait at least 10 minutes before attempting a procedure or manipulation.

Hematoma Block (See Figure 8.4)

- The hematoma block is another effective means of achieving analgesia for the purpose of manipulation. The procedure is generally used for closed fractures, is relatively safe to do, and is straightforward to apply.
- When a bone is broken, a hematoma forms and surrounds the broken ends of bone. This block takes advantage of the hematoma by adding an anesthetic agent to the hematoma "bath". The anesthetic agent should be injected as close to the broken bone as possible to allow rapid absorption. Within about 10 minutes, adequate anesthesia can be achieved. The procedure is described below.

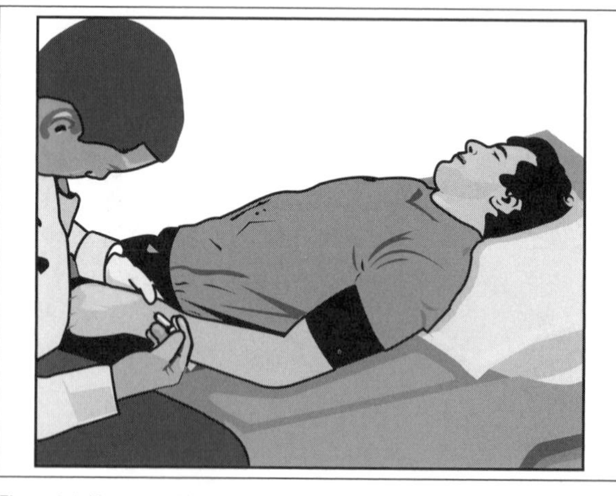

Figure 8.4 Hematoma block.

- Identify the fracture site and cleanse the skin overlying the fracture with an antiseptic solution such as chlorhexidine, alcohol, or iodine.
- Using a 10 cc syringe of 1%–2% lidocaine (or about 5 cc of bupivacaine) without epinephrine, and a fresh 20- or 22-gauge needle, aspirate around the hematoma to confirm proper placement of the needle within the hematoma. The hematoma can be evacuated to decrease volume and pressure at the fracture site.
- Instill up to 10 ml of lidocaine into the hematoma or close to the periosteoum. Wait 10 minutes before attempting manipulation. The hematoma block can be effective for hours.

Wrist Block (See Figure 8.5a)

A wrist block provides anesthesia in the distribution of the median, radial, and ulnar nerves at the wrist. With a successful wrist block, nearly any procedure in the hand can be performed without pain. For more localized procedures, blocking one or two of these three nerves may be adequate.

Median Nerve (See Figure 8.5b and 8.5c)

- Palpate the volar aspect of the distal forearm to identify the flexor carpi radialis (FCR) tendon (the most radial of the wrist flexor tendons) and the palmaris longus (PL) tendon just ulnar to the FCR.

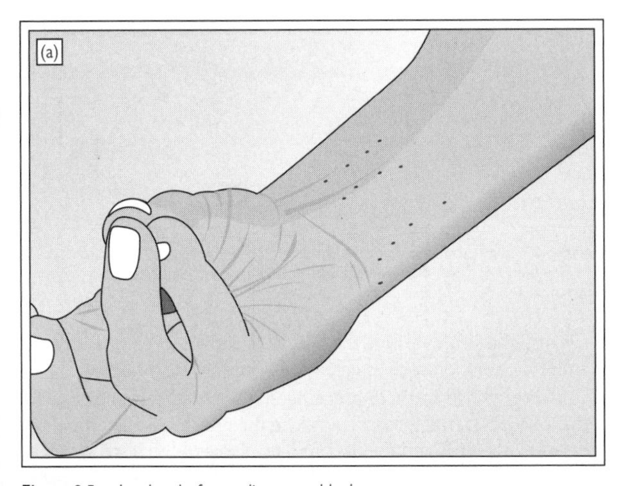

Figure 8.5 a. Landmarks for median nerve block.

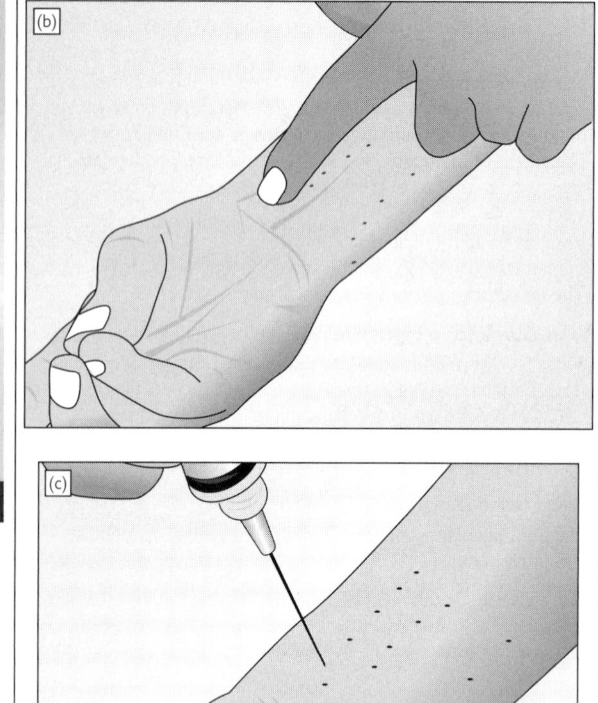

Figure 8.5 (Continued) b. Site for median nerve block. Dotted lines indicate the following tendons: flexor carpi radialis, palmaris longus, and flexi carpi ulnaris (from radial to ulnar side). c. Median nerve block.

- Cleanse the skin overlying this area at the distal forearm with an antiseptic solution such as chlorhexidine, alcohol, or iodine.
- With the syringe and a 25-gauge needle oriented perpendicular to the long axis of the forearm, advance the needle through the skin between the FCR and the PL 5 mm proximal to the distal wrist crease. You may feel a small "pop" as the needle passes into the carpal canal.

- Aspirate to ensure needle is not in a blood vessel and then inject 5–10 cc of anesthetic agent.

Ulnar Nerve (See Figure 8.6)

- Palpate the flexor carpi ulnaris (most ulnar of the wrist flexors) at the distal forearm.
- Cleanse the skin overlying this area with an antiseptic solution such as chlorhexidine, alcohol, or iodine.
- With the syringe and a 25-gauge needle oriented in a transverse plane parallel to and 1 cm proximal to the distal wrist crease, advance the needle through the skin and deep to the FCU tendon.
- Withdraw to confirm the needle is not in the ulnar artery and then inject 5–10 cc of anesthetic agent.

Radial Sensory Nerve (See Figure 8.7)

- Cleanse the skin in the dorsoradial aspect of the hand and wrist area with an antiseptic solution such as chlorhexidine, alcohol, or iodine.
- Using your nondominant index finger and thumb, elevate a pinch of skin at the dorsoradial aspect of the hand and wrist.
- Advance the syringe with a 25-gauge needle subcutaneously in the area of skin that you have pinched and slowly inject local anesthetic as the needle is withdrawn.

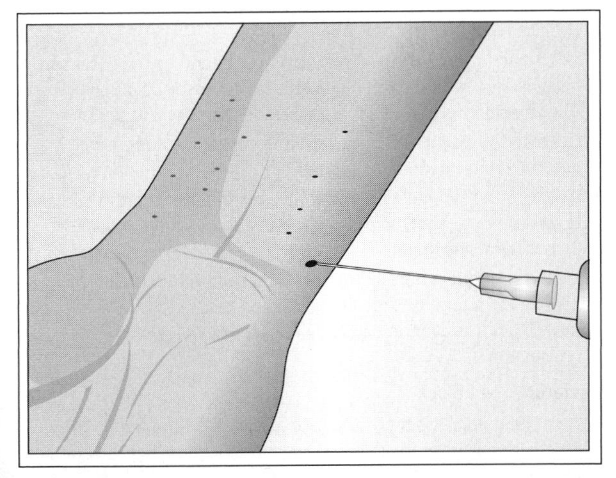

Figure 8.6 Ulnar nerve block.

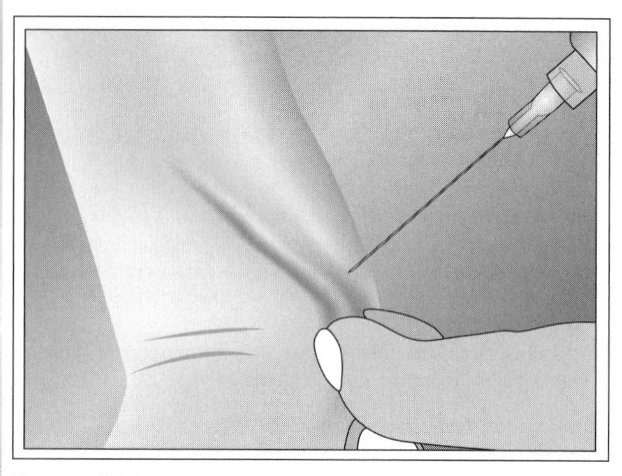

Figure 8.7 Radial sensory nerve block.

Ankle Block

Ankle blocks are an effective way to achieve analgesia for procedures involving the foot and ankle. The technique involves administering local anesthetic (lidocaine, bupivacaine or both) to the five nerves in the foot in a ring like distribution (see Figure 8.8).

Deep Peroneal Nerve

- With one finger, palpate the groove just lateral to the extensor hallucis longus (EHL) and medial to the extensor digitorum longus. The needle insertion site is just lateral to the dorsalis pedis artery.
- Cleanse the skin in this area with an antiseptic solution such as chlorhexidine, alcohol, or iodine.
- Advance the syringe with a 25-gauge needle through the skin and down to bone. Then, withdraw the needle 1–2 mm and inject 3 cc of local anesthetic.
- A fan technique whereby the needle is redirected 30 degrees medial and lateral to the initial direction can increase the success rate. Do not move the palpating finger; this will guide proper reinsertion.

Posterior Tibial Block

- Face the medial aspect of the foot and cleanse the skin in this area with an antiseptic solution such as chlorhexidine, alcohol, or iodine.

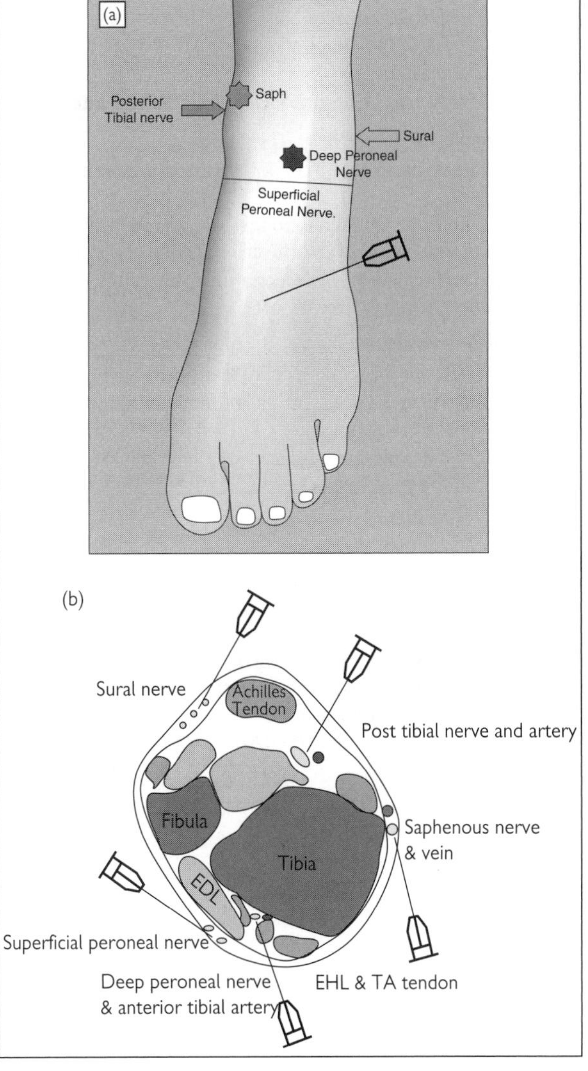

Figure 8.8 Ankle block. a. Skin anatomy for an ankle block. b. Cross-sectional anatomy for an ankle block. EDL, extensor digitorum longus; EHL, extensor hallucis longus; TA, tibialis anterior.

- Advance the syringe and 25-gauge needle in the groove behind the medial malleolus down to bone.
- The needle is then withdrawn 1–2 mm and 3 cc of local anesthetic is injected.
- A fan technique may be used to increase the success rate.

Sural Nerve Block

- The sural nerve passes 1–1.5 cm posterior to the lateral malleolus.
- This area should be identified and the skin cleansed with an antiseptic solution such as chlorhexidine, alcohol, or iodine.
- Nerve block is achieved by injecting a subcutaneous wheal with a 25-gauge needle at this level.

Superficial Peroneal Nerve Block

- Palpate the proximal lateral aspect of the foot.
- Cleanse the skin in this area with an antiseptic solution such as chlorhexidine, alcohol, or iodine.
- Nerve block is achieved by subcutaneous infiltration of local anesthetic at the lateral aspect of the foot.

Saphenous Nerve Block

- Palpate the medial malleolus.
- Cleanse the skin in this area and posteriorly to the Achilles tendon with an antiseptic solution such as chlorhexidine, alcohol, or iodine.
- Insert a 25-gauge needle at the level of the medial malleolus and inject local anesthetic subcutaneously from the medial malleolus posteriorly to the Achilles tendon and from the medial malleolus anteriorly to the tibial ridge. This may require two insertions of the needle.

Chapter 9

Miscellaneous Procedures

Joint Aspiration

Indications/Management

- Indicated to evaluate for septic joint in presentations of severe joint pain, limited ROM, exquisite pain with ROM, sometimes fever, erythema, effusion; also can evaluate for suspected crystal arthropathy
- PEARL—do not give antibiotics until after adequate aspirate obtained unless patient is septic
- Send aspirate for cell count with differential, gram stain to look for bacteria, crystals, aerobic, and anaerobic culture and sensitivity, and often also test for fungal or viral infection
- PEARL—do not aspirate through area of overlying cellulitis (due to risk of introducing infection into joint)
- PEARL—if patient has joint arthroplasty and infection is suspected, the orthopaedic surgeon should be consulted PRIOR to attempting any aspiration
- Aspirate is positive for infection if WBC > 100,000 with >75% polymorphonuclear cells (PMNs) and/or culture positive.
- Crystals suggest gout or calcium pyrophosphate deposition disease. Gout can cause WBC to be >100 K.
- If aspirate positive for infection, requires urgent (same day or early next AM) surgical I+D, IV antibiotics, splint immobilization

Gear Guide

- Chlorhexidine gluconate–based preparations or 10% povidone-iodine or 70% alcohol preparation swabs
- Sterile gloves
- Sterile fenestrated drape
- Sterile syringe (usually 20 cc, 10 cc for wrist or ankle) and needle (usually 20 gauge or larger)
- Tubes needed for lab tests on the joint fluid (cell count and differential, gram stain, culture and sensitivity, crystals)
- Lidocaine 1% as needed, 5–10 cc
- Sterile gauze pads and bandage

Aspiration of Specific Joints

Shoulder

- Patient sitting with arm resting comfortably at his or her side and forearm and hand so that shoulder is externally rotated
- Identify and palpate humeral head, coracoid process, and acromion.
- Prepare the area using sterile technique with chlorhexidine or betadine.

- ANTERIOR: Place needle just medial to humeral head and 1 cm lateral to coracoid process. The needle is directed posteriorly and slightly superolaterally. If the needle hits bone, it should be pulled back and redirected slightly.
- POSTERIOR (preferred): Place needle 2–3 cm inferior and 2 cm medial to the posterolateral corner of the acromion and direct anteriorly in the direction of the coracoid (see Figure 9.1).
- Aspirate and send joint fluid to lab for appropriate studies.
- Apply a clean, dry bandage.

Elbow

- Patient sitting with arm resting comfortably on table or Mayo stand
- Identify and palpate the triangle formed by the lateral olecranon, radial head, and lateral epicondyle (Figure 9.2). There is a soft spot in the center of this triangle.
- Prepare area using sterile technique with chlorhexidine or betadine.
- Insert needle into soft tissue between the lateral epicondyle, radial head, and lateral olecranon (Figure 9.3).
- If the needle hits bone, it should be pulled back and redirected slightly.
- Aspirate and send joint fluid to lab for appropriate studies.
- Apply a clean, dry bandage.

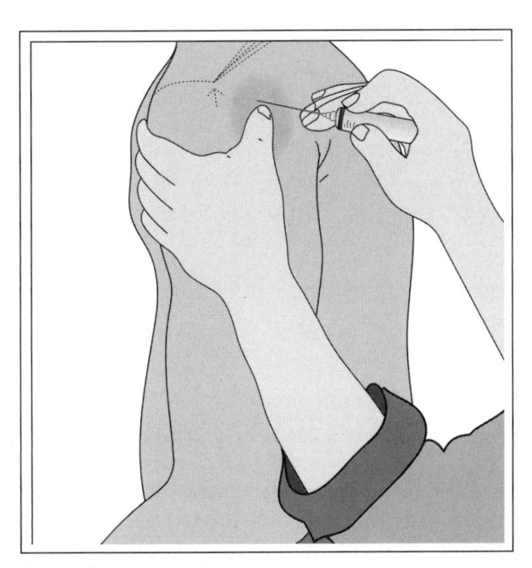

Figure 9.1 Posterior shoulder aspiration.

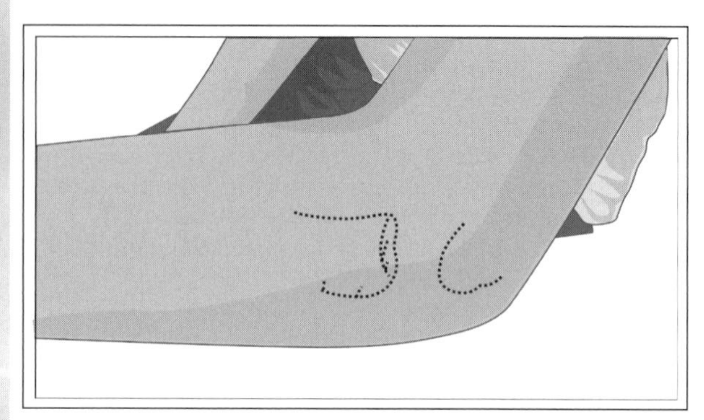

Figure 9.2 Elbow aspiration landmarks.

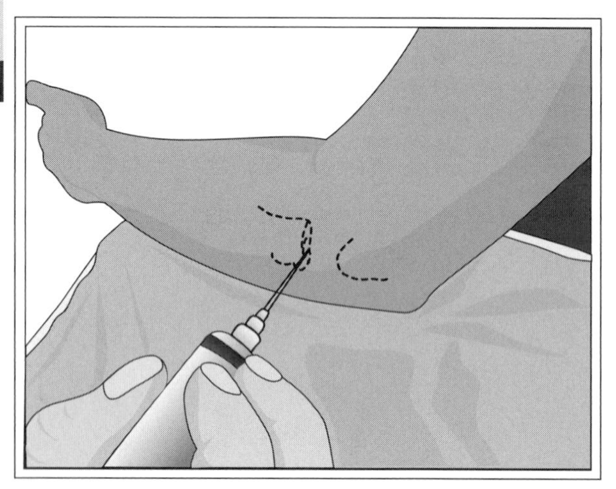

Figure 9.3 Elbow aspiration.

Wrist

- Patient sitting with arm resting comfortably on table or Mayo stand
- Identify and palpate Lister's tubercle at dorsoradial distal radius.
- Prepare area using sterile technique with chlorhexidine or betadine.
- Insert needle into soft spot 1 cm distal to Lister's tubercle—between the EPL and EDC tendons (Figure 9.4, Figure 9.5)

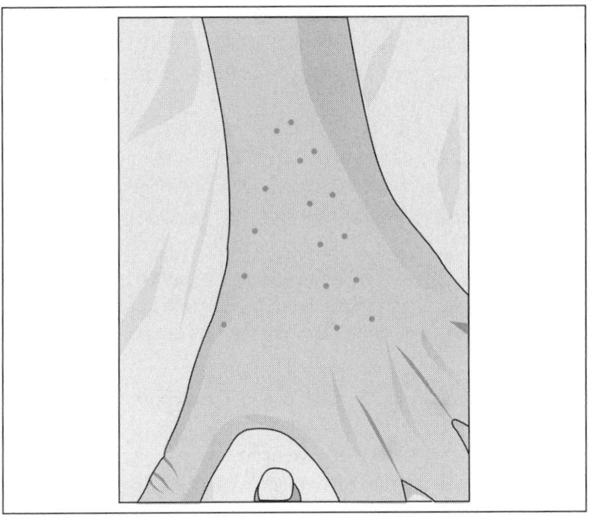

Figure 9.4 Dorsal wrist landmarks—EPL (third compartment) is radial to extensors to IF and MF (fourth compartment).

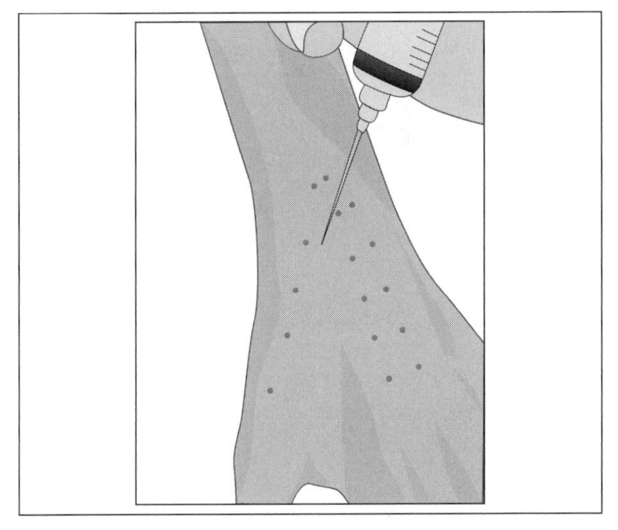

Figure 9.5 Wrist aspiration.

- Needle should be angled approximately 10 degrees proximally (to parallel distal radius)
- If the needle hits bone, it should be pulled back and redirected slightly.
- Aspirate and send joint fluid to lab for appropriate studies.
- Apply a clean, dry bandage.
- PEARL—May start with 1 cc sterile saline in syringe and inject and then aspirate to ensure enough fluid for analysis.

Hip

- Hip aspiration should be performed under fluoroscopic guidance by a radiologist or an orthopaedist if radiologist not available at your center and then joint fluid sent for appropriate studies.

Knee

- SUPRAPATELLAR (preferred if patient most comfortable with knee extended)
- Patient supine with the knee extended
- Identify and palpate superolateral pole of patella (Figure 9.6)
- Prepare area using sterile technique with chlorhexidine or betadine
- Insert needle 1 cm above and 1 cm lateral to this site. Needle should be directed 45 degrees distally and 45 degrees tilted below the patella (into the knee, toward intercondylar notch of femur) (Figure 9.7).

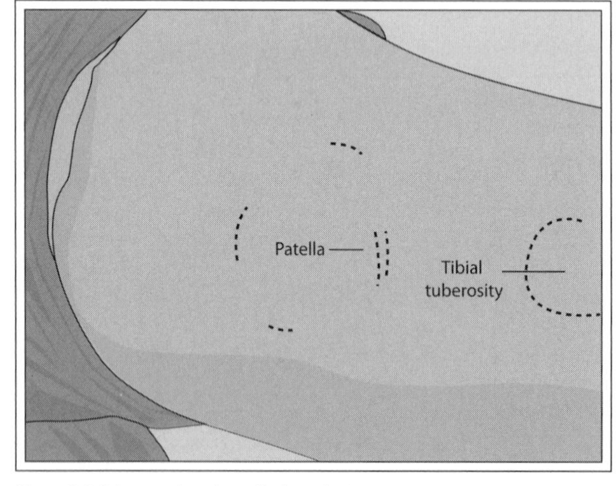

Figure 9.6 Palpate and mark patellar boundaries.

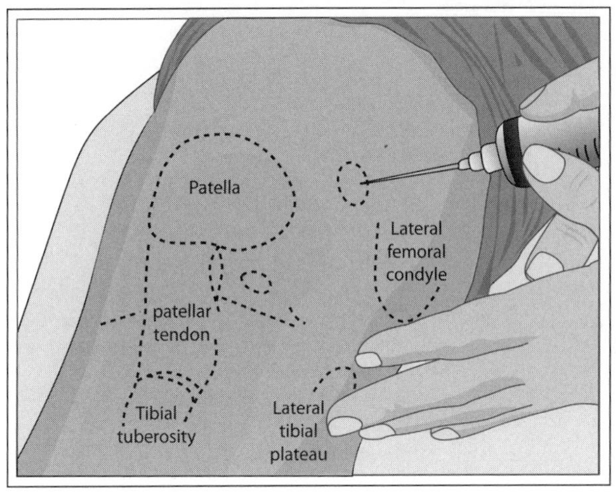

Figure 9.7 Suprapatellar knee aspiration.

- If the needle hits bone, it should be pulled back and redirected slightly
- Aspirate and send joint fluid to lab for appropriate studies
- Apply a clean, dry bandage.
- INFRAPATELLAR (preferred only if patient cannot straighten knee because fat pad may make aspiration difficult)
- Patient sitting upright with knee flexed 90 degrees
- Identify either side of inferior border of patella and the patellar tendon (Figure 9.8).
- Prepare area using sterile technique with chlorhexidine or betadine.
- Insert needle 5 mm below the inferior border of the patella and just lateral to the edge of the patellar tendon. Angle toward the center of the notch (center of knee) (Figure 9.9).
- If the needle hits bone, it should be pulled back and redirected slightly.
- Aspirate and send joint fluid to lab for appropriate studies.
- Apply a clean, dry bandage.
- PEARL—Insert needle through stretched skin (decreases pain of needle penetration through skin).
- PEARL—Use nondominant hand to compress opposite side of the joint or patella to help control the patella in the suprapatellar approach.

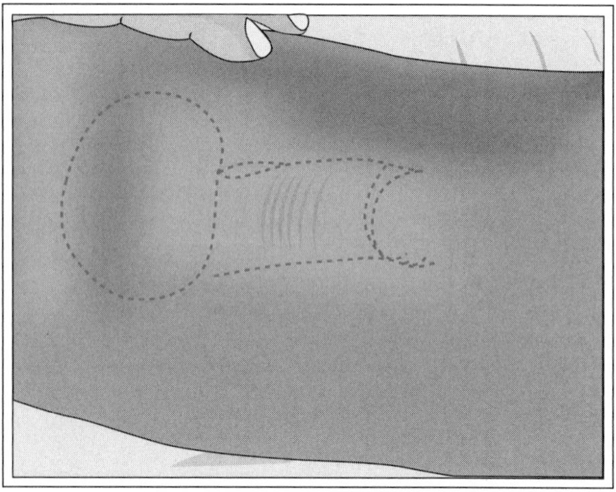

Figure 9.8 Knee landmarks.

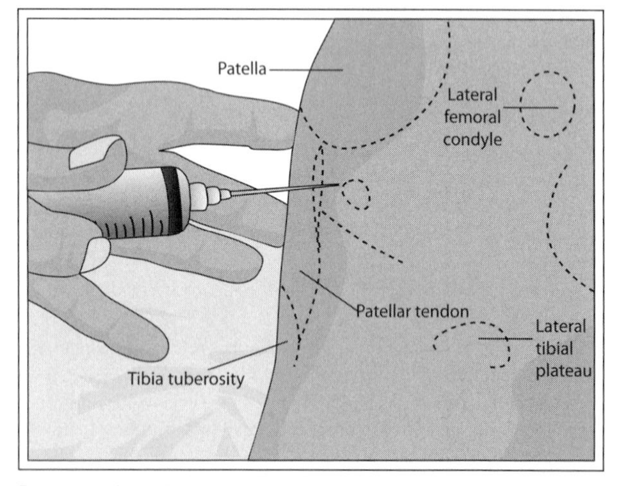

Figure 9.9 Infrapatellar knee aspiration.

- PEARL—For suprapatellar approach, can also start at midpoint of superomedial border of patella with needle directed to intercondylar notch of femur.

Ankle

- Patient supine or sitting with the knee flexed 90 degrees and with ankle relaxed
- ANTEROLATERAL (preferred)
  - Identify the ankle joint line, lateral malleolus, and lateral border of the extensor digitorum longus.
  - Active flexion/extension by the patient may help the physician identify the space between the base of the lateral malleolus and the lateral border of the extensor digitorum longus.
  - Prepare area using sterile technique with chlorhexidine or betadine.
  - Insert needle at the joint line midway between the base of the lateral malleolus and the lateral border of the extensor digitorum longus, advancing the needle perpendicular to the fibular shaft (Figure 9.10).
  - If the needle hits bone, it should be pulled back and redirected slightly.
  - Aspirate and send joint fluid to lab for appropriate studies.
  - Apply a clean, dry bandage.
- ANTEROMEDIAL

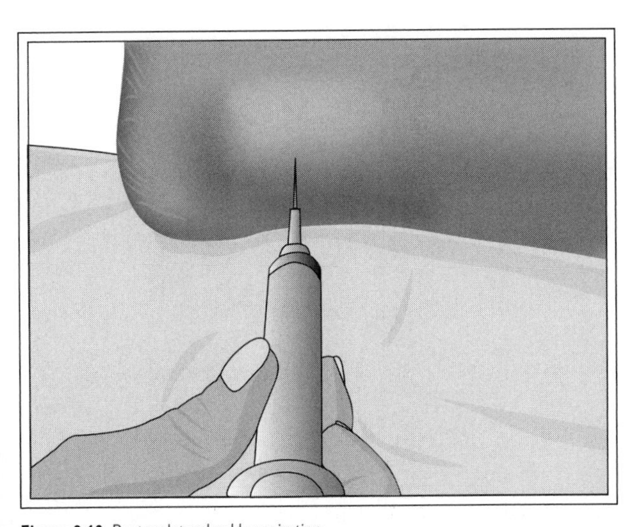

Figure 9.10 Posterolateral ankle aspiration.

- Identify and palpate the anterior border of the medial malleolus and the medial border of the tibialis anterior tendon and palpate in the space between these to identify the articulation between the tibia and the talus (Figure 9.11).

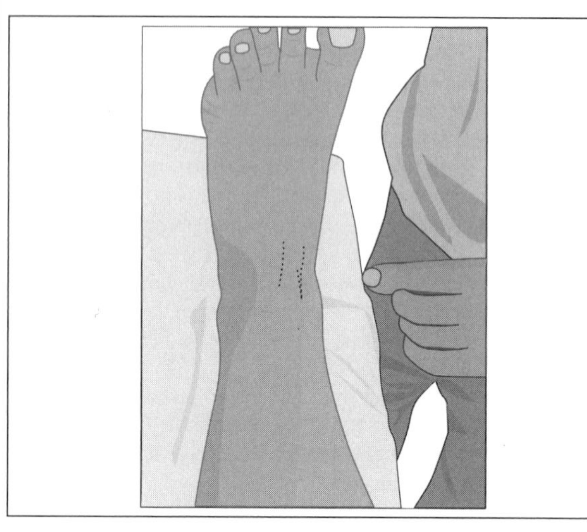

Figure 9.11 Anterior ankle landmarks.

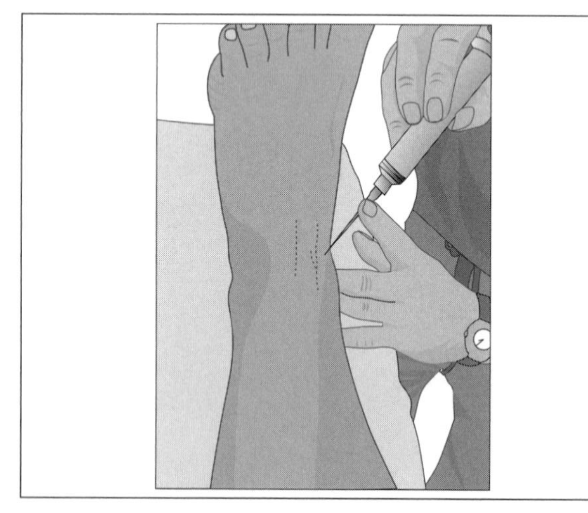

Figure 9.12 Anteromedial ankle aspiration.

- Prepare area using sterile technique with chlorhexidine or betadine.
- Insert needle into the joint space directed posterolaterally (Figure 9.12).
- If the needle hits bone, it should be pulled back and redirected slightly.
- Aspirate and send joint fluid to lab for appropriate studies.
- Apply a clean, dry bandage.

- PEARL—Exercise care with anteromedial approach to avoid injury to dorsalis pedis vessels or deep peroneal nerve (immediately below and lateral to extensor hallucis longus)

Nail Bed Repair (Figure 9.13a)

Gear Guide

- Betadine or betadine swabs or chlorhexidine
- Sterile gloves
- Sterile fenestrated drape
- Two sterile syringes (usually 10 cc)
- Sterile needle (usually 25 gauge)
- Lidocaine (no epinephrine)
- Sterile freer elevator if available
- Sterile narrow clamp—such as a Halsted
- Sterile hemostat

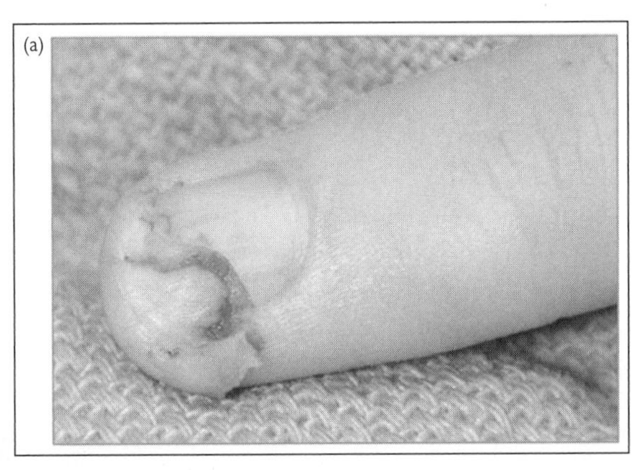

Figure 9.13 a. Nail bed injury.

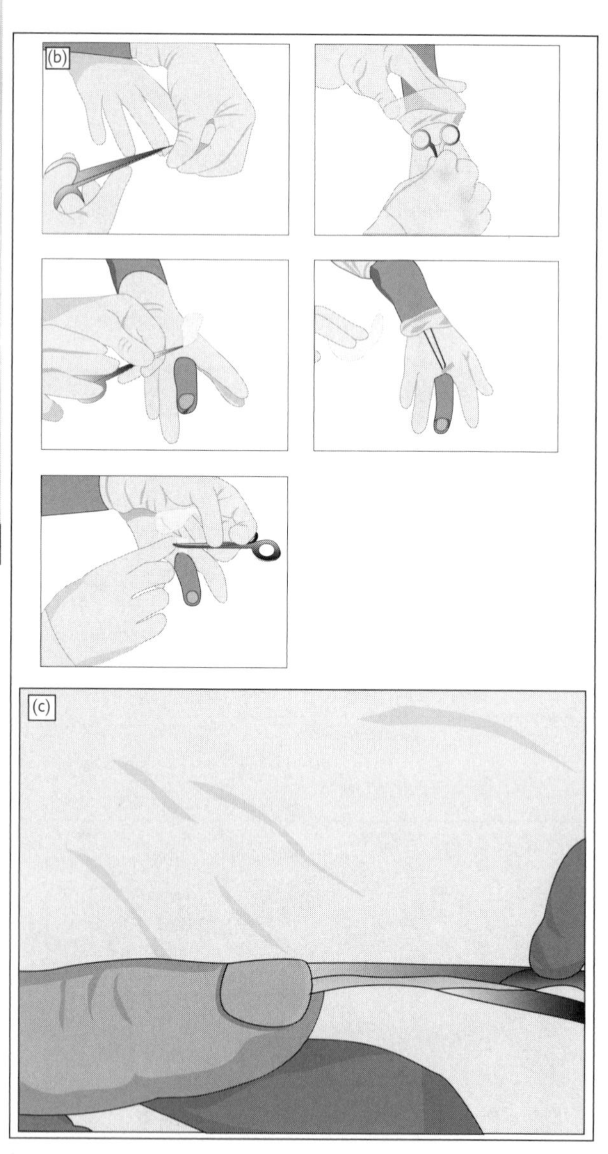

Figure 9.13 (Continued) b. Applying sterile glove as digital tourniquet. c. Elevating nail from nail bed with scissor blade.

- Sterile needle driver and forceps and suture scissor
- 6–0 or 7–0 chromic suture
- 5–0 nylon suture if nailfold laceration
- Sterile saline
- Nonadherent bandage—xeroform or adaptic
- Tube gauze/dressing materials

Technique

- Prepare the hand using sterile technique with betadine or chlorhexidine.
- Place patient's hand through sterile fenestrated drape.
- Draw up lidocaine in 10 cc syringe.
- Place digital block (see Chapter 8) with lidocaine (no epinephrine) at base of digit.
- Digital tourniquet (Figure 9.13b) will provide a bloodless field; place a slightly oversized sterile glove on the patient's hand.
- Pull distally on the glove tip on the operative finger and cut the tip with scissors.
- Roll the cut glove finger to the base of that digit (exsanguinates) and apply a hemostat to it near the base; the hemostat can be rotated to increase amount of compression as desired.
- Stabilize the handle of the hemostat under the glove cuff.

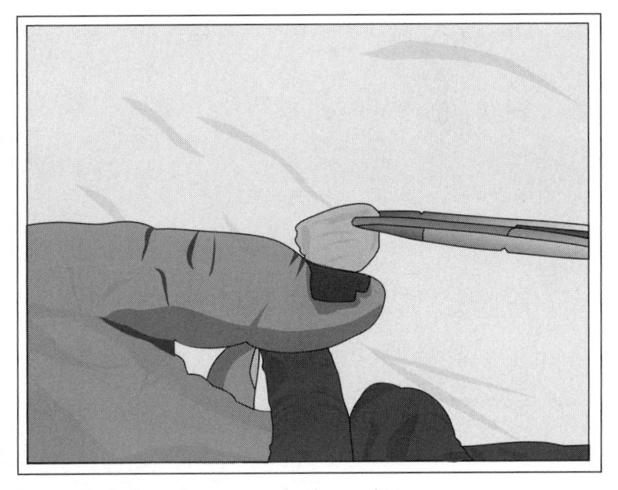

Figure 9.14 Rolling nail to remove nail without avulsion.

- Elevate nail from nailbed with freer elevator or small scissor blade (Figure 9.13c).
- Grasp nail distally with narrow clamp and apply distal traction to remove nail (can free from more proximal attachments with freer elevator or gentle blunt dissection with scissors); can roll nail with clamp to avoid avulsion of nailfold (Figure 9.14)
- Maintain nail on sterile drape/table.
- Irrigate nailbed with sterile saline using 10 cc syringe.
- Must ensure nailfold is maintained; if lacerated, repair with 5–0 nylon simple suture.
- Once nailbed laceration is well visualized, reapproximate nailbed with 6–0 or 7–0 chromic simple interrupted sutures (Figure 9.15); take care not to pull with excess force as nailbed is very friable.
- Irrigate with normal saline or sterile water.
- Maintain space beneath nailfold to prevent adhesions by placing xeroform, foil from suture wrapper (cut to appropriate size/shape), or deactivated fingernail (scrape deep side of the removed nail with sterile needle) with single hole at base (made with sterile needle, to avoid hematoma) under nailfold.
- Remove the hemostat to release the tourniquet
- Apply sterile nonadherent bandage with xeroform or adaptic, followed by sterile gauze and then tube gauze if available (or coban or tape).

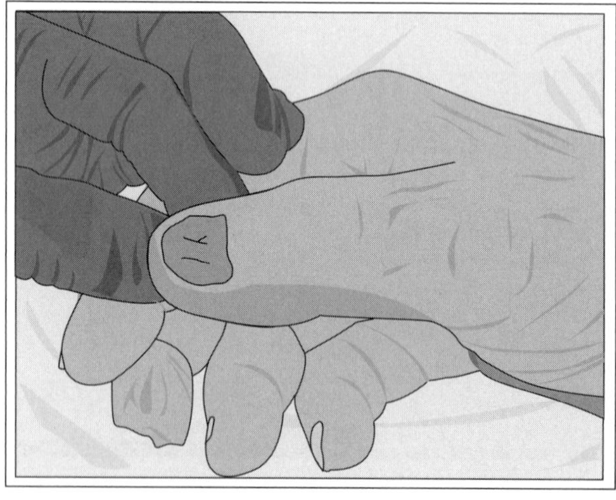

Figure 9.15 Repair of nail bed.

- PEARL—give IV ancef (or clindamycin if penicillin [PCN] allergy) before procedure and discharge with 5–7 days oral antibiotic (Keflex if not allergic)
- PEARL—always obtain radiographs of finger to evaluate for fracture.
- PEARL—apply fingertip splint if associated tuft fracture.
- PEARL—if fracture unstable, may need fixation with k-wire(s)
- PEARL—recent study shows dermabond effective in management of acute nailbed lacerations.

Superficial Lacerations

Gear Guide

- Betadine or betadine swabs
- Sterile gloves
- Sterile drape
- Two sterile syringes (usually 10 cc)
- Sterile needle (usually 25 gauge)
- Betadine swabs or chlorhexidine
- Sterile needle driver and forceps and suture scissor
- 3–0, 4–0, or 5–0 nylon suture (4–0 or 5–0 for hand/wrist, foot/ankle, forearm and 3–0 for leg or arm)
- Sterile saline
- Nonadherent bandage—xeroform or adaptic
- Dressing sponges

Technique

- Prepare the affected area using sterile technique with betadine.
- Place sterile drape(s) as appropriate.
- Draw up lidocaine in 10 cc syringe.
- Place local anesthetic (lidocaine) at area of laceration—can use field block—inject with 25-gauge needle subcutaneously in area of laceration.
- Irrigate laceration with sterile saline (can use 10 cc syringe as irrigator) and debride any foreign material or poorly viable tissue.
- Once adequately irrigated and debrided, proceed with reapproximation of skin edges with nylon simple interrupted sutures—enough tension to directly approximate skin edges but not too much tension, which would compromise tissue viability.
- Suture needle should be placed at 90 degrees to outer surface of skin and needle advanced through skin following contour of

curve of needle with pronation to supination motion of the hand advancing the needle driver.

- Enter the skin on the opposite side of the laceration with needle at 90 degrees to the skin as well.
- Needle placement in the skin should be the same distance from the laceration on both sides.
- Clean laceration site and dress with sterile nonadherent dressing with xeroform or adaptic and sterile gauze.
- PEARL—Consider splint for soft tissue rest if laceration at or close to a joint.
- PEARL—Discharge with 5–7 days oral antibiotic (Keflex if not allergic).

Tendon Lacerations (Except Hand Flexor Tendon Injuries, which Should Be Referred to a Hand Specialist)

Gear Guide

- Chlorhexidine or betadine or betadine swabs
- Sterile gloves
- Sterile fenestrated drape
- Two sterile syringes (usually 10 cc)
- Sterile needle (usually 25 gauge)
- Lidocaine
- Sterile needle driver and forceps and suture scissor
- 3–0 or 4–0 ticron or ethibond suture
- 4–0 nylon suture
- Sterile saline
- Nonadherent bandage—xeroform or adaptic
- Dressing sponges
- Sterile Sof-Roll
- Plaster for splint
- Ace bandage

Technique

- Prepare the affected area using sterile technique with chlorhexidine or betadine.
- Place sterile drape(s) as appropriate.
- Draw up lidocaine in 10 cc syringe.
- Place local anesthetic (lidocaine) at area of laceration—can use field block—inject with 25-gauge needle subcutaneously in area of laceration.

- Irrigate laceration with sterile saline (can use 10 cc syringe as irrigator) and debride any foreign material or poorly viable tissue.
- Retract skin edges to facilitate visualization of tendon laceration.
- Identify proximal and distal cut ends of tendon; debride sharply to clean edges.
- Position limb or digit to minimize tension across repair site; for example, extend wrist and fingers if extensor tendon laceration on hand.
- Place 3–0 or 4–0 ticron or ethibond suture as a core suture at one side of cut end of tendon and advance longitudinally to exit at side of tendon, then advance transversely across tendon at that level to other side, then reenter tendon on this side and advance longitudinally as a core suture to exit at laceration site (see Figure 9.16).
- Holding tendon approximated at laceration site, advance needle/suture as a core suture at same side of second half of cut tendon, cross tendon transversely at that level to opposite side, then reenter tendon on this side and advance longitudinally as a core suture to exit at laceration site.
- With tendon edges approximated, tie suture across repair site (see Figure 9.16).
- For larger tendons, can augment with horizontal mattress suture or running epitendinous suture (see Figure 9.16).

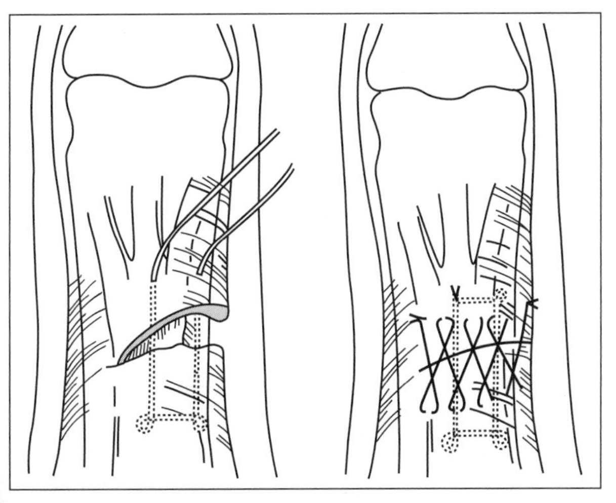

Figure 9.16 Extensor tendon repair.

- Avoid tension across repair site once suture is tied.
- Irrigate with sterile saline (can use 10 cc syringe as irrigator).
- Close skin with 4–0 nylon simple interrupted suture.
- Clean laceration site and dress with sterile nonadherent dressing with xeroform or adaptic and sterile gauze.
- Apply sterile Sof-Roll and then plaster splint to minimize tension on tendon repair site; for example, for wrist extensor tendon repair, splint with wrist in 30 degrees dorsiflexion.
- PEARL—As finger extensors share a common muscle belly, should splint all fingers in extension after finger extensor tendon repair
- PEARL—Additional options for core suture configuration are illustrated in Figure 9.17.

Stitches and Knots (See Suture Table in Chapter 10)

- Modified Kessler suture
- Bunnell suture
- Horizontal mattress suture

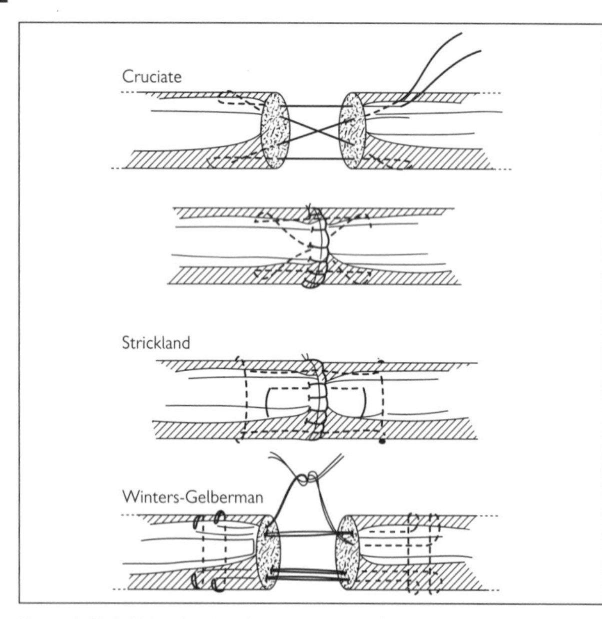

Figure 9.17 Additional options for core suture configuration.

Chapter 10

Quick Reference Guide

Tables

Table 10.1 Indications for Splints

Splint	Joints immobilized	Injuries	Page
Coaptation	Shoulder, elbow	Humeral shaft fracture	22
Posterior slab with buttress	Elbow, wrist	Wrist and forearm fractures	32
Long arm cast	Elbow, wrist, CMC	Elbow and forearm fractures	51
Intrinsic plus	Wrist, CMC, MCP, PIP, DIP	Carpal, metacarpal, and phalanx fractures	37
Radial gutter +/– thumb spica	Wrist, 1st MCP, PIP	Carpal, metacarpal, and phalanx fractures	42, 44
Ulnar gutter	Wrist, MCP, PIP	Carpal, metacarpal, and phalanx fractures	40
Short arm cast	Wrist, CMC	Distal forearm and wrist fractures	46
Volar slab/wrist cock-up	Wrist, can do MP, PIP, DIP depending on length	Carpal, metacarpal, central finger injuries	33
Bulky Jones splint for the ankle	Ankle, hindfoot, midfoot	Ankle, calc, talus, midfoot, forefoot injuries	60
Long leg bulky	Knee, ankle, hindfoot, midfoot	Tibia, distal femur, knee injuries	55
Long leg cast	Knee, ankle, hindfoot,	Tibia, distal femur, knee injuries in children	62
Short leg cast	Ankle, hindfoot, midfoot	Ankle, calc, talus, midfoot, forefoot injuries in children	67

Table 10.2 Upper Extremity Injuries , Recommended Radiographs and Splints

Injury	Radiographs	Splint	Page
Clavicle fractures	AP and 15 degree cephalad oblique	Sling	75
Acromioclavicular separations	AP chest, scapular-Y, and axillary	Sling	76
Sternoclavicular separations	Serendipity view, cephalic tilt AP	Sling	78
Scapular fractures	AP shoulder, scapular-Y, and axillary	Sling	79

Table 10.2 (Continued)

Injury	Radiographs	Splint	Page
Glenoid fractures	AP shoulder, Grashey, and axillary	Sling or shoulder immobilizer	80
Proximal humerus fractures	AP shoulder, Grashey, and axillary	Sling or shoulder immobilizer	85
Shoulder dislocations	AP shoulder, Grashey, and axillary	Sling or shoulder immobilizer	82
Physeal injuries of the proximal humerus	AP shoulder, Grashey, and axillary; ultrasound in newborns	Sling	87
Humeral shaft fractures	AP and lateral of the humerus	Coaptation	88
Distal humerus fractures in adults	AP and lateral of the elbow	Posterior long arm splint	91
Supracondylar elbow fractures in children	AP and lateral of the elbow	Bivalved long arm cast if no surgery; posterior long arm splint if surgery	94
Medial condyle fractures in children	AP and lateral of the elbow	Bivalved long arm cast if no surgery; posterior long arm splint if surgery	97
Lateral condyle fractures in children	AP and lateral of the elbow	Bivalved long arm cast if no surgery; posterior long arm splint if surgery	99
Medial epicondylar apophyseal fractures in children	AP and lateral of the elbow	Posterior long arm splint	100
Elbow dislocations	AP and lateral of the elbow	Posterior long arm splint	101
Olecranon fractures	AP and lateral of the elbow	Posterior long arm splint	104
Radial head fractures in adults	AP and lateral of the elbow	Posterior long arm splint	105
Radial head and neck fractures in children	AP and lateral of the elbow	Bivalved long arm cast if no surgery; posterior long arm splint if surgery	108
Radial head subluxations	Clinical diagnosis	No immobilization needed	109

Table 10.2 (Continued)

Injury	Radiographs	Splint	Page
Proximal ulna fracture and radial head dislocation	AP and lateral of the elbow	Posterior long arm splint	112
Shaft fractures of the radius and/or ulna in adults	AP and lateral of the forearm	Sugar tong splint	115
Shaft fractures of the radius and/or ulna in children	AP and lateral of the forearm	Bivalved long-arm cast	118
Distal-third radius fractures and radioulnar dislocation	AP and lateral of the forearm	Sugar tong splint	120
Distal radius fractures in adults	AP, lateral, and oblique of the wrist	Sugar tong splint	124
Distal radius fractures in children	AP, lateral, and oblique of the wrist	Bivalved short or long-arm cast	129
Scaphoid fractures	Scaphoid series	Thumb spica splint	132
Other carpal fractures	AP, lateral, and oblique of the wrist	Short arm intrinsic plus splint or short arm volar resting splint	135
Carpal dislocations and ligamentous injuries of the wrist	AP, lateral, and oblique of the wrist and clenched-fist PA view	Sugartong splint +/– thumb spica (if scaphoid fracture)	137
Thumb fractures and dislocations	AP, lateral, and oblique of the thumb	Short arm thumb spica splint	141
Metacarpal fractures	AP, lateral, and oblique of the hand	Short arm intrinsic plus splint	144
Phalanx fractures	AP, lateral, and oblique of the finger		147
Finger dislocations	AP, lateral, and oblique of the finger		149

Table 10.3 Lower Extremity Injuries, Recommended Radiographs and Splints

Injury	Radiograph	Splint	Page
Anterior and posterior hip dislocations	AP/lateral hip	Bedrest, traction	169
Femoral head fractures	Ap/lateral hip (CT commonly)	Bedrest	172
Slipped capital femoral epiphysis	AP/frog leg lateral hip	Bedrest	174

Table 10.3 (Continued)			
Injury	**Radiograph**	**Splint**	**Page**
Femoral neck fractures	AP/lateral hip	Bedrest	176
Intertrochanteric fractures	AP/lateral hip	Bedrest	179
Subtrochanteric fractures	AP/lateral femur	Traction	181
Femoral shaft fracture	AP/lateral femur	Traction	183
Distal femoral shaft (supracondylar) fractures	AP/lateral knee or femur	Long leg bulky splint	186
Quadriceps tendon rupture	AP/lateral knee (clinical dx [diagnosis])	Knee immobilizer	189
Patella fracture	AP/lateral knee	Knee immobilizer	191
Patella dislocation	AP/lateral knee (clinical dx or MRI if already reduced)	Knee immobilizer	193
Patellar tendon rupture	AP/lateral knee	Knee immobilizer	195
Acute ACL, PCL, PLC (posterior lateral corner), MCL or LCL rupture	AP/lateral knee AP/ lateral knee (clinical dx or MRI if unsure)	Knee immobilizer	196
Knee dislocations	AP/lateral knee, MRI	Long leg bulky splint	197
Acute meniscal tear vs. bucket handle, locked knee	AP/lateral knee, MRI	Knee immobilizer	200
Tibial spine avulsion fracture	AP/lateral knee	Knee immobilizer vs. extension long leg casting	201
Tibial tuberosity avulsion Fracture	AP/lateral knee	Knee immobilizer vs. extension long leg casting	202
Salter-Harris fracture of proximal tibia	AP/lateral knee	Long leg cast	205
Tibial plateau fractures	AP/lateral knee, CT	Long leg bulky splint	206
Tibial shaft fractures	AP/lateral tib-fib	Long leg bulky splint	209
Plafond or pilon fractures	AP/lateral/mortise ankle	Short leg bulky splint	212

Table 10.3 (Continued)			
Injury	**Radiograph**	**Splint**	**Page**
Fibular shaft fractures	AP/lateral tib-fib	None vs. boot	214
Fibular head dislocation	AP/lateral knee	Long leg bulky splint vs. knee immobilizer	216
High ankle sprains	AP/lateral/mortise ankle + AP/lateral knee	Short leg bulky splint	217
Maisonneuve	AP/lateral/mortise ankle + AP/lateral knee	Short leg bulky splint	219
Ankle fractures	AP/lateral/mortise ankle	Short leg bulky splint	220
Triplane fractures	AP/lateral/mortise ankle	Short leg cast	225
Tillaux fractures	AP/lateral/mortise ankle	Short leg cast	225
Ankle dislocations	AP/lateral/mortise ankle	Short leg bulky splint	227
Ligamentous injuries of the ankle	AP/lateral/mortise ankle	Aircast vs. short leg bulky splint	230
Talus fractures	AP/lateral/oblique foot	Short leg bulky splint	233
Talus/subtalus dislocations	AP/lateral/oblique foot	Short leg bulky splint	236
Calcaneus fractures	AP/lateral/oblique foot + Harris heel view	Short leg bulky splint	237
Tarsal bone fractures	AP/lateral/oblique foot	Short leg bulky splint	241
Navicular fractures	AP/lateral/oblique foot	Short leg bulky splint	241
Lisfranc fractures	AP/lateral/oblique foot	Short leg bulky splint	244
Metatarsal fractures (1–4)	AP/lateral/oblique foot	Short leg bulky splint	246
Fifth metatarsal fractures	AP/lateral/oblique foot	Short leg bulky splint	248
Great toe injuries	AP/lateral/oblique foot	Postop shoe	250
Lesser toe injuries	AP/lateral/oblique foot	Postop shoe	252
Sesamoid fractures	AP/lateral/oblique foot	Postop shoe	253

Table 10.4 Radiographs for Pelvic Injuries, Recommended Radiographs and Splints

Injury	Radiograph	What the X-ray shows	Page
Pelvic ring injury	AP pelvis	Pubic symphysis, bony anatomy, some SI joint	263
Pelvic ring injury	Inlet	SI, sacrum	263
Pelvic ring injury	Outlet	Sacral foramen, pubic symphysis	263
Acetabulum	AP pelvis	Bony anatomy, femoral head	266
Acetabulum	Judet internal (obturator) oblique	Anterior column, posterior wall	266
Acetabulum	Judet external (iliac) oblique view	Posterior column, anterior wall	266

Table 10.5 Radiographs for Spine Injuries.

C-spine	AP	Bony anatomy C2–T1
C-spine	Open mouth odontoid	Bony anatomy C1–C2
C-spine	Lateral	Bony anatomy cranium to T1
C-spine	Swimmers	Improved visualization C7–T1
C-spine	Flexion extension lateral	Dynamic stability
T-spine	AP	Bony anatomy T1–T12
T-spine	Lateral	Bony anatomy T1–T12
T-spine	Flexion extension (rarely done)	Dynamic stability T-spine
L-spine	AP	Bony anatomy L1-sacrum
L-spine	Lateral	Bony anatomy L1-sacrum
L-spine	Spot	L4–S1 magnified view
L-spine	Flexion extension	Dynamic stability of L-spine

Table 10.6 Types of Sutures

Suture	Material	Type	Use	No. of Knots
Prolene	Polypropylene	Nonabsorbable monofilament	Subcuticular closure, infected or contaminated wounds	4–5
Ethilon	Nonbraided nylon	Nonabsorbable monofilament	Skin, nerve repair	4–5
Silk	Silk	Nonabsorbable braided polyfilament	Vessel ligation	4–5
Ethilon	Braided nylon	Nonabsorbable braided polyfilament	Skin	4–5
Monocryl	Polyglecaprone 25	Absorbable monofilament	Subcuticular closure, soft tissue, vessel ligation	4–5
PDS	Polydioxanone	Absorbable monofilament	Soft tissue esp. in pediatrics, plastics	4–5
Plain gut	Mucosa of sheep intestines	Absorbable	Skin, superficial vessel ligation	4–5
Chromic gut	Plain gut tanned with chromic salt	Absorbable	Nailbed repair, securing skin grafts	4–5
Vicryl, ethicon	Polyglactin-910	Absorbable braided polyfilament	Subcutaneous soft tissue approximation, vessel ligation, tendon repair	3–4
Dexon	Polyglycolic acid	Absorbable braided polyfilament	Subcutaneous soft tissue approximation, vessel ligation, tendon repair	3–4
Ethibond/ticron	Polyester fiber	Nonabsorbable multifilament	Vessel anastomoses, tendon repair	4–5

Monofilament—Relatively more resistant to harboring microorganisms, so preferred in potentially contaminated tissues

Polyfilament braided—Holds knots better. More susceptible to harboring infection.

Table 10.7 **Antibiotic Recommendations. (*Tetanus vaccination for Everyone Who Has Not Had a Documented Booster*)**

Injury	Antibiotic recommendation	Duration
Type I and II open fracture	Cefazolin 1g IV Q8H	24 h after closure
Type III open fracture	Same as Type I and II + Gentamicin 3–5 mg/kg/day or Levofloxacin 500 mg/day	72 h after closure
Open fracture with barnyard exposure	Same as Type III + penicillin 2 million units Q4H or Metronidazole 500 mg Q6H	72 h after closure
Laceration	Many lacerations do not require antibiotics. Clinical judgment of need based on injury pattern/mechanism, consider abx for large wounds, crush wounds or infected, or contaminated wounds; Keflex 500 mg PO QID or clindamycin 600 mg PO TID	3–7 days
Nailbed injury	Clinical judgment of need based on injury pattern/mechanism; Keflex 500 mg PO QID or clindamycin 600 mg PO TID	3–7 days
Human bite	Amoxicillin/clavulanate 875/125 mg PO BID or moxifloxacin 400 mg PO QD if penicillin allergic	7 days
Animal bite	Rabies prophylaxis if needed + amoxicillin/clavulanate 875/125 mg PO BID or moxifloxacin 400 mg PO QD if penicillin allergic	7 days
Tendon laceration	Keflex 500 mg PO QID or clindamycin 600 mg PO TID	3–7 days/until follow-up
Shoe puncture injury	Ciprofloxacin 750 mg PO BID or TMP/SMX 15–20 mg/kg/day for pediatric patients	3 days past inflammation resolution/until follow-up

Table 10.8 Opiate Characteristics				
Medication	**Strength**	**Equivalent oral dose to 5 mg Oxycodone**	**Freq.Q-hr**	**t1/2(hr)**
Morphine	Strong	10 mg	3–4	2–3
Hydromorphone	Very strong	1.25	3–4	2–4
Fentanyl	Very strong	2.5 mcg/hr	1–2	3–4
Methadone	Strong	3.33 mg	6–8	21–25
Oxycodone	Moderate	5 mg	4–6	3–4
Hydrocodone	Mild–mod	5.55 mg	3–4	3–4
Codeine	Weak	20 mg	3–4	2.5–4

Glossary

- Angulation: The degree and direction of the distal segment compared to the proximal segment
- Apex: The tip of the fractured bone on the convex side (opposite to the concave side of the fracture). Apex is used to describe a fracture's deformity, for example, if the tip of the fracture is pointing anterior, the fracture would be described as apex anterior.
- Bimalleolar: Fracture of both the medial and lateral malleoli in an ankle fracture. This is an unstable ankle fracture by definition as there is widening of the ankle mortise.
- Bimalleolar equivalent: Fracture of the lateral malleolus and disruption of the syndesmotic ligament between the tibia and the fibula, resulting in widening of the ankle mortise. This is an inherently unstable fracture.
- Bivalve: Cut a cast on two sides to allow the cast to be opened and permit swelling.
- Buckle (fracture): Fracture type most commonly seen in children whereby the bone breaks on one side but buckles outward on the opposite side without fully breaking; synonymous with torus fracture
- Butterfly: A triangle or wedge that has broken off at the site of the fracture
- Comminuted: Multiple fracture lines
- Compression: Side of the fracture where the bone has been pushed together or compressed; opposite of tension
- Diaphyseal/Diaphysis: Relating to the shaft of a long bone
- Diastasis: Separation of two normally attached bones between which there is no joint, for example, separation of the epiphysis from a long bone
- Dislocation: Complete displacement of a joint from normal anatomic position
- Displacement: The amount of translation one segment has compared to the other. Fractures can be nondisplaced (should be hard to see on radiograph), minimally displaced, or displaced. Amount of displacement can be described by the percent of the width of the bone in question.
- Dorsal: Relating to the back or posterior of a structure; opposite of ventral
- Epiphyseal/Epiphysis: Relating to the end of a long bone; area of growth
- External fixation (Ex-fix): Mechanism of fracture or joint immobilization using pins into the bone attached to each other by means of a frame

- Extra-articular: Outside of the joint
- Fat pad sign: Seen when a fat pad is elevated off the bone by swelling of the periosteum due to fracture, such that a more radiolucent structure is seen adjacent to the bone, indicating a fracture that may not otherwise be seen
- Greenstick: Incomplete fracture of a bone whereby one side of the bone is broken while the other side is bent
- Intra-articular: Extending into or involving a joint
- Involucrum: Layer of new bone that forms around an area of pyogenic osteomyelitis
- Malunion: Incomplete or faulty union of a fracture
- Metaphyseal/Metaphysis: Wider part of the long bone between the diaphysis and the epiphysis
- Mortise: Refers to a rectangular cavity in a piece of wood; in orthopaedics, used to describe the box formed by the distal tibia around the talus. Radiologically, used to describe the specific X-ray view in which the articular space between tibia and talus are best visualized without overlap of the bones.
- Non-Union: A fracture that has failed to heal appropriately; may be hypertrophic (associated with new bone growth) or atrophic (lacking new bone growth)
- Oblique: Slanted, sloped, or on an incline
- Overwrap: Application of a new layer of fiberglass over an existing bivalved cast; seals the bivalve of the cast
- Physeal/Physis: Area of bone between the metaphysis and the epiphysis in which cartilage grows
- Plastic deformation: Permanent change in the shape or size due to a deforming force exerted on an object
- Pronation: Rotation of the hands or feet so that the palm or sole faces downward
- Radiolucent: Permits the penetration of X-rays; shows up on radiographs as black (e.g., air)
- Radio-opaque: Prohibits the penetration of X-rays; shows up on radiographs as white (e.g., bone)
- Reduction: Correction of a fracture into its anatomic position
- Salter-Harris: Fracture involving the growth plate or physis; classification scheme for physeal fractures
- Segmental: Fractures above and below a segment of a bone such that segment is free floating
- Simple: One fracture line
- Sprain: Stretch or tearing injury to the ligaments
- Strain: Stretch or tearing injury to muscles or tendons
- Subluxation: Partial displacement of a joint from normal anatomic position

- Sequestrum: Piece of dead bone separated from surrounding bone by a wall of new bone or the involucrum
- Splint: A rigid device designed to immobilize; noncircumferential to allow for swelling
- Supination: Rotation of the hands or feet so that the palm or sole faces upward
- Tension: Side of the fracture where the bone has been pulled apart and is under tension; opposite of compression
- Tolerance: The limit of acceptable reduction; reduction to a position less than a fracture's tolerance (e.g., 20 degrees of angulation where the tolerance is 10 degrees) requires additional treatment for optimal result. Additional treatment may be operative or nonoperative.
- Torus: Fracture type most commonly seen in children whereby the bone breaks on one side but buckles outward on the opposite side without fully breaking; synonymous with buckle
- Trimalleolar: Fracture that involves the medial, lateral, and posterior malleoli of the ankle
- Varus: Displacement toward the midline of the body
- Valgus: Displacement away from the midline of the body
- Ventral: Relating to the front or anterior of a structure; opposite of dorsal
- Volar: Relating to the palm of the hand or the sole of the foot

Eponyms

- Baby Bennett's fracture: Articular fracture of the fifth metacarpal base; also called a reverse Bennett fracture
- Bankart lesion: Avulsion of the anteroinferior glenoid labrum, usually secondary to an anterior shoulder dislocation; if there is avulsion of labrum and the glenoid bone, the lesion is termed a "bony Bankart."
- Barton fracture: Displaced articular fracture of the distal radius typically associated with carpal dislocation; may be volar or dorsal
- Bennett's fracture: Oblique intra-articular fracture of the volar lip of the first metacarpal base
- Boxer's fracture: Fracture of the fifth metacarpal neck with volar displacement of the metacarpal head
- Buckle fracture: Impaction fracture of a child's bone whereby the bone buckles or bends but does not break completely; also called a Torus fracture
- Burst fracture: Fracture of the anterior and middle column of a vertebral body due to an axial load; the fragments of the vertebral

body are typically displaced; may occur in cervical, thoracic, or lumbar vertebrae.

- Chance fracture: Transverse fracture of the vertebral body and spinous process with horizontal disruption and flexion displacement; most commonly occurs in the thoracolumbar vertebrae
- Chauffeur's fracture: Oblique fracture of the radial styloid; also called a Hutchinson's fracture
- Chopart's fracture: Fracture dislocation of the talonavicular and calcaneocuboid joints (a.k.a. Chopart's joints) of the foot
- Clay shoveler's fracture: Spinous process fracture of the lower cervical or upper thoracic vertebrae
- Colles fracture: Dorsally displaced distal radius fracture
- Compression Fracture: Fracture of the anterior column caused by an axial compressive force
- Cotton's fracture: Trimalleolar ankle fracture—fracture of the medial malleolus, lateral malleolus, and posterior malleolus
- Die punch fracture: Intra-articular distal radius fracture with dorsal impaction of the distal radius at the lunate fossa
- Essex-Lopresti's fracture: Radial head fracture and dislocation of the distal radioulnar joint
- Galeazzi fracture: Distal radial shaft fracture with dislocation of the distal radioulnar joint; also called a Piedmont fracture
- Gamekeeper's thumb: Disruption of the ulnar collateral ligament of the thumb; may have an associated avulsion fracture
- Greenstick fracture: Incomplete fracture of a bone in a child with fracture of the tension side of the bone and intact cortex and periosteum on the compression side of the fracture
- Hangman's fracture: Fracture through the pars interarticularis or neural arch of the axis (C2)
- Hill-Sachs lesion: Cortical depression on the posterolateral surface of the humeral head caused by the humeral head hitting the glenoid in a shoulder dislocation
- Hoffa fragment: Fracture through the coronal plane of the femoral condyle
- Holstein-Lewis fracture: Fracture of the distal third of the humerus with entrapment of the radial nerve
- Hutchinson's fracture: Oblique fracture of the radial styloid; also called a Chauffeur's fracture
- Jefferson's fracture: Fracture of the atlas ring due to axial compression; fractures are usually anterior and posterior to the lateral facet joints.
- Jersey finger: Rupture of the flexor digitorum profundus attachment to the distal phalanx

- Jones fracture: Diaphyseal fracture of the base of the fifth metacarpal
- Lisfranc's: Fracture dislocation of the tarsometatarsal joint (Lisfranc joint) of the foot
- Little League elbow: Medial epicondyle apophysitis due to repeated valgus overload
- Maisonneuve's fracture: Fracture of the proximal fibula with syndesmotic disruption, resulting in medial malleolus or deltoid ligament rupture
- Malgaigne's fracture: Vertical pelvic fractures anterior and posterior to the hip joint
- Mallet finger: Flexion deformity of the distal phalanx due to avulsion of the extensor tendon from the distal phalanx
- Monteggia's fracture: Proximal third ulna fracture and dislocation of the radial head
- Nightstick fracture: Isolated ulna shaft fracture
- Nursemaid's elbow: Radial head subluxation due to a sudden force on an extended, pronated arm
- Piedmont fracture: Distal radial shaft fracture with dislocation of the distal radioulnar joint; also called a Galeazzi fracture
- Reverse Bennett's fracture: Articular fracture of the fifth metacarpal base; also called a Baby Bennett fracture
- Reverse Colles' fracture: Volarly displaced distal radius fracture; also called a Smith's fracture
- Reverse Galeazzi fracture: Distal ulnar shaft fracture with dislocation of the distal radioulnar joint
- Rolando's fracture: Y-shaped intra-articular fracture of the base of the thumb metacarpal
- Segond's fracture: Avulsion fracture of the bony insertion of the iliotibial band from the lateral tibial condyle
- Smith's fracture: Volarly displaced distal radius fracture; also called a reverse Colles' fracture
- Straddle fracture: Bilateral fractures of the superior and inferior pubic rami
- Teardrop fracture: Flexion fracture/dislocation of the anterior fragment of the vertebral body; most commonly occurs in the cervical vertebrae
- Tillaux fracture: Fracture of the lateral side of the distal tibial physes after the medial side of the physes has already closed
- Torus fracture: Impaction fracture of a child's bone whereby the bone buckles or bends but does not break completely; also called a Buckle fracture

Abbreviations

- ABI: ankle-brachial index
- ACL: anterior cruciate ligament
- ADI: atlanto-dens interval
- AIN: anterior interosseous nerve
- AP: anterior-posterior
- APC: anterior-posterior compression
- ATFL: anterior talofibular ligament
- AVN: avascular necrosis
- CC: coracoclavicular
- CMC: carpometacarpal
- CR: closed reduction
- CRPP: closed reduction percutaneous pinning
- CTEV: congenital tales equino-varus
- DDH: developmental dysplasia of the hip
- DIP: distal interphalangeal
- DISI: dorsal intercalated segmental instability; indicated by a dorsal scapholunate angle > 70 degrees
- DJD: degenerative joint disease
- DP: dorsalis-pedis
- DRUJ: distal radioulnar joint
- ECR: extensor carpi radialis
- ED: extensor digitorum
- EHL: extensor hallucis longus
- ER: external rotation
- FCR: flexor carpi radialis
- FDS: flexor digitorum superficialis
- FDP: flexor digitorum profundus
- FFWB: foot flat weight-bearing
- FHL: flexor hallucis longus
- GS: gastrocnemius-soleus
- GSW: gunshot wound
- HO: heterotopic ossification
- HS: hamstrings
- IF: index finger
- IMN/IMR: intramedullary nail/intramedullary rod
- INT: interossei
- IP: interphalangeal (joint)
- IP: iliopsoas
- IR: internal rotation

- IT: iliotibial
- LAC: long arm cast
- LC: lateral compression
- LCL: lateral collateral ligament
- LF: long finger
- LLC: long leg cast
- MC: metacarpal
- MCL: medial collateral ligament
- MCP: metacarpal-phalangeal
- MFC: medial femoral condyle
- MP: metacarpophalangeal
- MT: metatarsal
- MTP: metatarsophalangeal
- NWB: non-weight-bearing
- ORIF: open reduction internal fixation
- PA: posterior-anterior
- PCL: posterior collateral ligament
- PIN: posterior interosseous nerve
- PIP: proximal interphalaneal
- SAC: short arm cast
- SC: sternoclavicular
- SF: small finger
- SH: Salter-Harris
- SLAP: superior labrum anterior-posterior
- SLC: short leg cast
- SLR: straight leg raise
- TA: tibialis anterior
- TFCC: triangular fibrocartilage complex
- THR/THA: total hip replacement/total hip arthroplasty
- TKR/TKA: total knee replacement/total knee arthroplasty
- TMT: tarsometatarsal
- TSR/TSA: total shoulder replacement/total shoulder arthroplasty
- TP: tibialis posterior
- TTWB: toe-touch weight-bearing
- UCL: ulnar collateral ligament
- VISI: volar intercalated segmental instability; indicated by a volar scapholunate angle < 35 degrees
- VS: vertical shear
- WBAT: weight-bearing as tolerated

Bibliography

American College of Surgeons: American College of Surgeons Committee on Trauma. *Advanced Trauma Life Support: ATLS*, 8th ed. Chicago, IL, American College of Surgeons, 2008.

Anderson PA, Gugala Z, Lindsey RW, Schoenfeld AJ, Harris MB. Clearing the cervical spine in the blunt trauma patient. *Journal of the American Academy of Orthopaedic Surgeons* 2010;18:149–59.

Browner BD, Jupiter JB, Levine AM, Trafton PG, Krettek C. *Skeletal Trauma: Basic Science, Management and Reconstruction*, 4th ed. St. Louis, MO: W.B. Saunders Company, 2008.

Bucholz RW, Heckman JD, Court-Brown CM, Tornetta P. Rockwood and Green's Fractures in Adults, 7th ed. Philadelphia, PA: Lippincott Williams & Wilkins, 2010.

Denis F. Spinal Instability as Defined by the Three-Column Spine Concept in Acute Spinal Trauma. *Clinical Orthopaedics and Related Research* 1984;189:65–76.

Egol KA, Koval KJ, Zuckerman JD. Handbook of Fractures, 4th ed. Philadelphia, PA: Lippincott Williams & Wilkins, 2010.

Elliott KGB, Johnstone AJ. Diagnosing Acute Compartment Syndrome. *Journal of Bone and Joint Surgery (Br)* 2003;85:625–632.

Gustilo RB, Anderson JT. Prevention of infection in the treatment of one thousand and twenty-five open fractures of long bones: retrospective and prospective analyses. *Journal of Bone and Joint Surgery Am* 1976;58:453–458.

Hak DJ, Smith WR, Suzuki T. Management of Hemorrhage in Life-threatening Pelvic Fracture. *Journal of the American Academy of Orthopaedic Surgeons* 2009;17:447–457.

Hoffman JR, Mower W et al. Validity of a set of clinical criteria to rule out injury to the cervical spine in patients with blunt trauma. *The New England Journal of Medicine* 2000;343:94–99.

Hoppenfeld S. Chapter 4: Physical Examination of the Cervical Spine and Temporomandibular Joint. Physical Examination of the Spine and Extremities. Norwalk, CT: Appleton and Lange, 1976.

Hoppenfeld S. Chapter 9: Physical Examination of the Lumbar Spine. Physical Examination of the Spine and Extremities. Norwalk, CT: Appleton and Lange, 1976.

McKinney BI, Nelson C, Carrion W. Apophyseal avulsion fractures of the hip and pelvis. *Orthopedics* 2009;32:42–48.

Mehta S, Auerbach JD, Born CT, Chin KR. Sacral Fractures. *Journal of the American Academy of Orthopaedic Surgeons* 2006; 4:656–665.

Morrissy RT, Weinstein SL. Lovell and Winter's Pediatric Orthopaedics, 7th ed. Philadelphia, PA: Lippincott Williams & Wilkins, 2005.

Okike K, Bhattacharyya T. Trends in the Management of Open Fractures. A Critical Analysis. *Journal of Bone and Joint Surgery Am* 2006;88:2739–2748.

Olson SA, Glasgow RR. Acute Compartment Syndrome in Lower Extremity Musculoskeletal Trauma. *Journal of the American Academy of Orthopaedic Surgeons* 2005;13:436–444.

Stiell IG, Wells GA, Vandemheen KL, Clement CM, Lesiuk H, De Maio VJ, Laupacis A, Schull M, McKnight RD, Verbeek R, Brison R, Cass D, Dreyer J, Eisenhauer MA, Greenberg GH, MacPhail I, Morrison L, Reardon M, Worthington J. The Canadian C-spine rule for radiography in alert and stable trauma patients. *Journal of the American Medical Association* 2001;286:1841–1848.

Tornetta III, P. Displaced Acetabular Fractures: Indications for Operative and Nonoperative Management. *Journal of the American Academy of Orthopaedic Surgeons* 2001;9:18–28.

Volgas DA, Stannard JP, Alonso JE. Current orthopaedic treatment of ballistic injuries. *Injury* 2005;36:380–386.

White CE, Hsu JR, Holcomb JB. Haemodynamically unstable pelvic fractures. *Injury*, 2008.

Zalavras CG, Marcus RE, Levin LS, Patzakis MJ. Management of Open Fractures and Subsequent Complications. *Journal of Bone and Joint Surgery Am* 2007;89:884–895.

Zalavras CG, Patzakis MJ. Open Fractures: Evaluation and Management. *Journal of the American Academy of Orthopaedic Surgeons* 2003;11:212–219.

Zernicke RF, Garhammer J, Jobe FW. Human patellar-tendon rupture. *J Bone Joint Surg Am* 1977;59(2):179–183.

Index

A

B

C

D

E

F

G

H

I

J

K

L

M

N

O

P

Q

R

S

T

U

V